RED-EARED SLIDER TURTLES live throughout the south-central United States.

BEHAVIOR Slider turtles, like all reptiles, are cold-blooded, so they spend a lot of time sunning themselves on rocks or logs.

Habitat Slider turtles prefer to live near slow-moving water, such as marshes or ponds.

Shell The turtle's shell is actually made from its ribs, which grow together, covered with a layer of skin.

HSP Illinois Science

SCHOOL PUBLISHERS

Visit *The Learning Site!*
www.harcourtschool.com

HSP Illinois Science

SCHOOL PUBLISHERS

Copyright © 2009 by Harcourt, Inc.

All rights reserved. No part of this publication may be reproduced or transmitted in any form or by any means, electronic or mechanical, including photocopy, recording, or any information storage and retrieval system, without permission in writing from the publisher.

Requests for permission to make copies of any part of the work should be addressed to School Permissions and Copyrights, Harcourt, Inc., 6277 Sea Harbor Drive, Orlando, Florida 32887-6777. Fax: 407-345-2418.

HARCOURT and the Harcourt Logo are trademarks of Harcourt, Inc., registered in the United States of America and/or other jurisdictions.

Printed in the United States of America

ISBN – 13: 978-0-15-363827-5
ISBN – 10: 0-15-363827-3

2 3 4 5 6 7 8 9 10 48 16 15 14 13 12 11 10 09 08

If you have received these materials as examination copies free of charge, Harcourt School Publishers retains title to the materials and they may not be resold. Resale of examination copies is strictly prohibited and is illegal.

Possession of this publication in print format does not entitle users to convert this publication, or any portion of it, into electronic format.

Excursion Photo Credits

Page Placement Key: (t) Top, (b) bottom, (l) left, (r) right, (c) center, (bg) background, (fg) foreground

Life: 41 (bg) Bill Beatty/Earth Scenes; 41 (t) Missouri State University; 41 (b) Visuals Unlimited; 42 (c) Nicole Lafevre; 43 (tr) Jim & Mary Whitmer; 43 (b) Cosley Zoo, Wheaton Park District; 44 (bg) Chicago Botanic Garden. The Chicago Botanic Garden is owned by the Forest Preserve District of Cook County; 45 (t) Don Smetzer/PhotoEdit; 46 (b) Just Our Pictures

Earth: 235 (bg) Karina Wang/ The Image Bank/Getty Images; 235 (b) James Shaffer/PhotoEdit; 235 (inset) Tzu-Fang Sheu; 236 (bg) STScI/NASA/Corbis; 236 (b) Herbert Trackman Planetarium—Joliet College; 237 (c) Bettmann/Corbis; 238 (bg) Roger Townsend; 239 (t) James Killam; 239(B) Larry Backe

Physical: 445 (all) Magic Waters Waterpark, Rockford Park District, Illinois; 446 (bg) Joseph Sohm/Visions of America/Corbis; 447 (t) Ron Vesely/MLB Photos via Getty Images; 447 (b) Tannen Maury/EPA/Corbis; 448 (l) Discovery Center Museum; 449 (t) Discovery Center Museum

Consulting Authors

Michael J. Bell
Associate Professor of Early Childhood Education
College of Education
West Chester University of Pennsylvania
West Chester, Pennsylvania

Michael A. DiSpezio
Curriculum Architect
JASON Academy
Cape Cod, Massachusetts

Marjorie Frank
Former Adjunct Professor, Science Education
Hunter College
New York, New York

Gerald H. Krockover
Professor of Earth and Atmospheric Science Education
Purdue University
West Lafayette, Indiana

Joyce C. McLeod
Adjunct Professor
Rollins College
Winter Park, Florida

Barbara ten Brink
Science Specialist
Austin Independent School District
Austin, Texas

Carol J. Valenta
Senior Vice President
St. Louis Science Center
St. Louis, Missouri

Barry A. Van Deman
President and CEO
Museum of Life and Science
Durham, North Carolina

Senior Editorial Advisors

Napoleon Adebola Bryant, Jr.
Professor Emeritus of Education
Xavier University
Cincinnati, Ohio

Tyrone Howard
Associate Professor
Graduate School of Education and Information Studies
University of California, Los Angeles
Los Angeles, California

Robert M. Jones
Professor of Education Foundations
University of Houston–Clear Lake
Houston, Texas

Mozell P. Lang
Former Science Consultant
Michigan Department of Education
Science Consultant, Highland Park Schools
Highland Park, Michigan

Jerry D. Valadez
K–12 Science Coordinator
Fresno Unified School District
Fresno, California

Contents

Introductory Chapter
Getting Ready for Science x

Essential Questions

- **Lesson 1** What Are Some Science Inquiry Tools? 2
- **Lesson 2** What Are Some Science Inquiry Skills? 14
- **Lesson 3** What Is the Scientific Method? 26
- **Chapter Review and Test Preparation** 36
- **Safety in Science** 38

Big Idea
Science is raising questions about Earth and the universe and seeking answers by careful observation and investigation.

LIFE SCIENCE

Illinois Excursions and Projects 40

Unit A — Living Things in Our World 47

Chapter 1 Types of Living Things 48

Essential Questions

- **Lesson 1** What Are Some Types of Living Things? 50
- **Lesson 2** How Do Living Things Grow and Change? 60
- **People in Science**
  - Akira Okubo 70
  - Charles Henry Turner 71
- **Chapter Review and Test Preparation** 72

Big Idea
Cells are the building blocks of all the many types of living things.

Chapter 2 Types of Plants 74

Essential Questions

- **Lesson 1** What Do Plants Need to Live? 76
- **Lesson 2** What Are Some Types of Plants? 86
- **Lesson 3** How Do Plants Make Food? 98
- **People in Science**
  - Marie Clark Taylor 106
  - Rosa Ortiz 107
- **Chapter Review and Test Preparation** 108

Big Idea
Plants have parts that help them meet their needs.

Chapter 3 Types of Animals.......... 110
Essential Questions
Lesson 1 What Do Animals Need to Live?...... 112
Lesson 2 What Are Vertebrates?............ 122
Lesson 3 What Are Invertebrates?........... 134

Science Spin Weekly Reader Technology
Better Than Nature Can Make?......... 146

Chapter Review and Test Preparation........ 148
Unit A Visual Summary................ 150

> **Big Idea**
> Animals can be classified by their traits.

UNIT B — Living Things Interact — 151

Chapter 4 Where Living Things Are Found........ 152
Essential Questions
Lesson 1 What Are Ecosystems?............ 154
Lesson 2 What Are Some Types of Ecosystems?...... 164
Lesson 3 How Do Living Things Survive in Ecosystems?..... 174
Lesson 4 How Do Ecosystems Change?........ 184

Science Spin Weekly Reader Technology
Hiding in Plain Sight......... 192

Chapter Review and Test Preparation........ 194

> **Big Idea**
> All the living, once-living, and nonliving things in an environment interact in an ecosystem.

Chapter 5 Living Things Depend on One Another.... 196
Essential Questions
Lesson 1 How Do Plants and Animals Interact?...... 198
Lesson 2 What Are Food Chains?........... 208
Lesson 3 What Are Food Webs?............ 218

People in Science
Margaret Morse Nice................ 228
Rodolfo Dirzo................... 229

Chapter Review and Test Preparation........ 230
Unit B Visual Summary................ 232

> **Big Idea**
> All living things need energy to survive and grow.

EARTH SCIENCE

Illinois Excursions and Projects 234

UNIT C: Earth's Land — 241

Chapter 6 Minerals and Rocks 242

Essential Questions

Lesson 1 What Are Minerals and Rocks? 244
Lesson 2 What Are the Types of Rocks? 254
Lesson 3 What Are Fossils? 266

Science Spin Weekly Reader **Technology**
Stuck in the Muck 276

Chapter Review and Test Preparation 278

Big Idea
Rocks are made up of different minerals, and they go through changes during the rock cycle.

Chapter 7 Forces That Shape the Land 280

Essential Questions

Lesson 1 What Are Landforms? 282
Lesson 2 How Do Landforms Change Slowly? 292
Lesson 3 How Do Landforms Change Quickly? 302

People in Science
Marisa Quinones 312
Charles Richter 313

Chapter Review and Test Preparation 314

Big Idea
Processes on Earth can change Earth's landforms. Some of these changes happen slowly, while others happen quickly.

Chapter 8 Conserving Resources 316

Essential Questions

Lesson 1 What Are Some Types of Resources? 318
Lesson 2 What Are Some Types of Soil? 328
Lesson 3 How Do People Use and Impact the Environment? . 338
Lesson 4 How Can Resources Be Used Wisely? 348

Science Spin Weekly Reader **Technology**
The Great Dam 358

Chapter Review and Test Preparation 360
Unit C Visual Summary 362

Big Idea
Living things use Earth's resources to meet their needs. Some of these resources can be recycled or reused.

UNIT D — Weather and Space 363

Chapter 9 The Water Cycle 364

Essential Questions

Lesson 1 Where Is Water Found on Earth? 366
Lesson 2 What Is the Water Cycle? 376
Lesson 3 What Is Weather? 386
People in Science
    Adriana Ocampo 398
    Bin Wang 399
Chapter Review and Test Preparation 400

Big Idea
Water is important to all living things in many different ways.

Chapter 10 Earth's Place in the Solar System 402

Essential Questions

Lesson 1 What Causes Earth's Seasons? 404
Lesson 2 How Do Earth and the Moon Interact? 414
Lesson 3 What Is the Solar System? 424

Science Spin Weekly Reader — Technology
    Ancient Planet Found 438
Chapter Review and Test Preparation 440
Unit D Visual Summary 442

Big Idea
The sun and everything that orbits it make up the solar system. Other objects, such as constellations, are found outside the solar system.

PHYSICAL SCIENCE

Illinois Excursions and Projects 444

UNIT E — Investigating Matter and Energy — 451

Chapter 11 Properties of Matter 452

Essential Questions

- **Lesson 1** What Is Matter? 454
- **Lesson 2** What Are States of Matter? 466
- **Lesson 3** How Does Matter Change? 476

People in Science
- Steven Chu 486
- Tara McHugh 487

Chapter Review and Test Preparation 488

Big Idea
Matter has properties that can be observed, described, and measured.

Chapter 12 Energy 490

Essential Questions

- **Lesson 1** What Is Energy? 492
- **Lesson 2** How Can Energy Be Used? 502
- **Lesson 3** Why Is Energy Important? 510

People in Science
- Enrico Fermi 518
- Evangelista Torricelli 519

Chapter Review and Test Preparation 520

Big Idea
You use many forms of energy every day to grow and live.

Chapter 13 Electricity and Magnets 522

Essential Questions

- **Lesson 1** What Is Electricity? 524
- **Lesson 2** What Are Magnets? 532
- **Lesson 3** How Are Electricity and Magnets Related? 540

Science Spin Weekly Reader — Technology
Batteries Included 548

Chapter Review and Test Preparation 550

Big Idea
Electricity and magnetism are related and are part of things you use every day.

Chapter 14 Heat, Light, and Sound 552

Essential Questions

Lesson 1 What Is Heat? 554
Lesson 2 What Is Light? 562
Lesson 3 How Are Light and Color Related? 570
Lesson 4 What Is Sound? 580

Science Spin Weekly Reader Technology
A New Source of Energy? 588

Chapter Review and Test Preparation 590
Unit E Visual Summary 592

Big Idea
Heat, light, and sound are different forms of energy that can move from place to place.

UNIT F — Exploring Forces and Motion — 593

Chapter 15 Forces and Motion 594

Essential Questions

Lesson 1 What Is Motion? 596
Lesson 2 What Are Forces? 606
Lesson 3 How Do Waves Move? 616

People in Science
Percy Spencer 624
Christine Darden 625

Chapter Review and Test Preparation 626

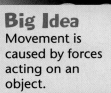

Big Idea
Movement is caused by forces acting on an object.

Chapter 16 Work and Machines 628

Essential Questions

Lesson 1 What Is Work? 630
Lesson 2 What Are Some Simple Machines? 640
Lesson 3 What Are Some Other Simple Machines? 652

Science Spin Weekly Reader Technology
Say Hello to ASIMO 662

Chapter Review and Test Preparation 664
Unit F Visual Summary 666

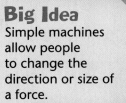

Big Idea
Simple machines allow people to change the direction or size of a force.

REFERENCES

Picture Glossary R1
Index R31

Getting Ready for Science

What's the Big Idea?

Science is raising questions about Earth and the universe and seeking answers by careful observation and investigation.

Essential Questions

Lesson 1

What Are Some Science Inquiry Tools?

Lesson 2

What Are Some Science Inquiry Skills?

Lesson 3

What Is the Scientific Method?

 Student eBook www.hspscience.com

Students preparing to investigate

What do you wonder?

Do you ever wonder why plants are green or what causes the seasons? If so, you are already thinking like a scientist! Scientists ask questions. How are these students preparing to answer questions? How does this relate to the Big Idea?

LESSON 1

Essential Question
What Are Some Science Inquiry Tools?

Investigate to find out how to observe objects.

Read and Learn what science inquiry tools are and how they are used.

Fast Fact

Measuring Up
The first tools used by people to measure things were not rulers and measuring cups, but body parts! In fact, the foot measurement was based on the length of a person's foot. In the Investigate, you will learn more about measurement.

2

Using a graduated cylinder

Vocabulary Preview

inquiry [IN•kwer•ee] A question that is asked about something, or a close study of something (p. 6)

forceps [FAWR•seps] A tool used to pick up and hold on to objects (p. 7)

3

Investigate

Making Bubbles

Guided Inquiry

Start with Questions

The children in this photo are washing a dog in water.

- How is this water different from plain water?
- What do you think causes bubbles to form in the water?
- What do bubbles look like up close?

Investigate to find out. Then read and learn to find out more.

Prepare to Investigate

Inquiry Skill Tip

A **measurement** includes a number and a unit. Suppose the width of a bubble is 6 centimeters. This measurement includes the number 6 and the unit *centimeters*. Other units are ounces, pounds, inches, miles, seconds, and liters.

Materials

- safety goggles
- large container
- straw
- water
- metric measuring cup
- dishwashing soap
- small containers
- stirring stick
- hand lens

Make an Observation Chart

Draw what you see when you look at the bubbles with a hand lens.

Follow This Procedure

CAUTION: **Put on safety goggles.**

1. Use the measuring cup to **measure** 1 L (1,000 mL) of water. Pour the water into a large container.

2. Then **measure** 50 mL of dishwashing soap. Add the soap to the container of water, and then stir.

3. Pour some of the soap-and-water solution into small containers. Use the straw to blow air into the solution. Be careful not to blow too hard or to spill some of the solution. Bubbles should form. **Observe** the bubbles with a hand lens. **Record** your **observations**.

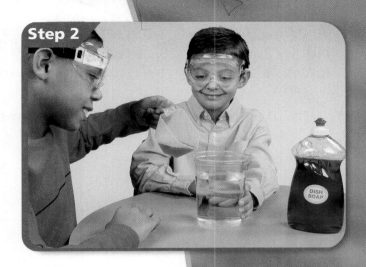
Step 2

Step 3

Draw Conclusions

1. What did you observe about the bubbles?

2. **Inquiry Skill** Scientists use many different tools to **measure** things. In this Investigate, you used a measuring cup to **measure** both water and soap. What kind of measuring tool could you use to **measure** the size of the bubbles you made? Explain your answer.

Independent Inquiry

Add 60 mL of glycerine and 8 mL of sugar to the solution. Blow bubbles. **Compare** these bubbles to the first bubbles.

Read and Learn

VOCABULARY
inquiry p. 6
forceps p. 7

SCIENCE CONCEPTS
▶ what some measurement tools are
▶ how measurement tools are used

MAIN IDEA AND DETAILS
Look for details about tools used for measuring things.

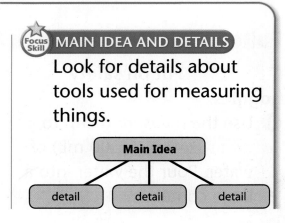

Tools Used for Inquiry

Have you ever asked questions about something? If so, you made an inquiry. An **inquiry** happens when someone asks a question or studies something closely. In almost every science inquiry, tools must be used to observe, measure, and compare the objects that are being studied.

Hand Lens

What It Is: A hand lens is a tool used to magnify, or enlarge, an object's features.

How to Use It: Hold a hand lens up to an object that you want to observe closely. Look through the clear part of the hand lens. You can move the hand lens closer to or farther from the object to make the object clearer.

Safety: Some hand lenses have a glass lens. Be careful not to drop a hand lens. If a hand lens breaks, do not try to clean up the broken pieces. Ask an adult for help.

Forceps

What They Are: Forceps are used to pick up and hold on to objects. They are similar to tongs and tweezers.

How to Use Them: Place the tips of the forceps around the object you want to pick up. Apply pressure to the forceps' handles, and lift the object.

Safety: The tips of the forceps can be sharp. Keep forceps away from your face. Always clean forceps after you use them.

Magnifying Box

What It Is: Like a hand lens, a magnifying box is a tool used to magnify, or enlarge, an object's features.

How to Use It: Place the magnifying box on top of a flat surface. Look through the clear part of the box.

Safety: Some magnifying boxes have a glass lens. Be careful not to drop the box. Do not use a magnifying box that has a cracked or damaged lens.

Dropper

What It Is: A dropper is a tool that can be used to pick up and release small amounts of liquid. Some droppers have marks on them that can be used to measure the liquid.

How to Use It: Squeeze the dropper's bulb. Place the end of the dropper in a liquid and release the bulb. Some of the liquid will move up into the dropper. To release the liquid, squeeze the bulb again.

Safety: Droppers should be cleaned after each use.

Use a Magnifying Box

Place a magnifying box on your textbook. Use it to look at the letters on this page. What do you observe? Now look at other objects, such as a leaf.

Thermometer

What It Is: A thermometer is a tool that measures temperature, or how hot or cold something is.

How to Use It: Put the thermometer in the place where you would like to measure the temperature. Wait about five minutes. Then see where the liquid in the thermometer's tube is. Use the markings along the side of the tube to read the temperature.

Safety: If a thermometer breaks, do not touch it. Ask an adult for help.

Ruler

What It Is: A ruler is a tool used to measure length, width, height, or depth.

How to Use It: Place the ruler against the object you would like to measure. Use the markings on the ruler to see how long, wide, high, or deep the object is.

Safety: Many rulers are made of plastic or wood. Do not use rulers to measure warm objects. This may cause the plastic to melt or the wood to catch on fire.

Measuring Tape

What It Is: Like a ruler, a measuring tape is used to measure length, width, height, or depth. A measuring tape is useful for measuring a curved object.

How to Use It: Place the measuring tape along the object you would like to measure. Use the markings on the measuring tape to see how long, wide, high, or deep the object is.

Safety: Measuring tapes are often made of plastic. Do not use measuring tapes to measure warm objects.

Measuring Cup

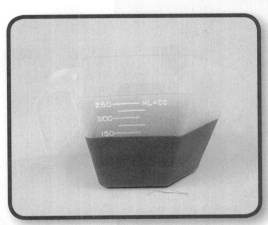

What It Is: A measuring cup measures volume, or the amount of space that something takes up. Measuring cups are usually used for liquids and loose solids such as powders.

How to Use It: Pour the substance you need to measure into the measuring cup. Use the marks on the outside of the cup to see how much of the substance is in the cup.

Safety: Some measuring cups are made of glass. Be careful not to drop the cup, or it could break.

Spring Scale

What It Is: A spring scale is a tool that measures an object's weight.

How to Use It: Attach the object you want to weigh to the hook at the bottom of the spring scale. Allow the object to hang as you hold the spring scale up. The weight of the object will be displayed on the scale's readout.

Safety: Only weigh objects on a spring scale when told to do so by your teacher.

Forceps, rulers, spring scales, and more can all be used during science inquiry. Each of these tools is used for different things. Part of learning about science is learning how to choose which of these tools you can use to help you answer your questions.

 **MAIN IDEA AND DETAILS**

What are two tools used to observe an object closely?

Some Other Tools Used in Science

Many science tools have similar uses. For example, a measuring cup, measuring spoons, and graduated cylinders can all be used to measure how much space something takes up. You choose which of these three tools to use by finding out how much you have to measure.

Microscopes are used to magnify an object. They are helpful to see things that you can't see with your eyes alone.

Balances are used to measure the mass of objects. By placing the object in one pan and weights in the other pan, you are able to "balance" out the object's mass.

MAIN IDEA AND DETAILS

When might you use a microscope?

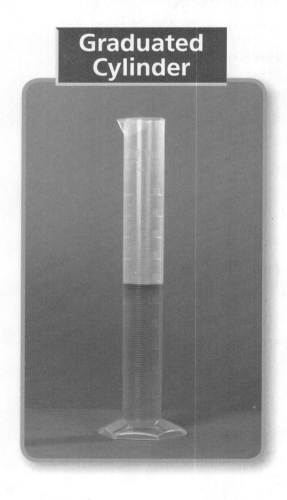

Graduated Cylinder

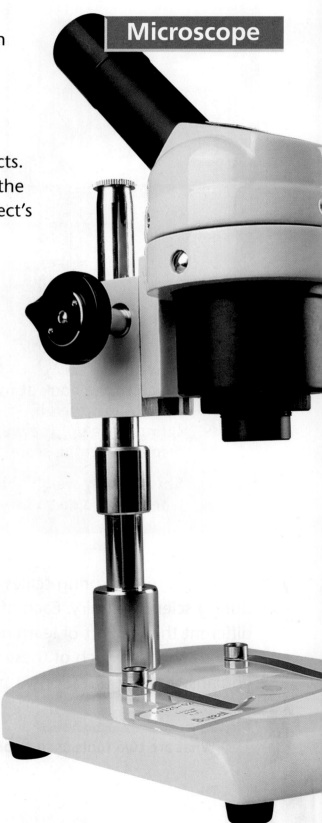

Microscope

Pan Balance

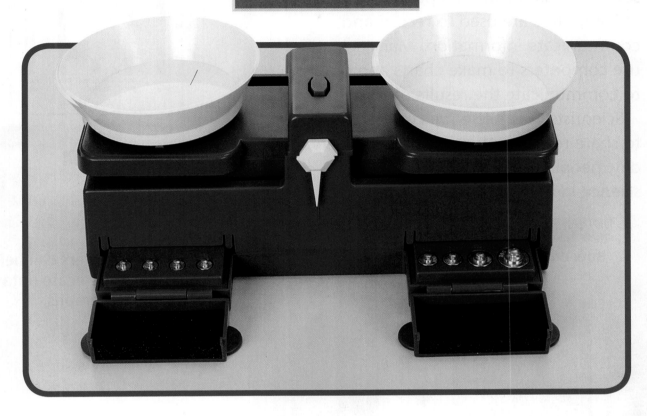

Measuring Spoon

Computers as Tools

Computers can be tools for science inquiry. You can use them to gather research from various sources. The Internet and many libraries allow you to search for data with computers.

Computers are used to record and communicate information. Many scientists use computers to make charts and graphs to communicate the results of an inquiry.

Scientists may use e-mail or Web sites to share results with other people. Every day, people use computers to make science inquiry better and easier.

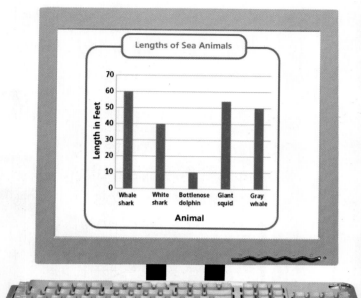

▲ Computers can help you communicate data by using graphs.

How do computers help scientists communicate?

◄ This student is using a computer to gather information.

Lesson Review

Essential Question

What are some science inquiry tools?

In this lesson, you learned about tools that can help you observe and measure things. Scientists use these tools in inquiries.

1. **MAIN IDEA AND DETAILS** Draw and complete a graphic organizer for this main idea: There are many tools that can be used in science inquiry.

   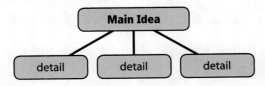

2. **SUMMARIZE** Write a four-sentence summary of this lesson. Tell about all the tools you learned about along with their uses.

3. **DRAW CONCLUSIONS** You need to look closely at a soil sample. Which tool would you use?

4. **VOCABULARY** Write a sentence describing how you can use forceps during a science inquiry.

Test Prep

5. Which tool should you use to add two drops of food coloring to a bowl of water?
   A. dropper
   B. forceps
   C. measuring cup
   D. measuring spoons

Make Connections

 Writing

Expository
Write two paragraphs that **explain** the way you use technology compared to the way scientists use technology.

 Math

Measure Elapsed Time
Timers and stopwatches can be used to measure how much time passes. Use one of these two tools to measure how long it takes five of your classmates to each run 50 meters. Record the results in a table.

13

LESSON 2

Investigate to find out how scientists predict.

Read and Learn what inquiry skills scientists use and how they use them.

Essential Question

What Are Some Science Inquiry Skills?

Fast Fact

Blowing Bubbles
The longest bubble ever blown and measured was about 32 m (105 ft) long! What shape do you think this huge bubble was? Do the Investigate to find out the different shapes that bubbles can take.

Bubble maker

Vocabulary Preview

infer [in•FER] To draw a conclusion about something (p. 19)

variable [VAIR•ee•uh•buhl] The one thing that changes in a science inquiry or experiment (p. 22)

formulate [FOR•myoo•layt] To come up with a plan for something (p. 23)

Investigate

Shapes of Bubbles

Guided Inquiry

Start with Questions

The boy in this picture is using an interesting wand to make a bubble.

- In the photo, what is the shape of the wand that the boy is using?
- Do bubble wands that have different shapes make bubbles that have different shapes?

Investigate to find out. Then read and learn to find out more.

Prepare to Investigate

Inquiry Skill Tip

Good predictions are not random guesses. They are educated guesses based on past experience. Before you predict, review any observations or data that suggest what might happen.

Materials

- safety goggles
- wire hangers
- pie pan
- bubble solution or dishwashing soap

Make an Observation Chart

Wand	Sketch of Wand Shape	Bubble Shape	Other Observations
1			
2			
3			

Follow This Procedure

CAUTION: Put on safety goggles.

1. Use wire hangers to make bubble wands of different shapes. For example, you could make a round, a square, and a triangular wand.

2. **Predict** the shape of the bubbles that will be made by each wand.

3. Pour some of the bubble solution into the pie pan. Dip one of your wands into the solution. Use the wand to make bubbles. **Observe** the bubbles' shapes. Repeat this activity with all of the wands that you made.

Step 1

Step 3

Draw Conclusions

1. What did you predict about the shape of the bubbles? Were your predictions correct?

2. **Inquiry Skill** Scientists use **observations** of the natural world to make **predictions**. Use your **observations** to **predict** the shape of a bubble blown with a heart-shaped wand.

Independent Inquiry

What wand shape would make the biggest bubble? Blow bubbles with different wands. **Measure** and **compare** the bubbles' sizes.

Read and Learn

VOCABULARY
infer p. 19
variable p. 22
formulate p. 23

SCIENCE CONCEPTS
▶ what some inquiry skills are
▶ how inquiry skills are used

MAIN IDEA AND DETAILS
Look for details about skills for science inquiry.

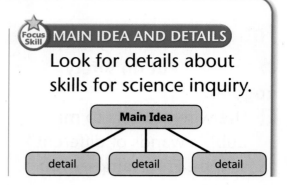

Skills Used for Inquiry

Scientists raise questions about Earth and the universe. When scientists try to find an answer to those questions, they use thinking tools called inquiry skills. You have already used many of these skills while doing the Investigates. You have measured, observed, compared, and made predictions. All of these are inquiry skills.

Think about how you used inquiry skills to answer questions about bubbles. You used some of the skills at the same time. That's because many inquiry skills work well together.

These students are setting up an investigation. They will need to use inquiry skills to complete the investigation. ▼

Use Numbers
Scientists use numbers when they collect and display their data. Understanding numbers and using them to show the results of investigations are important skills.

Measure
You use numbers when you measure something. To make measurements, you can use tools such as thermometers, timers, rulers, measuring tapes, spring scales, and measuring cups.

Gather, Record, Display, or Interpret Data
When you make measurements, you are gathering data. Data are pieces of information. Data can be displayed in charts, tables, graphs, and so on. Recording data can help you interpret, or understand, what the data are telling you.

Infer
When you infer, you draw conclusions about something by using gathered data or previous knowledge. Often you must make inferences to interpret data.

This student is using a ruler to measure yarn.

MAIN IDEA AND DETAILS
List two inquiry skills and state how they are connected.

This student is making a bar graph to display his data.

More Skills Used for Inquiry

You will use more than one inquiry skill in every science inquiry you do. You have already learned how some inquiry skills are used together. On the next few pages, you will read about more skills that can be used during an inquiry.

This student is classifying different writing items. What characteristics is the student using to classify the items? ▼

Compare
When you compare, you identify characteristics of things or events to find out how they are alike or different.

Predict
When you predict something, you use what you know to tell what may happen in the future.

Classify/Order
To classify something, you group or organize it into categories based on specific characteristics. To order things, you place them in the correct sequence.

This student is using a stopwatch to time a process. Timing something involves using a time/space relationship. ▼

Use Time/Space Relationships
Where were you at noon yesterday? By answering this question, you are using a time/space relationship. In fact, whenever you notice where something is at a certain time, you classify that object according to the time and space that it is in.

Variables

Suppose you want to know what color of flower bees like best. To find out, you could plant roses of different colors in a garden and count how many bees visit each rose. In this inquiry, only one thing is different—the color of the roses. The flower's color is a variable. A **variable** is one thing in a science inquiry that is different. You also need to make sure that all the other parts of the inquiry stay the same.

What is the variable in the inquiry below? ▼

These students are building a model of a volcano.

Formulate or Use Models
Formulate means to come up with a plan. Formulating plans can help you stay organized and on track. Models are often used in science to study things that are too big or too small to see easily in real life.

Other inquiry skills include observing, communicating, planning, and more. Throughout this book, you will get the chance to practice each of these skills.

Focus Skill — MAIN IDEA AND DETAILS

What inquiry skills would you use to place events in the correct order on a time line?

Insta-Lab

Make a Model
Use modeling clay to make a model of a plant or an animal that you are interested in. How does your model compare to the real plant or animal?

23

Fact and Opinion

A fact is something that can be proven. You know something is a fact if you can prove it based on evidence. An opinion is a personal belief that is not based on evidence.

Scientists do not draw conclusions based on opinions. They use facts and evidence to understand the world.

MAIN IDEA AND DETAILS Give an example of an opinion and an example of a fact about an apple.

Ashley records that the bug is pretty. Erin records that the bug has six legs. Which student is recording her opinion? Which student is recording a fact? ▼

Lesson Review

Essential Question

What are some science inquiry skills?

In this lesson, you learned why inquiry skills, or thinking skills, are important. Scientists use these skills to answer questions about Earth and the universe.

1. **MAIN IDEA AND DETAILS** Draw and complete a graphic organizer for this main idea: There are many skills that can be used in science inquiry.

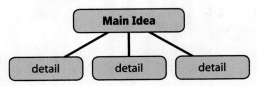

2. **SUMMARIZE** Write a paragraph to summarize this lesson. Include the five inquiry skills and explain how they might be used.

3. **DRAW CONCLUSIONS** Can two objects be in the same space at the same time? Explain.

4. **VOCABULARY** Choose one vocabulary word from this lesson, and explain its meaning.

Test Prep

5. **Critical Thinking** You need to find out how the heights of third graders differ from the heights of second graders. What inquiry skills should you use?

Make Connections

Writing

Persuasive
What do you think the three most useful inquiry skills are? Write a paragraph listing the three skills that in your **opinion** are the most useful.

Language Arts

Solving a Mystery
Inquiry skills are not only useful for scientists—they are also useful for detectives. Write a short story about a detective who uses inquiry skills to solve a mystery.

LESSON 3

Essential Question

What Is the Scientific Method?

Investigate to find out how to compare the results of an experiment.

Read and Learn about the steps of the scientific method and how it is used.

Fast Fact

Bubble Art
These students added tempera paint to bubble solution to make art. But paint doesn't need to be added to bubbles for them to be colorful. Do the Investigate to find out why.

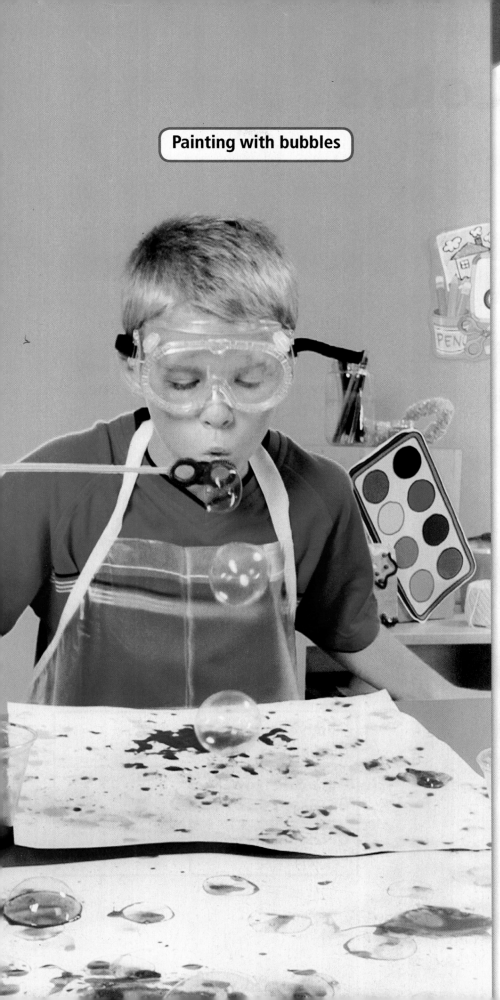

Painting with bubbles

Vocabulary Preview

scientific method
[sy•uhn•TIF•ik METH•uhd]
An organized plan that scientists use to conduct a study (p. 30)

investigation
[in•ves•tuh•GAY•shuhn]
A study that a scientist does (p. 30)

hypothesis
[hy•PAHTH•uh•sis] A possible answer to a question that can be tested to see if it is correct (p. 31)

experiment
[ek•SPAIR•uh•muhnt] A test done to see if a hypothesis is correct or not (p. 31)

Investigate

Bubble Colors

Guided Inquiry

Start with Questions

This bubble is colorful. Why do you think bubbles have different colors?

- Can you blow bubbles in different colors?
- Can you make bubbles change color?

Investigate to find out. Then read and learn to find out more.

Prepare to Investigate

Inquiry Skill Tip

Compare means to tell how things are alike and different. When you compare, observe things and tell at least one way they are alike and at least one way they are different.

Materials

- safety goggles
- cotton ball
- bubble solution or dishwashing soap
- clear tape
- clear plastic lid
- spoon
- flashlight
- straw

Make an Observation Chart

Observations Before Blowing	
Observations While Blowing	
Observations During Popping	

28

Follow This Procedure

CAUTION: Put on safety goggles.

1. Tape a plastic lid over the part of a flashlight that light shines from.
2. Hold the flashlight so the light will shine straight up. Dip a cotton ball in the bubble solution. Wipe the cotton ball over the whole top of the lid. Then put a spoonful of the solution on the lid.
3. Use a straw to blow one big bubble. Turn off the lights, and hold the flashlight so that the attached lid is about even with your eyebrows.
4. **Observe** the bubble. Dip the end of the straw in bubble solution, and put the straw inside the big bubble. Blow very gently. **Observe** what happens.

Draw Conclusions

1. Communicate your observations by drawing what happened.
2. **Inquiry Skill** Use your observations to compare the colors in the bubble, when you first watched it, to the colors you saw just before the bubble popped.

Independent Inquiry

Predict how adding some tempera paint to bubbles will change the Investigate. Plan and conduct a simple experiment to test your predictions.

29

Read and Learn

VOCABULARY
scientific method p. 30
investigation p. 30
hypothesis p. 31
experiment p. 31

SCIENCE CONCEPTS
▶ what the scientific method is
▶ how to use the scientific method

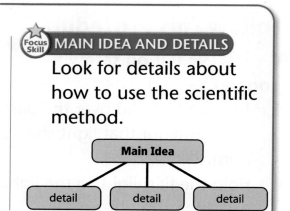

MAIN IDEA AND DETAILS
Look for details about how to use the scientific method.

Scientific Method

How do scientists answer a question or solve a problem? They use an organized plan called the **scientific method** to conduct a study. The study that a scientist does is called an **investigation**. In this lesson, you will learn how the scientific method can be used to plan an investigation to study bubbles.

MAIN IDEA AND DETAILS
What do scientists use to help them answer questions?

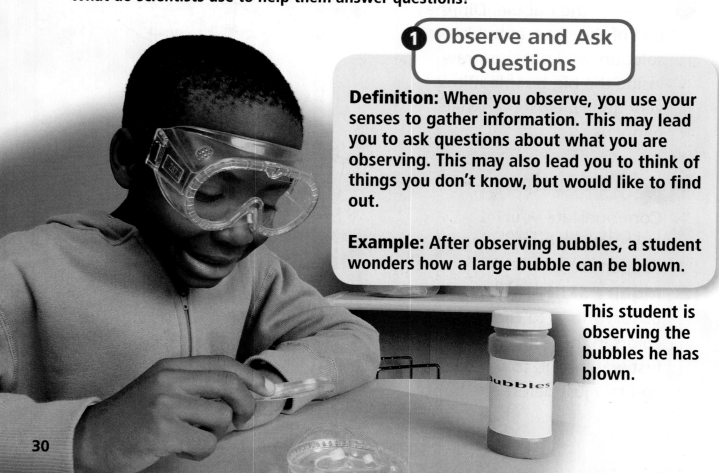

① Observe and Ask Questions

Definition: When you observe, you use your senses to gather information. This may lead you to ask questions about what you are observing. This may also lead you to think of things you don't know, but would like to find out.

Example: After observing bubbles, a student wonders how a large bubble can be blown.

This student is observing the bubbles he has blown.

② Form a Hypothesis

Definition: Write a possible answer to your question. A possible answer to a question is a hypothesis. A hypothesis can be tested to see if it is correct.

Example: The student thinks that the best way to make a big bubble is to use a wand that is flexible instead of a wand that is rigid.

This student is writing his hypothesis in a complete sentence.

③ Plan an Experiment

Definition: An experiment is a test done to find out if a hypothesis is correct or not. When you plan an experiment, you need to describe the steps, identify the variables, list the equipment you will need, and decide how you will gather and record your data.

Example: The student will test his ideas about bubbles by making a *rigid* bubble wand out of chenille sticks and a *flexible* bubble wand out of string.

Safety: Consider your safety when planning an experiment. Study the steps of the experiment and include any safety equipment needed.

This student is planning an experiment to study bubble size.

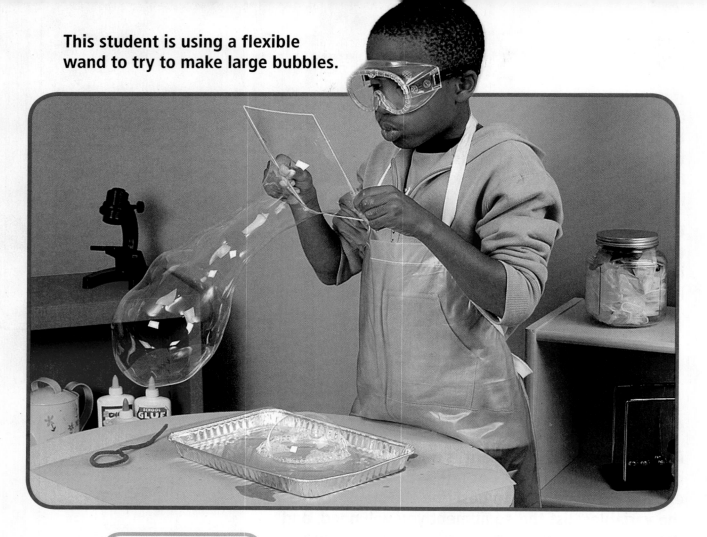

This student is using a flexible wand to try to make large bubbles.

④ Conduct an Experiment

Definition: Follow the steps of the experiment you planned. Observe and measure carefully. Record everything that happened. Organize your data so you can study it more easily.

Example: The student makes two different wands of the same size and blows bubbles with both. He has a ruler to measure each bubble.

Safety: Follow all of the safety instructions in the experiment's plans.

Blow a Super Bubble! Thread a 90-cm string through two straws. Tie the ends of the string together to make a wand. Hold the straws so that the string is tight. Dip the wand in bubble solution and make bubbles. What did your bubbles look like?

5. Draw Conclusions and Communicate Results

Definition: Analyze the data you gathered. Make charts, tables, or graphs to show your data. Write a conclusion. Describe the evidence you used to determine whether your test supported your hypothesis. Decide whether your hypothesis was supported. Communicate your results.

Example: The student looks at the data he collected and realizes that the biggest bubbles were blown by the more flexible wand.

This student is analyzing the data he collected.

Often the observations you make during an experiment will lead you to ask new questions and form a new hypothesis. Remember, you can learn something even if your hypothesis is not supported.

MAIN IDEA AND DETAILS
What part of the scientific method is often done after a scientist asks a question?

Independent Inquiry

Definition: If your hypothesis was supported, you may want to ask another question about your topic that you can test. If your hypothesis was not supported, you may want to form another hypothesis and do a test of a different variable.

Example: The student decides to test how different movements of his flexible wand affect bubble size.

Safety: Make sure your new investigation will keep you and others safe.

Comparing Results

It is important to make sure your results from an experiment are accurate. The best way to check your accuracy is to repeat the experiment. You might repeat it yourself. Someone else could also repeat it. This helps you make sure you get similar results each time.

What happens when the same experiment gives different results? You may choose to do the experiment again. The more times you perform an experiment, the more accurate your observations will be.

You may ask questions about the experiment. *Why did I get different results each time? What caused the difference?* Incorrectly recording your observations can cause you to get different results. Using different tools and recording different observations can also lead to different results.

What might have caused these students to get different results?

MAIN IDEA AND DETAILS

What can you do if you get different results from the same experiment?

Lesson Review

Essential Question

What is the scientific method?

In this lesson, you learned the steps for planning and conducting an investigation. These steps, called the scientific method, make up an organized plan that you can follow to conduct a study or answer a question.

1. **MAIN IDEA AND DETAILS** Draw and complete a graphic organizer for this main idea: There are five main parts to the scientific method.

   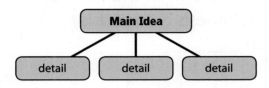

2. **SUMMARIZE** Write a summary to tell how the steps of the scientific method are related.

3. **DRAW CONCLUSIONS** A scientist hypothesizes that cats prefer eating fish to eating chicken. He does an experiment and finds that his hypothesis is not supported. Did the scientist learn anything? Explain your answer.

4. **VOCABULARY** Explain how the terms *hypothesis* and *experiment* are related.

Test Prep

5. Tina did an experiment and drew conclusions. What should she do next?
   - **A.** ask a question
   - **B.** communicate her results
   - **C.** plan an experiment
   - **D.** form a hypothesis

Make Connections

 Writing

Expository Writing
Write a **how-to** booklet about the scientific method. You may want to include illustrations and examples in your booklet.

 Math

Construct a Bar Graph
Graphs are often used to communicate data. Go through this lesson and count the number of times the words *hypothesis* and *experiment* appear. Make a bar graph to show your results.

Review and Test Preparation

Vocabulary Review

Use the terms below to complete the sentences. The page numbers tell you where to look in the chapter if you need help.

inquiry p. 6

forceps p. 7

infer p. 19

variable p. 22

scientific method p. 30

investigation p. 30

experiment p. 31

hypothesis p. 31

1. Steps that are called the _____ are used to plan and conduct a scientific study.

2. When you _____, you draw conclusions about something by using gathered data or previous knowledge.

3. When someone asks a question or closely studies something, an _____ is done.

4. The one thing in a science inquiry that is different is a _____.

5. A tool used to pick up and hold on to objects is called _____.

6. A study that a scientist does is called an _____.

7. A possible answer to a question is a _____.

8. A test done to find out if a hypothesis is correct is an _____.

Check Understanding

Write the letter of the best choice.

9. **MAIN IDEA AND DETAILS** Which of the following tools would you use to measure the height of a book?
   A. measuring cups
   B. balance
   C. ruler
   D. spring scale

10. Identify the tool in the picture.
    F. forceps
    G. hand lens
    H. magnifying box
    J. microscope

36

11. Which of the following tools is **not** used for measurement?
  A. graduated cylinder
  B. hand lens
  C. ruler
  D. thermometer

12. Which inquiry skill would most likely require you to use this tool?

  F. formulate models
  G. infer
  H. interpret data
  J. use time/space relationships

13. MAIN IDEA AND DETAILS Which inquiry skill do you use when you identify how things are alike or different?
  A. compare
  B. formulate models
  C. control variables
  D. predict

14. When you measure, which other inquiry skill would you most likely use?
  F. classify   H. predict
  G. infer   J. use numbers

15. Which of the following is an example of a good hypothesis?
  A. Plants grow best in sunlight.
  B. Some apples are red.
  C. What is inside the sun?
  D. Why are bubbles round?

16. During which scientific method step do you observe and make measurements?
  F. Conduct an experiment.
  G. Draw conclusions.
  H. Form a hypothesis.
  J. Plan an experiment.

Inquiry Skills

17. Infer why it is important for scientists to communicate their experiments' results with one another.

18. Form a hypothesis and plan a simple experiment that could test the hypothesis.

Critical Thinking

19. What inquiry skills are necessary for forming a hypothesis?

20. What tools would be best to use to measure the milk and book? Explain your answers.

Safety in Science

Doing investigations in science can be fun, but you need to be sure you do them safely. Here are some rules to follow.

1. **Think ahead.** Study the steps of the investigation so you know what to expect. If you have any questions, ask your teacher. Be sure you understand any caution statements or safety reminders.

2. **Be neat.** Keep your work area clean. If you have long hair, pull it back so it doesn't get in the way. Roll or push up long sleeves to keep them away from your experiment.

3. **Oops!** If you spill or break something, or if you get cut, tell your teacher right away.

4. **Watch your eyes.** Wear safety goggles anytime you are directed to do so. If you get anything in your eyes, tell your teacher right away.

5. **Yuck!** Never eat or drink anything during a science activity.

6. **Don't get shocked.** Be especially careful if an electric appliance is used. Be sure that electrical cords are in a safe place where you can't trip over them. Never pull a plug out of an outlet by pulling on the cord.

7. **Keep it clean.** Always clean up when you have finished. Put everything away and wipe your work area. Wash your hands.

8. **Play it Safe.** Always know where safety equipment, such as fire extinguishers, can be found. Be familiar with how to use the safety equipment around you.

LIFE SCIENCE

ILLINOIS

Illinois Excursions and Projects
White Oak 40
Cosley Zoo 42
Chicago Botanic Garden 44
Needs of Plants 46

UNIT A
LIVING THINGS IN OUR WORLD 47

CHAPTER 1
Types of Living Things. 48

CHAPTER 2
Types of Plants. 74

CHAPTER 3
Types of Animals 110

UNIT B
LIVING THINGS INTERACT 151

CHAPTER 4
Where Living Things Are Found. 152

CHAPTER 5
Living Things Depend on
   One Another 196

Illinois Excursions

White Oak

The white oak tree is the state tree of Illinois. It has long, wide branches. It has pale gray bark and big leaves. In the fall, the leaves turn red and purple. A white oak can grow to be 100 feet tall. That's as tall as a 10-story building!

Oak trees begin life as acorns. After the acorns begin to sprout, they form seedlings. It takes many years for an oak seedling like this to grow as big as the tree shown at the right.

Tree Growth

Do you know how the life of this tree began? The tree started out as an acorn. Every fall, acorns drop from white oak trees. Squirrels and birds eat some of them. Other acorns can grow into tall white oaks.

Once acorns are in the ground, they sprout roots and stems. The roots grow into the ground to draw in water and minerals. Soon tree stems start poking up from the ground and reaching toward the sun. The leaves on the plants start making food. The tiny new plants are called seedlings.

Every year, the trees will continue to grow, as long as they get the nutrients they need. When a white oak is about 20 years old, it starts making acorns that can grow into new trees. The next time you are sitting under a mighty oak tree, remember that it started out as an acorn!

The acorn of the white oak tree is about 1 inch long.

Think And Write

❶ **Scientific Thinking** How might a young oak tree be similar to its parents? In what ways could it be different? Write a paragraph comparing and contrasting the two plants. Tell how the parent plants may be similar to and different from their offspring.

❷ **Scientific Thinking** All living things go through life cycles. Research Illinois' state bird, the cardinal, and write a paragraph describing its life cycle.

Illinois Excursions

Wheaton

Cosley Zoo

The Cosley Zoo, in Wheaton, can be a fun and interesting place to visit. There are many types of animals at the zoo. You can learn many different things as you observe the animals here.

Notice the sheep at the zoo. How did they get their wool coats and long ears? Notice the horse. How did it get its yellow hair and long mane? Both the sheep and the horse were born with these traits. They inherited them from their parents. All animals, including zoo animals, inherit features from their parents.

The Cosley Zoo is a good place to learn about animals. You can see many traits of animals that they inherited from their parents. You may also notice some traits and behaviors the animals learned on their own.

Learned Behavior

There are also some things animals at the zoo have learned on their own. Many of the animals know when it is time to be fed. They might know food is coming when they hear the zookeeper's keys or when the cage door opens. Knowing when it is time for a meal is not something animals are born with. Zookeepers sometimes teach animals behaviors for certain times. For example, animals might be given special treats, such as food snacks, praise, or being petted during health check-ups. This makes the animals more willing to have health check-ups later.

Think And Write

1. **Scientific Thinking** Do research about an animal found at the zoo. Describe some traits it inherited from its parents. Tell about some things it learned on its own.

2. **Scientific Inquiry** Research what zookeepers at the Cosley Zoo do. Write a paragraph telling how you would train one of the animals to be calm during a health check-up.

Illinois Excursions

Chicago Botanic Garden

Do you like to see and smell beautiful flowers? Do you like to look at different plants and watch wild animals in their natural environments? You can do all this and more at the Chicago Botanic Garden.

While you are there, study the animals closely. Observe the birds as they fly. They have beaks to obtain their food. The big heron by the water has a beak like a dagger to spear fish and frogs. Other animals also have traits that help them survive. The beaver building its dam has webbed feet to help it swim. Watch out for the skunk! It has special glands that spray bad-smelling oil when it is scared.

The Chicago Botanic Garden is filled with plants and wild animals. Most plants and animals have special traits that help them survive in their environments.

Animals aren't the only living things that have special traits that help them. Plants have special traits, too. In one of the greenhouses, you can see a golden barrel cactus. The cactus has a waxy coating to help hold in water. The water lilies on the pond have wide leaves to help them float. Be careful not to touch the rose! It has sharp thorns to protect it from animals.

Think And Write

1. **Scientific Inquiry** Research other wild animals you might see at the Chicago Botanic Garden. Write a paragraph describing some of the special traits that help these animals live in their environment.

2. **Scientific Thinking** The Chicago Botanic Garden has grouped its plants in 23 different display gardens. What reasons might there be for displaying the plants in separate groups instead of in one large display garden?

How do a swan's webbed feet help it survive in its environment?

Illinois Projects and Activities

Project: Needs of Plants

Materials
- 1 marigold plant
- 1 aloe plant
- 2 paper cups
- potting soil
- water

Procedure
1. Gently shake the soil from the roots of each plant. Pour a layer of soil into each paper cup. Put a plant in each cup. Add soil until the roots are covered. Give each plant the same amount of water. Put the cups near a sunny window.

2. Do not water the plants again. Observe the plants every other day for two weeks.

3. Record any changes you observe.

Draw Conclusions
1. Which plant looks healthier after two weeks? How would you explain this?

2. What type of environment do you think would be good for each plant?

Living Things in Our World

UNIT A — LIFE SCIENCE

CHAPTER 1
Types of Living Things..........48

CHAPTER 2
Types of Plants.................74

CHAPTER 3
Types of Animals..............110

Unit Inquiry

Body Coverings

Living things respond to the cold in different ways. You might put on a jacket, a hat, and gloves to stay warm. How do wild animals stay warm in cold weather? Does hair help keep a mammal warm? Plan and conduct an experiment to find out.

CHAPTER 1
Types of Living Things

What's the Big Idea?

Cells are the building blocks of all the many types of living things.

Essential Questions

Lesson 1

What Are Some Types of Living Things?

Lesson 2

How Do Living Things Grow and Change?

Student eBook
www.hspscience.com

What do you wonder?

This sloth is a kind of living thing. Sloths spend most of their lives hanging upside down in trees. How is a sloth different from other living things? How does this relate to the **Big Idea?**

Three-toed sloth

LESSON 1

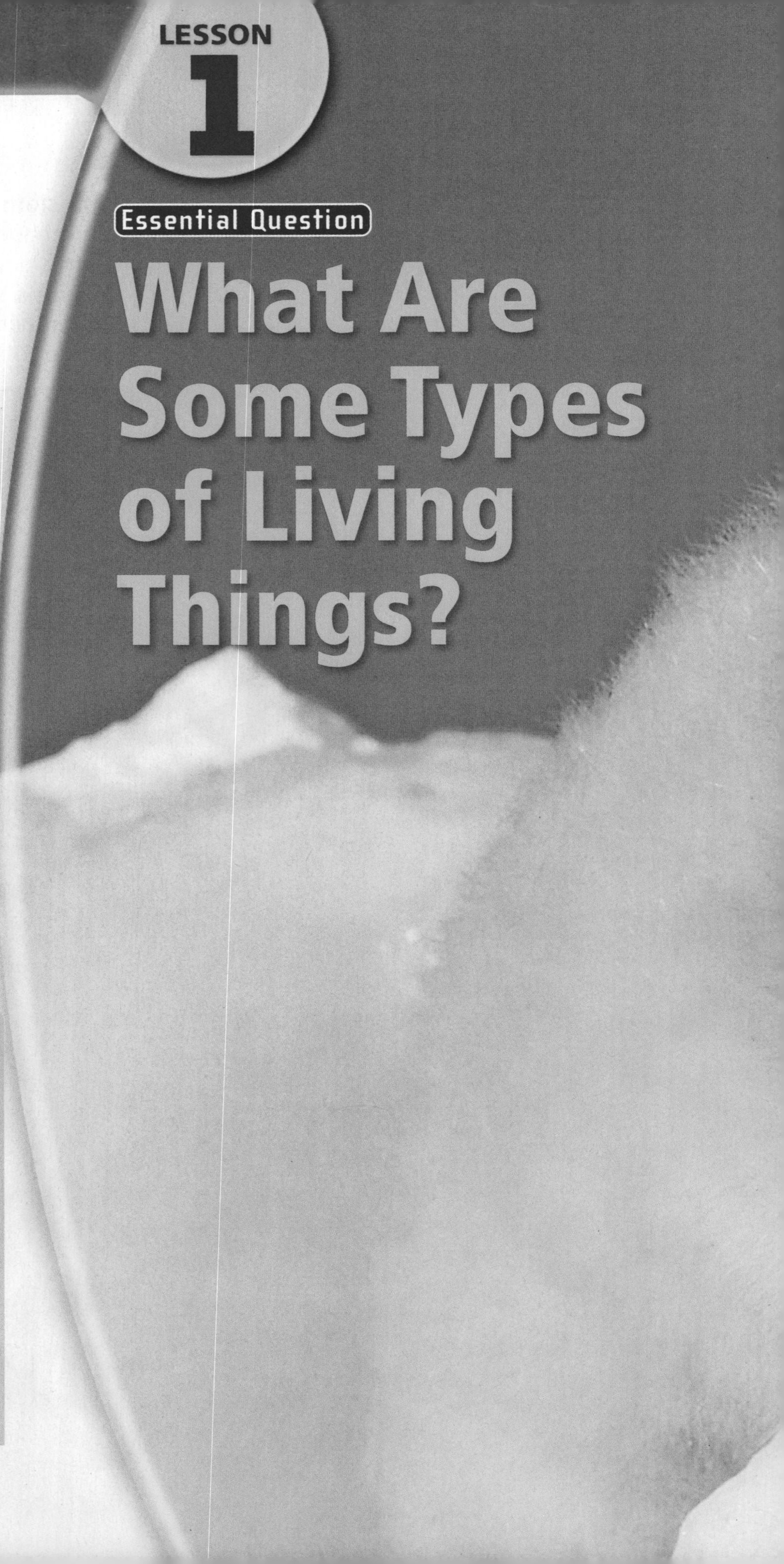

Investigate to find out how living and nonliving things are different.

Read and Learn about parts of living things and how to observe them.

Essential Question

What Are Some Types of Living Things?

Fast Fact

Home Sweet Home
Most living things have a home. Baby seals begin their lives on rocks near the ocean. When they grow larger, they live in the ocean, where they can catch food. In the Investigate, you'll learn why animals live where they do.

Vocabulary Preview

organism
[AWR•guh•niz•uhm] Any living thing (p. 54)

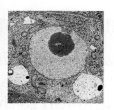

cell [SEL] A tiny building block that makes up every part of an organism (p. 56)

Young harp seal

Investigate

Homes for Living Things

Guided Inquiry

Start with Questions

These birds are living things. The nest is a home for the young bird until it's old enough to live on its own.

- How is it like your home?
- Do nonliving things need homes?
- Do all living things need homes?

Investigate to find out. Then read and learn to find out more.

Prepare to Investigate

Inquiry Skill Tip

When you make an inference, you are making a logical statement about what you have observed. You can test your inference by using it to finish the sentence, "It makes sense that...."

Materials
- picture sorting cards
- crayons or colored pencils
- paper

Make an Observation Chart

What makes the animal a living thing?	Where does the animal live?

Follow This Procedure

1. Look at the picture sorting cards your teacher has given you. Sort the cards into two piles—one for living things and one for things that are not living.

2. Choose an animal from the stack of cards that show living things.

3. Describe what makes the animal a living thing. Then describe where the animal lives. **Record** your **observations**.

4. Suppose you work at a zoo. Use crayons or colored pencils to draw a habitat for your animal. Draw everything your animal might need.

5. **Compare** the home for your animal with the homes that your classmates drew.

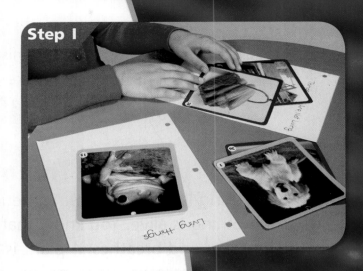

Draw Conclusions

1. What did you **observe** about the home where your animal lives?

2. **Inquiry Skill** **Infer** what an animal gets from its home. Why does that animal live where it does?

Independent Inquiry

Look at the cards. **Compare** the body coverings of several animals. **Classify** the animals by the types of body coverings they have.

Read and Learn

VOCABULARY
organism p. 54
cell p. 56

SCIENCE CONCEPTS
▶ what an organism is
▶ how cells make up organisms

COMPARE AND CONTRAST
Compare living and nonliving things.

alike — different

Living and Nonliving Things

Any living thing is an **organism**. Sometimes it is hard to tell if something is alive. You know that a cat is alive because it moves and breathes. A plant is also alive, but it doesn't move and breathe the same way a cat does.

There are many differences between living things and nonliving things. One difference is that living things reproduce. When a living thing reproduces, it makes more of its kind. A cat reproduces when it has kittens. Many plants reproduce by making seeds. New plants grow from the seeds.

The rooster in this picture isn't alive. It cannot reproduce.

Living things react to changes around them. When a mouse sees a cat, it reacts by running away. Many plants react to cold weather in the fall by losing their leaves. How do you react to cold weather?

All living things use energy to grow. Animals eat food to get the energy they need to stay healthy. Plants use energy from the sun to make their own food.

 **COMPARE AND CONTRAST** **What are three differences between living and nonliving things?**

This rooster is alive. It needs energy to grow. It reacts to changes that happen around it. This rooster can reproduce.

Parts of Living Things

Your body has many parts. Inside, you have a heart and a brain. Outside, you have a nose and ears. All living things have parts. A tiger has claws and a tail. A butterfly has wings. A tree has leaves, stems, and roots.

The parts of living things are made of cells. A **cell** is the tiny building block of an organism. Many cells make up each part of an organism. The cells of each part of an organism, such as a plant, do a job for the organism. Leaf cells make food for the plant, and root cells take in water from the soil. All the cells in an organism work together to help the organism survive.

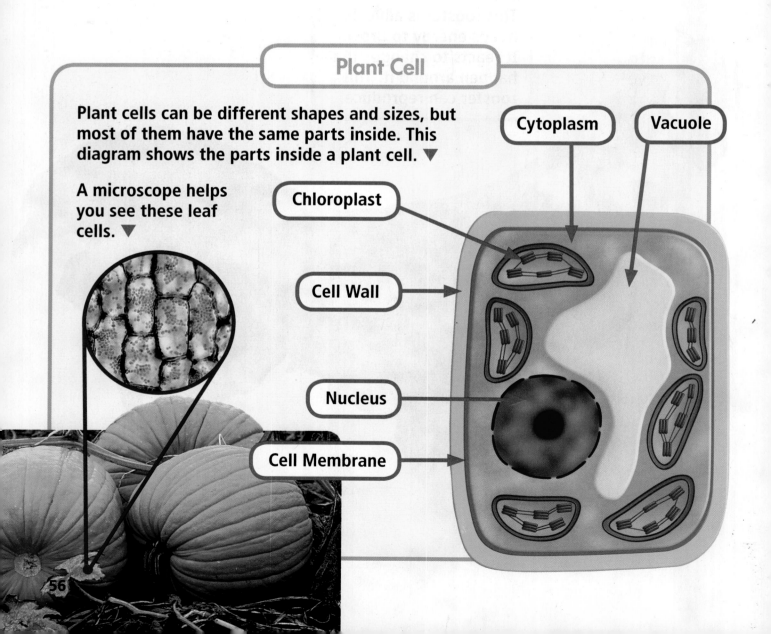

Plant Cell

Plant cells can be different shapes and sizes, but most of them have the same parts inside. This diagram shows the parts inside a plant cell. ▼

A microscope helps you see these leaf cells. ▼

- Cytoplasm
- Vacuole
- Chloroplast
- Cell Wall
- Nucleus
- Cell Membrane

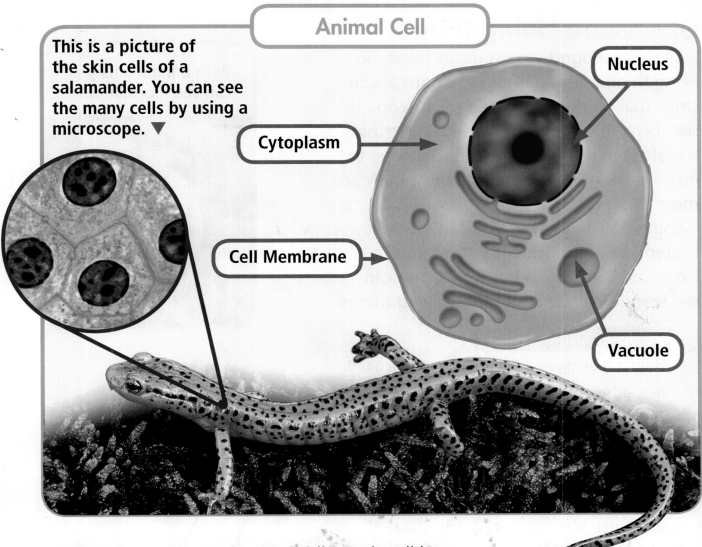

Animal Cell

This is a picture of the skin cells of a salamander. You can see the many cells by using a microscope. ▼

- Cytoplasm
- Cell Membrane
- Nucleus
- Vacuole

Animals are also made up of cells. Each cell has a job to do. An animal's skin cells protect it from the outside. Muscle cells join together to make muscles. These help an animal move.

Animal cells are different from plant cells. An animal cell doesn't have a cell wall. It doesn't have chloroplasts.

Most cells in organisms are very small. They are too small to see with just your eyes. You need to use a microscope to see cells.

COMPARE AND CONTRAST

How are plant cells different from animal cells?

Observing Living Things

Life is all around you. You can see some living things easily. Other organisms are so small that you need to use a microscope to see them. You might have seen a lake that has algae in it. If you used a microscope to observe the algae, you would most likely see more living organisms.

Using only your eyes, you can see butterfly wings, cricket legs, and flowers. However, if you use a hand lens, you can see many small details of these living things.

 COMPARE AND CONTRAST

How is looking at a living thing with a hand lens different from not using the lens?

With a hand lens you can see more details.

You can use binoculars to see the alligator from a safe distance.

What Is a Feather?

Use a hand lens to examine a feather. Describe what you see. Make a drawing of the feather. Look at a different kind of feather. How would you compare the two feathers?

Lesson Review

Essential Question

What are some types of living things?

In this lesson, you learned that living things have cells that help the organism survive. Animal and plant cells are different.

1. **COMPARE AND CONTRAST** Draw and complete a graphic organizer to compare organisms and nonliving things.

2. **SUMMARIZE** Write a three-sentence summary of this lesson. Tell how living and nonliving things are different.

3. **DRAW CONCLUSIONS** Why does a scientist often use a microscope or hand lens when studying living things?

4. **VOCABULARY** Write one or two sentences that tell how an organism and a cell are related.

Test Prep

5. Which part does a plant cell have that an animal cell does **not** have?
   A. cell membrane
   B. cell wall
   C. nucleus
   D. cytoplasm

Make Connections

 Writing

Expository
Plant cells and animal cells are alike, but they are also different. Write two paragraphs that **compare and contrast** plant cells and animal cells.

 Math

Ordering Numbers
Choose four different types of living things. Find the average life span of each organism. Order the life spans from shortest to longest.

LESSON 2

[Essential Question]

How Do Living Things Grow and Change?

Investigate to find out how quickly different seeds grow.

Read and Learn about how living things grow and change.

Fast Fact

Fast Starters
Young plants in a forest clearing must grow quickly. If they don't grow fast enough, other plants will block the sunlight. In the Investigate, you'll see how fast some seeds grow.

Young evergreen

Vocabulary Preview

life cycle [LYF SY•kuhl] The changes that happen to an organism during its life (p. 64)

metamorphosis [met•uh•MAWR•fuh•sis] A series of changes in appearance that some organisms go through (p. 66)

larva [LAHR•vuh] The stage of complete metamorphosis after an organism hatches from its egg (p. 67)

pupa [PYOO•puh] The stage of complete metamorphosis where an organism is wrapped in a cocoon or a chrysalis (p. 67)

inherit [in•HAIR•it] To have a trait passed on from parents (p. 68)

Investigate

How Fast Do Seeds Grow?

Guided Inquiry

Start with Questions

These cherries are growing on a cherry tree. The cherries have seeds inside them.

- If these seeds grow, what will they become?
- How long do you think it will take them to grow?
- How fast do other types of seeds grow?

Investigate to find out. Then read and learn to find out more.

Prepare to Investigate

Inquiry Skill Tip

To make an inference, make a list of all your data. Then write a sentence to explain what your data show.

Materials

- paper towels
- tape
- 3 small zip-top bags
- 3 kinds of seeds
- water
- ruler

Make a Data Table

Seeds	Day1	Day2	Day3	Day4	Day5
Bag #1					
Bag #2					
Bag #3					

Follow This Procedure

1. Fold three paper towels to fit inside a plastic bag. Place one paper towel into each bag with a small amount of water. Be careful not to make the paper towel too wet.

2. Number the bags 1, 2, and 3. Put your name on each bag.

3. Place seeds of one kind in bag 1, another kind in bag 2, and the third kind in bag 3. Label the type of seed in each bag. Zip the bags shut, and place them near a sunny window.

4. For five days, use a ruler to **measure** how much each seed has grown each day. **Record** your **observations** in a table like the one shown.

5. Add water to the bags if you notice the paper towels drying out.

Draw Conclusions

1. What changes did you **observe** in each bag?
2. **Inquiry Skill** **Infer** what will happen if you leave the seeds near the window longer than five days.

Independent Inquiry

Will seeds sprout without light? **Predict** what would happen if you placed new bags of seeds in a dark closet instead of near a sunny window. Try it.

63

Read and Learn

VOCABULARY
life cycle p. 64
metamorphosis p. 66
larva p. 67
pupa p. 67
inherit p. 68

SCIENCE CONCEPTS
▶ how living things grow and change over time

 **SEQUENCE**
Look for the sequence in which living things change.

Plants Grow and Change

Organisms grow and change during their lives. In the Investigate, you saw seeds grow into young plants. The changes that happen to an organism during its life make up its **life cycle**. For example, the life cycle of a bean starts with a seed. Then the organism grows. An adult plant may reproduce. In time, it dies.

The picture below shows how bean seeds begin to grow. Bean plants grow from seeds into tall vines that have flowers. They make pods containing seeds that can grow into new bean plants. A new life cycle begins when one of these new seeds begins to grow.

This garden is full of plants.

Plant Life Cycle

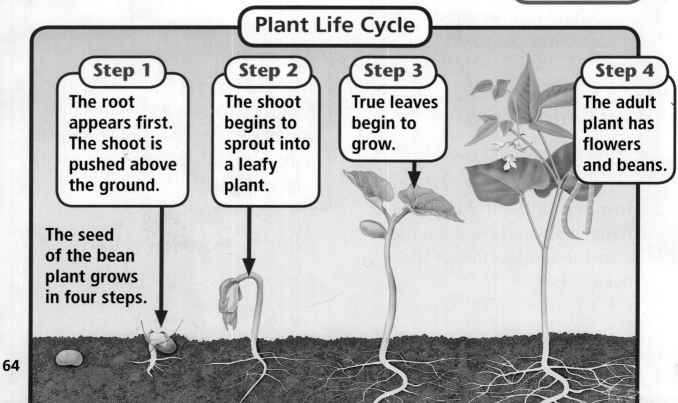

Step 1 The root appears first. The shoot is pushed above the ground.

Step 2 The shoot begins to sprout into a leafy plant.

Step 3 True leaves begin to grow.

Step 4 The adult plant has flowers and beans.

The seed of the bean plant grows in four steps.

64

Plants have seeds of different shapes and sizes. But not all plants begin their life cycles as seeds. Tulips and daffodils grow from underground stems called bulbs. The year before, the bulbs stored water and food for the following year's young plants. Tubers are another type of underground stem. A white potato is an example of a tuber.

With some plants, a new plant can grow from a part of the plant that breaks off. If you put a piece of a leaf's stem into moist soil, it may grow roots. No matter how plants begin their life cycles, the young plants will be the same kind of plants as their parents.

SEQUENCE How does a plant grow from a seed?

Math in Science
Interpret Data

Seed Lengths
Seeds have many different sizes and shapes. Some are very large. Some are very small.

Seed	Length (mm)
Apple	7 mm
Corn	10 mm
Butter bean	18 mm
Avocado	38 mm
Coconut	300 mm

The rings of a tree trunk tell how old the tree was. One ring forms each year as the tree grows. How many rings can you count in this trunk?

Plants continue to grow throughout their lives. The giant sequoia can grow to be 91 meters (300 ft) tall.

Animals Grow and Change

Like plants, animals grow and change during their *life spans*, or the time from birth to death. You have grown and changed since you were born. You will continue to grow and change.

The way you look changes as you become older. For some animals, this series of changes in appearance is called **metamorphosis** (met•uh•MAWR•fuh•sis).

For most animals, the life cycle starts with an egg. After the animal is hatched or born, it grows and develops. It becomes an adult and reproduces. In time, it dies.

SEQUENCE What events happen in the life cycle of an animal?

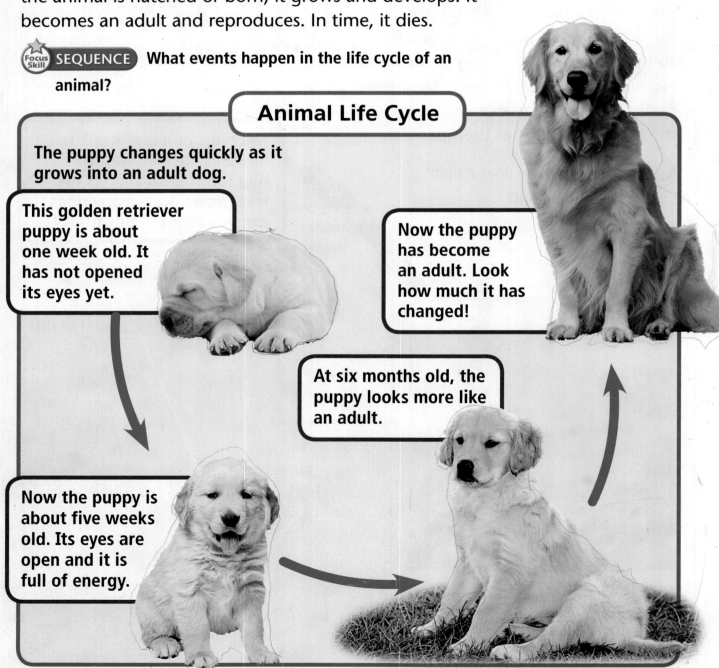

Animal Life Cycle

The puppy changes quickly as it grows into an adult dog.

This golden retriever puppy is about one week old. It has not opened its eyes yet.

Now the puppy is about five weeks old. Its eyes are open and it is full of energy.

At six months old, the puppy looks more like an adult.

Now the puppy has become an adult. Look how much it has changed!

Science Up Close

For more links and animations, go to www.hspscience.com

Metamorphosis

Like other animals, insects change during their life cycle. There are two kinds of metamorphosis insects can go through. Some insects change their entire appearance. Others don't.

Incomplete Metamorphosis

A grasshopper goes through incomplete metamorphosis. The insect looks about the same during its life cycle.

1. When a grasshopper hatches from an egg, it looks like a tiny adult grasshopper.

2. In each step, the grasshopper gets slightly larger.

3. The adult grasshopper has wings and can reproduce. The life cycle can begin again.

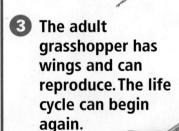

Complete Metamorphosis

Butterflies and moths go through complete metamorphosis. In this kind of metamorphosis, the insect changes the way it looks during its life cycle.

1. This life cycle begins with a tiny egg laid on a leaf.

2. The egg hatches into a caterpillar. In this stage of its life cycle, the organism is a **larva**. It eats leaves.

3. Next the caterpillar becomes a **pupa**. The pupa is wrapped in a chrysalis. The pupa doesn't eat or move.

4. An adult insect comes out of the chrysalis. The butterfly looks nothing like the larva. The adult butterfly finds a mate and lays eggs. The life cycle begins again.

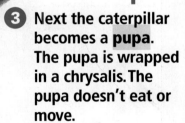

Heredity

Do you know someone who looks like his or her mom or dad? They share *traits,* or features. You **inherited** some traits from your parents. This means that your parents have passed on the traits to you. All organisms pass on traits. Red roses make more red roses when they reproduce.

Some traits aren't inherited. An organism learns them during its life cycle. You learned how to talk. Birds must learn to fly.

Explain how roses get their colors.

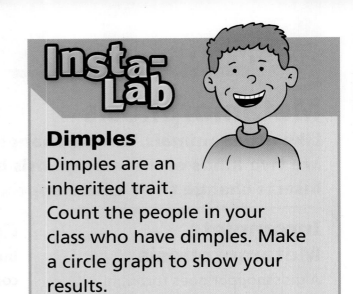

Dimples
Dimples are an inherited trait. Count the people in your class who have dimples. Make a circle graph to show your results.

▼ This father horse looks different from the mother horse.

◀ Notice this mother horse's markings and coloring.

What traits has this foal inherited from its parents?

Lesson Review

Essential Question

How do living things grow and change?

In this lesson, you learned how plants and animals grow and change. Some plants grow from seeds. Some insects go through metamorphosis. Organisms inherit traits from their parents.

1. **SEQUENCE** Draw and complete a graphic organizer that shows the steps of complete metamorphosis in sequence.

2. **SUMMARIZE** Write a summary of this lesson. Tell about the stages of a plant's life cycle.

3. **DRAW CONCLUSIONS** You find an insect that is in a chrysalis. What kind of metamorphosis is this?

4. **VOCABULARY** Write a paragraph about the life of an organism. Use the words *life cycle*, *inherit*, and *metamorphosis*.

Test Prep

5. **Critical Thinking** You and a friend are walking through a forest. You find dozens of trees covered with chrysalises. How will this forest change in a couple of weeks?

Make Connections

 Writing

Expository
Write an essay **comparing** the similarities and differences between plants and their offspring.

 Art

Make a Model
Make a model of an insect that goes through incomplete metamorphosis. Include all stages of the life cycle.

People in Science

Akira Okubo

As he was growing up in Japan, a country surrounded by water, Akira Okubo became interested in the ocean. His interest led him to study oceanography.

Questions that Okubo has studied include why and how fish live in schools. Fish gather in schools for protection. A school may break up at night to feed, but the fish gather again the next morning. A school may have as few as two dozen fish or as many as several million. All the fish in a school are about the same size. Adult fish and young fish are never in the same school. Some fish form schools when they are young and stay together all their lives. Other species of fish form schools for only a few weeks after hatching.

Okubo has also studied plankton—tiny animal-like and plant-like living things that float near the water's surface. Most living things that make up plankton are so small that they can be seen only with a microscope. Plankton is food for many other living things in the sea. Animal-like kinds of plankton eat plant-like kinds. Many plankton are eaten by fish. Some whales eat nothing but tons of plankton!

▶ AKIRA OKUBO
▶ Oceanographer

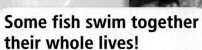
Some fish swim together their whole lives!

 Think and Write
1. Why is plankton important?
2. What is the advantage for fish of traveling in schools?

Charles Turner

▶ **CHARLES HENRY TURNER**
▶ Entomologist

Have you ever wondered if insects can hear? If so, you're not the only one. Charles Henry Turner studied insects for many years. He studied many kinds of insects, including ants, honeybees, wasps, and cockroaches. In addition to insects, Charles Henry Turner also studied spiders.

Charles Turner became the first scientist to prove that insects can hear. He also proved that honeybees can see colors and patterns.

 Think and Write
1. Why would it be helpful for honeybees to be able to see colors and patterns?
2. Which kind of insect would you like to study? Why?

Career Agricultural Extension Agent

Extension agents often work for state or local governments. These people offer help and information to farmers and gardeners. They often travel to different areas to teach people to use new technology in agriculture.

Chapter 1 Review and Test Preparation

Vocabulary Review

Use the terms below to complete the sentences. The page numbers tell where to look in the chapter if you need help.

organism p. 54

cell p. 56

life cycle p. 64

metamorphosis p. 66

larva p. 67

inherit p. 68

1. The changes that take place during an organism's life make up its _____.

2. To _____ a trait is to get it from your parents.

3. A living thing is an _____.

4. A caterpillar can also be called a _____.

5. A _____ is one of many small parts that make up an organism.

6. A _____ is a change in the way an organism looks as it grows and develops.

Check Understanding

7. Which of these can a nonliving thing do?
   A. breathe
   B. be moved
   C. reproduce
   D. respond

8. What kind of cell does the picture show?
   F. insect
   G. plant
   H. puppy
   J. rooster

9. You find a caterpillar. Which stage comes next?
   A. adult
   B. egg
   C. larva
   D. pupa

10. What is the correct order for complete metamorphosis?
    F. adult→pupa→egg→larva
    G. egg→larva→pupa→adult
    H. larva→egg→adult→pupa
    J. pupa→egg→adult→larva

11. What is Part C in the picture?
   A. cell membrane
   B. cytoplasm
   C. nucleus
   D. vacuole

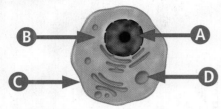

12. Which sentence about a cell is true?
   F. You can see a cell by using just your eyes.
   G. Each cell has a specific job.
   H. A plant cell doesn't have a cell wall.
   J. An animal cell has chloroplasts.

13. COMPARE AND CONTRAST What does a plant cell have that an animal cell doesn't have?
   A. cell wall
   B. cytoplasm
   C. nucleus
   D. vacuole

14. SEQUENCE Which sequence is correct for most animals?
   F. egg→reproduction→growth→death
   G. egg→growth→reproduction→death
   H. reproduction→egg→death→growth
   J. egg→death→growth→reproduction

15. What is the first stage of the life cycle for most animals?
   A. adult
   B. cocoon
   C. egg
   D. seed

16. Where do traits come from?
   F. They are only inherited.
   G. They are only learned.
   H. They are inherited or learned.
   J. They are found.

Inquiry Skills

17. You observe something that has wings and legs. It does not move or breathe. What can you infer about it?

18. A young horse has black hair. What can you infer about its parents?

Critical Thinking

19. How do you know a rock doesn't have muscle cells?

20. Why does the seed of a palm tree develop into a palm tree instead of an oak tree?

CHAPTER 2
Types of Plants

What's the Big Idea? Plants have parts that help them meet their needs.

Essential Questions

Lesson 1
What Do Plants Need to Live?

Lesson 2
What Are Some Types of Plants?

Lesson 3
How Do Plants Make Food?

Student eBook
www.hspscience.com

What do you wonder?

When people buy cut flowers like these, they often put the stems in water to help the flowers live longer. How do the stems of these flowers help the flowers live longer? How does this relate to the **Big Idea?**

LESSON 1

Investigate to find out what plants need.

Read and Learn about what plants need and how their parts help them meet those needs.

Essential Question

What Do Plants Need to Live?

Fast Fact

Colorful Tulips
Tulip comes from the Turkish word for a turban, which is a kind of hat. The hat and the flower are shaped very much alike. In the Investigate you will discover a need of most plants.

Tulip field

Vocabulary Preview

root [ROOT] The part of a plant that grows underground and takes water and nutrients from the soil (p. 82)

nutrients [noo•TREE•uhnts] The parts of the soil that help plants grow and stay healthy (p. 82)

stem [STEM] The part of a plant that grows above ground and helps hold the plant up. It also carries water and nutrients up to the leaves (p. 82)

leaf [LEEF] The part of a plant that grows out of the stem and is where a plant makes food (p. 82)

77

Investigate

Needs of Plants

Guided Inquiry

Start with Questions

Plants need water to grow and survive. This ghost plant stores water in its thick leaves.

- How would its leaves help this plant during dry weather?
- What else does this plant need?

Investigate to find out. Then read and learn to find out more.

Prepare to Investigate

Inquiry Skill Tip

You can make it easier to compare three objects by comparing them in pairs first. Compare 1 and 2, 1 and 3, and then 2 and 3.

Materials

- 3 small plants
- sand
- gravel
- 3 cups
- potting soil
- water

Make an Observation Chart

Day	Plant 1 (Sand)	Plant 2 (Gravel)	Plant 3 (Soil)
2			
4			
6			
8			
10			
12			
14			

Follow This Procedure

1. Gently shake the soil from each plant. Plant one in a cup of sand. Plant another in a cup of gravel. Plant the third in a cup of soil. Water each plant. Put the cups in a sunny window.

2. **Observe** the plants every other day for two weeks. Water the plants every few days with the same amount of water.

3. **Record** any changes you **observe**. Be sure to look for changes in plant size.

Draw Conclusions

1. Which plant looked the healthiest after two weeks? Explain why.

2. Which plant looked the least healthy after two weeks? What was different for this plant?

3. **Inquiry Skill** Scientists often **compare** the results they get in their experiments. How could you **compare** your findings?

Step 1

Step 2

Independent Inquiry

Predict how different amounts of water might affect the growth of plants. Try it!

Read and Learn

VOCABULARY
root p. 82
nutrients p. 82
stem p. 82
leaf p. 82

SCIENCE CONCEPTS
▶ what plants need to live
▶ how plants get what they need to live

MAIN IDEA AND DETAILS
Look for details about the things plants need to live.

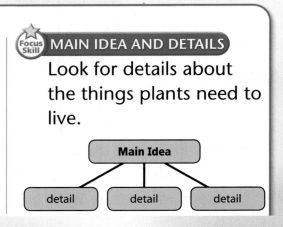

What Plants Need

Plants live and grow in many places. They can grow in a city and in the country. They grow in forests, in fields, and in parks. Some even grow underwater or in deserts! No matter where they grow, most plants need the same things to live.

Every living thing has basic needs. Organisms cannot survive unless those needs are met. Plants need air, water, light, and nutrients. Most plants get nutrients from the soil, but some plants can get them from the air. How are your basic needs the same as and different from a plant's?

◀ This girl is making sure her plants get air, light, water, and soil.

Think about places where plants don't grow. Plants don't grow deep in caves, because there is no light. Most plants don't grow on bare rock, because there is no soil. Why do you think there are more plants in a forest than in a desert?

Many people like to garden. They care for plants by giving them water and good soil. However, most plants grow without human care. They get what they need from the sun, the air, the rain, and the soil.

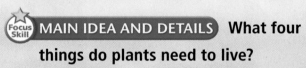 MAIN IDEA AND DETAILS What four things do plants need to live?

Make a Model
Observe how stems carry water to a plant's leaves. Put red food coloring into a cup of water. Add a stalk of celery that has leaves. Let it sit overnight. Measure how far the color traveled up the stalk. Did the color reach the leaves?

Some Plant Parts and What They Do

Compare the ivy on this page to the flower on the page on the right. The plants look very different, but they have the same parts.

The **roots** of most plants grow under the ground. They take in water and nutrients from the soil. **Nutrients** (NOO•tree•uhnts) help plants grow and stay healthy.

The **stem** grows above the ground and holds the plant up. It carries water and nutrients to the leaves. The stem also takes food from the leaves to the roots.

A **leaf** grows out from the stem. The leaves are where the plant makes food. Sunlight, air, and water are used inside a leaf to make the food the plant needs to live and grow.

MAIN IDEA AND DETAILS How does water from the soil travel to the leaves of a plant?

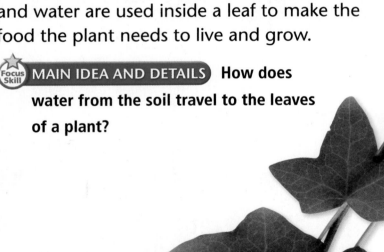

◀ Like other plants, ivy has roots, stems, and leaves. What does each of these parts do?

Plant Parts

The stem of the plant helps hold up the leaves. It also carries water and nutrients from the roots to the leaves.

Most plant leaves are green. The green coloring helps leaves collect sunlight. The coloring also helps the plant make its own food so it can live and grow. Some plants are food for animals, too.

The roots of the plant take in water and nutrients from the soil. They also help hold the plant in the ground.

Plants Live in Different Environments

Plants are found almost everywhere on Earth. They even grow in ponds and in deserts. In these places, it can be hard for plants to get what they need.

For example, the desert, where cactus plants grow, is very dry. You might wonder how a cactus gets water. It has a special stem and roots for getting the water it needs. A waterlily grows underwater. The roots gather nutrients from the soil and water. A long stem leads to a floating leaf that absorbs sunlight.

MAIN IDEA AND DETAILS How do plants get what they need from their environments?

Cactus
The roots of the cactus are close to the surface of the soil. They quickly take in rainwater before it dries up. The cactus's thick stem stores the water.

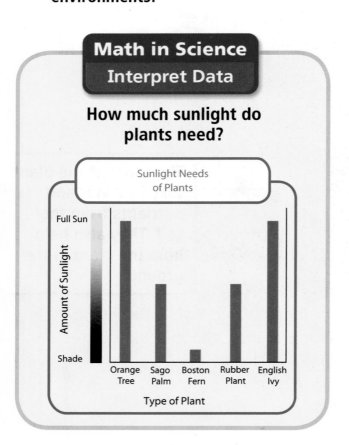

Math in Science
Interpret Data

How much sunlight do plants need?

Sunlight Needs of Plants

(Amount of Sunlight: Full Sun to Shade)

Type of Plant: Orange Tree, Sago Palm, Boston Fern, Rubber Plant, English Ivy

Paperwhites
This plant grows from a bulb. Until its roots grow into the ground, the bulb gives the plant the food it needs.

Lesson Review

Essential Question

What do plants need to live?

In this lesson, you learned that plants need nutrients, water, air, and light to live.

1. **MAIN IDEA AND DETAILS** Draw and complete a graphic organizer for this main idea: Plant parts help them meet their needs.

   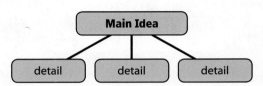

2. **SUMMARIZE** Write two sentences that summarize this lesson. Tell about three different parts of plants and how they help plants.

3. **DRAW CONCLUSIONS** Ryan notices that the lettuce plants in the garden are drooping. The soil is dry, and the sun is hot. What should Ryan do? How do you know?

4. **VOCABULARY** Write a paragraph explaining how roots and nutrients are related.

Test Prep

5. Which is **not** something most plants need to live and grow?
   A. nutrients
   B. shelter
   C. oxygen
   D. water

Make Connections

 Writing

Expository

Your neighbors want to plant a garden. Write a **friendly letter** to help them choose the best spot to put the garden. Be sure to tell them about the basic needs of plants.

 Art

State Symbols

Every state has a state tree and a state flower. What are your state's tree and flower? Draw or paint a picture of each.

85

LESSON 2

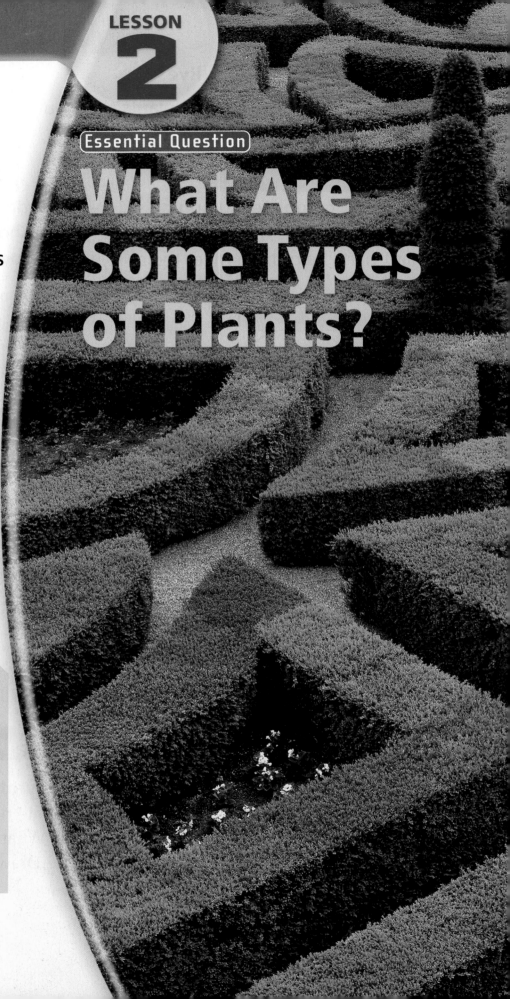

Investigate to find out what seeds need to begin growing.

Read and Learn about different types of plants.

Essential Question

What Are Some Types of Plants?

Fast Fact

Amazing Mazes

Garden mazes have been popular for hundreds of years. The longest one contains about 16,000 trees. In the Investigate, you will grow and observe some plants.

Hedge maze

Vocabulary Preview

seed [SEED] The first stage of life for many plants (p. 91)

deciduous [dee•SIJ•oo•uhs] Relating to plants that lose all their leaves at about the same time every year (p. 93)

evergreen [EV•er•green] A plant that stays green and makes food all year long (p. 93)

Investigate

Growing Lima Beans

Guided Inquiry

Start with Questions

Have you ever been to a place where there weren't any plants? Some deserts have almost no plants.

- Why do you think there are no plants in this picture?
- Do you think a seed would grow if you planted it in this area of the desert?

Investigate to find out. Then read and learn to find out more.

Prepare to Investigate

Inquiry Skill Tip

You draw conclusions from observations and facts. After an investigation, make a list of what you know, and then use the list to make and support a conclusion.

Materials

- 3 lima bean seeds
- 2 zip-top plastic bags
- 2 paper towels
- hand lens
- water

Make an Observation Chart

Day	Wet Seed	Dry Seed
1		
2		
3		
4		
5		
6		
7		

Follow This Procedure

1. Split one lima bean seed in half, and use the hand lens to look inside. Identify the new plant inside the seed. After you have finished, put the seed to the side.

2. Fold each paper towel in half. Moisten one of them with water, but not too much. Don't moisten the other towel. Then place each towel inside a plastic bag.

3. Place one seed in each bag. Label the bag with the moistened towel WET, and label the other bag DRY. Seal the bags, and place them where they won't be disturbed.

4. **Observe** the seeds for 10 school days. **Record** your **observations**.

Draw Conclusions

1. How does a new bean plant grow?

2. **Inquiry Skill** Use what you have observed to **draw conclusions** about what bean seeds need to grow. Why do you think the seeds didn't need soil to start growing?

Step 1

Step 2

Independent Inquiry

Predict how seeds will grow at different temperatures. **Design and conduct an experiment** to see if seeds grow faster in warm or cold weather.

Read and Learn

VOCABULARY
seed p. 91
deciduous p. 93
evergreen p. 93

SCIENCE CONCEPTS
▶ how plants can be grouped

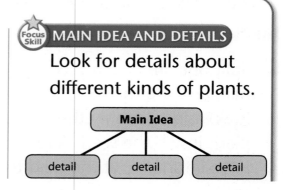

MAIN IDEA AND DETAILS
Look for details about different kinds of plants.

Trees, Shrubs, and Grasses

Suppose your teacher asked you to organize the classroom bookshelves. How would you group the books so that students could find what they needed? You might start by putting similar books together. All the science books could go together on one shelf, and all the math books could go on another.

◀ You would have to look closely to see the flowers of the oak tree. The small flowers grow into the tree's seed, the acorn.

There's no way to miss the flowers on the showy azalea. However, the shrub's seeds are tiny.

90

The flowers of these plants are hard to spot, but the seeds of one kind are easy to see. These palm trees produce the world's largest seeds—coconuts!

Just as you had a way to arrange the books, scientists have a way to arrange plants. They group similar plants. For example, plants that have seeds are put into two groups. One group of plants with seeds has flowers, and the other does not. A **seed** is the earliest stage of life for many plants.

The trees, shrubs, and grasses shown on these pages look very different. They may even grow in different places. Yet the trees, shrubs, and grasses are alike in two ways. They all have seeds and they all have flowers.

 **MAIN IDEA AND DETAILS**

How is grouping plants helpful to scientists?

Types of Leaves

You already know that leaves help plants get the light and air they need to make food. Scientists group some plants by the kind of leaves the plants have.

Leaves have many sizes and shapes. Many leaves, such as the leaves of maple and oak trees, are large and flat. Others, like those from the pine tree, are long and narrow. A leaf may have smooth edges, or it may be jagged. The way a leaf looks can help you identify what plant it came from.

Pine

The long, thin leaves of a pine tree look like needles. Pine trees are grouped as evergreens.

Maple

A maple leaf is broad and flat, with five lobes that have small notches. A maple tree loses its leaves each fall.

Oak

Most oaks are deciduous trees with flat leaves. In the winter, most oak trees have no leaves.

Fern

The fern is a plant with many fronds—the leafy branches are made up of smaller leaflets. ▶

All plants lose leaves and grow new ones. Scientists group plants based on how they lose their leaves.

Deciduous (dee•sɪj•oo•uhs) plants lose their leaves each year. This usually happens in the fall, when the days get shorter and cooler. During the cold winter months, deciduous plants do not make food.

Evergreen plants stay green and make food all year long. They lose leaves or needles a few at a time, not all at once.

How can leaves be used to group plants?

Magnolia
The flat leaves of the magnolia tree are different in shape from pine needles, yet both are evergreen trees.

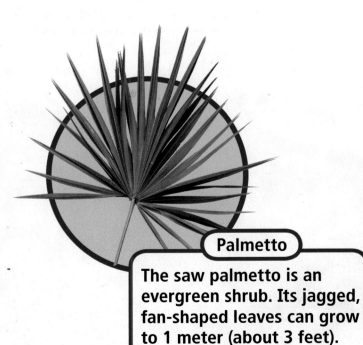

Palmetto
The saw palmetto is an evergreen shrub. Its jagged, fan-shaped leaves can grow to 1 meter (about 3 feet).

Compare Leaves
Look at the leaves on these pages. Make a list of each leaf's features. Which features are alike? Which are different? Group the leaves by their features.

Plants with Flowers

Think about a garden overflowing with beautiful flowers. Butterflies and bees go from flower to flower. Scientists often group plants by whether or not they have flowers.

Some flowers are large and very colorful. Others are small and hard to see. Not all plants even have flowers. In plants that do, the flowers have an important job. Special parts produce seeds.

MAIN IDEA AND DETAILS

What role do flowers play in a plant's life cycle?

Cherry Tree
Flowering cherry trees can be grouped as deciduous, as forming seeds, and as having flowers.

Lilies
Like other plants with flowers, day lilies make seeds.

Phlox
The flowers of these phlox (FLAHKS) plants produce seeds.

Plants Without Flowers

Some plants do not form flowers. They make seeds in their other parts. For example, many evergreen plants make their seeds inside cones. These evergreens are called conifers (KAHN•uh•ferz).

Other plants, such as ferns, make spores (SPORZ). Spores are usually brown and can be seen on the underside of a leaf. These spores grow into new fern plants.

MAIN IDEA AND DETAILS

How do plants without flowers form new plants?

Spruce Tree
This evergreen tree forms seeds in its cones.

Dwarf Juniper
The dwarf juniper is an evergreen plant that makes seeds in its cones.

Boston Fern
This Boston fern has spores on the undersides of its fronds.

Seeds

You've probably seen seeds inside apples or other fruits. These seeds are important because they contain the food to help a new plant grow. Like an apple seed, all seeds look very different from the plants they become.

Sometimes seeds need to spread to new places to grow. Some plants have seeds that stick to animals. The animals spread the seeds as they move about. Other plants have seeds that the wind can carry. Some seeds, such as coconuts, can float on water.

 MAIN IDEA AND DETAILS What does a seed do?

Peas
Each of these tasty peas is a seed that could grow into a new plant.

Sugar Maple
The sugar maple tree has seeds that can be blown to new places by the wind.

Oranges
The seeds of this orange can grow into orange trees.

Thistle
Downy fluff covers thistle seeds. The fluff helps the seeds float away in the wind.

96

Lesson Review

Essential Question

What are some types of plants?

In this lesson, you learned that types of plants can be classified in different ways.

1. **MAIN IDEA AND DETAILS** Draw and complete a graphic organizer for this main idea: Plants are classified in many different ways.

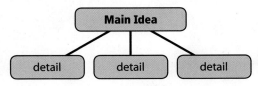

2. **SUMMARIZE** Write a four-sentence summary for this lesson. Tell about the different types of plants.

3. **DRAW CONCLUSIONS** You find a plant that has many small green leaves up and down each branch. There are brown spots under the leaves. What kind of plant might it be?

4. **VOCABULARY** Use the lesson vocabulary to write a sentence about trees.

Test Prep

5. **Critical Thinking** What can the shape, color, and size of a leaf help you know?

Make Connections

 Writing

Expository
Choose two kinds of plants that you have seen before. Write a paragraph **comparing and contrasting** the two plants.

 Language Arts

Generous Plants
Read *The Giving Tree*, by Shel Silverstein. Then write your own story about the things you get from plants.

LESSON 3

Essential Question

How Do Plants Make Food?

Investigate to find out if plants grow better with or without light.

Read and Learn about how plants make food, and how plants can be helpful and harmful.

Fast Fact

Shady Skyscrapers
The tall trees in a forest keep much of the sunlight from reaching the plants that grow on the ground. In the Investigate, you will see how important light is to a plant.

Redwood trees

Vocabulary Preview

photosynthesis [foht•oh•SIN•thuh•sis] The process that plants use to make sugar (p. 102)

chlorophyll [KLAWR•uh•fil] The green substance inside leaves that helps a plant use light energy to make food (p. 103)

Investigate

Lights, Plants, Action!

Guided Inquiry

Start with Questions

The plants below are sunflowers. They need energy to grow.

- Where do plants get energy to grow?
- How do you think sunlight is helpful to these sunflowers?

Investigate to find out. Then read and learn to find out more.

Prepare to Investigate

Inquiry Skill Tip

Organizing your observations makes it easier to compare. Make a table with a column for each object you want to compare. Make a row for each property you observe.

Materials
- 2 cups
- potting soil
- 2 small plants
- water

Make an Observation Chart

Day	Sunny	Dark

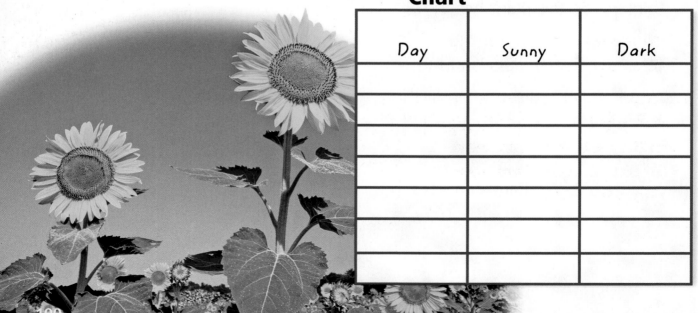

Follow This Procedure

1. Half-fill each cup with potting soil. Gently place a plant in each cup. Fill the cups with soil. Water each plant lightly. Use just enough water to make the soil damp.

2. Put one cup near a sunny window. Put the other cup in a dark place, such as a cabinet or closet.

3. **Observe** the cups for two weeks. Water the plants when necessary. **Record** the changes you observe. Make sure to look for changes in plant size and color.

Draw Conclusions

1. Which plant looked healthier? How did it look different from the other plant? What do you think made this plant healthy?

2. What was different for the plant that looked less healthy?

3. **Inquiry Skill** Scientists often compare the results they get in their experiments. How would you compare your findings?

Independent Inquiry

Predict how growing plants under different colored light might affect them. Now try it!

Read and Learn

VOCABULARY
photosynthesis p. 102
chlorophyll p. 103

SCIENCE CONCEPTS
▶ how plants make food

 **CAUSE AND EFFECT**
Look for effects of plants using water and light.

cause → effect

How Plants Make Food

Suppose you've had a long day at school and you still have a ball game ahead of you. You need energy, so you eat a healthful snack. To get energy, all animals need to eat food.

Plants are different from animals in that plants can make their own food. The food that plants make is sugar. Plants make sugar by a process called **photosynthesis** (foht•oh•SIN•thuh•sis). The plant uses the sugar it makes for energy.

 **CAUSE AND EFFECT** What is the result of photosynthesis?

These plants are part of a roof garden, where they make their own food. ▼

Science Up Close

For more links and animations, go to www.hspscience.com

Photosynthesis

During photosynthesis, water and carbon dioxide (dy•AHKS•yd) come together to make sugar. Sunlight provides the energy needed for this to happen. **Chlorophyll** (KLAWR•uh•fil) is a green substance inside leaves. It helps the plant use light energy.

Plants need light from the sun for photosynthesis. Chlorophyll helps plants use the sunlight.

Air is made up of different gases. One type of gas is called carbon dioxide. Plants need carbon dioxide to make sugar.

Oxygen is another kind of gas. It is made during photosynthesis. You need to breathe oxygen to stay alive. Plants need oxygen to use the energy in the sugar.

Water soaks into the soil, where the roots take it in. Stems carry the water to the leaves. Plants also need water to make sugar.

How Plants Are Helpful and Harmful

Plants are helpful to people in many ways. One way we use them is to eat them for food. We get energy and nutrients from eating certain parts of the plants. We get wood to make paper from trees, and we use other plants to make cloth and medicines.

However, some plants can be harmful. They may have poisons in their parts or give people allergy problems.

Focus Skill CAUSE AND EFFECT Why shouldn't you touch a plant that you have never seen before?

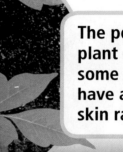

People get energy and nutrients from foods, such as strawberries.

Pollen from plants causes some people to sniffle, sneeze, and have itchy eyes.

The poison ivy plant can cause some people to have an itchy skin rash.

Grains feed many people around the world.

Insta-Lab

Make a Graph

Make a list of 20 foods that come from plants. Tell whether each food comes from roots, stems, leaves, flowers, fruits, or seeds. Make a bar graph from your list.

104

Lesson Review

Essential Question

How do plants make food?

In this lesson you learned that plants get energy from the sun through the process of photosynthesis.

1. 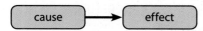 **CAUSE AND EFFECT** Draw and complete a graphic organizer that shows cause and effect for photosynthesis.

   cause → effect

2. **SUMMARIZE** Write a summary for this lesson. Tell about photosynthesis.

3. **DRAW CONCLUSIONS** How does photosynthesis help make the foods that people eat?

4. **VOCABULARY** Use lesson vocabulary to describe how the green color in leaves helps plants make food.

Test Prep

5. What do plants give off during photosynthesis?
   A. carbon dioxide
   B. oxygen
   C. sunlight
   D. water

Make Connections

 Writing

Expository
List the "ingredients" that are needed for photosynthesis. Then write a **how-to** "recipe" telling the ways the plant uses the ingredients to make its own food.

 Health

Natural Health
Use an encyclopedia to learn about a medicine that comes from a plant. Report to your class how the medicine helps people.

People in Science

Marie Clark Taylor

Have you ever thought about all the plants around you? Marie Clark Taylor did. She received her college degree in botany, the study of plants.

After she earned her degree, Marie Clark Taylor wanted to share what she had learned, so she taught college. She showed other science teachers how using plants in their science courses could help them. Since most plants are easy to obtain, many teachers could use them in their classes.

In 1956, during the Summer Institute of Botany for Teachers at Small Colleges, Marie Clark Taylor led a committee on educational problems. She helped find problems with textbooks, lab manuals and other teaching materials. One solution they found was to make a list of plant items they could share with each other throughout the year. Taylor later became the head of the Botany department at Howard University.

▶ **MARIE CLARK TAYLOR**
▶ Former Chair of Botany Department at Howard University

 Think and Write

1. Why might plant cells be easier to study than animal cells?
2. What kind of plant would you like to study? Why?

Rosa Ortiz

▶ **ROSA ORTIZ**
▶ Botanist

Rosa Ortiz is on the hunt for a plant. Ortiz is a botanist. A botanist is a scientist who studies plants. Ortiz travels to study one species, or kind, of plant. This plant is part of the moonseed family.

Moonseed is a vine found in many places. Some types of moonseed can be used as medicines. Ortiz is studying how the plant has changed over time.

Think and Write

1. Why might someone want to study only one species of plant?
2. Do you think it is important to know how plants have changed over time? Why or why not?

Career Nursery Worker

Nursery workers take care of plants. They know how much sun and water plants need. They also know how to protect plants from diseases, insects, and the weather.

Chapter 2 Review and Test Preparation

Vocabulary Review

Use the terms below to complete the sentences. The page numbers tell you where to look in the chapter if you need help.

root p. 82
nutrient p. 82
stem p. 82
leaf p. 82
deciduous p. 93
evergreen p. 93
photosynthesis p. 102
chlorophyll p. 103

1. The green substance that helps plants make food is _____.
2. The plant part in which its food is made is the _____.
3. A plant that loses its leaves each fall is said to be _____.
4. A substance that helps plants grow and stay healthy is a _____.
5. An underground plant part that takes in water is the _____.
6. The process that plants use to make food is _____.
7. Plants that do not lose their leaves in the fall are _____.
8. The part that connects a plant's roots to its leaves is the _____.

Check Understanding

Write the letter of the best choice.

9. How are roots like stems?
    A. Both are green.
    B. Both move water.
    C. Both make food.
    D. Both grow under the ground.

10. **CAUSE AND EFFECT** How can plants cause harm?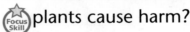
    F. We get food from them.
    G. We make cloth from them.
    H. They can make us have allergy problems.
    J. They can be used to make medicines.

11. Which are the four things that plants need to live and grow?
    A. water, soil, air, and light
    B. water, soil, air, and warmth
    C. water, soil, oxygen, and warmth
    D. water, soil, oxygen, and light

12. Which things do plants need to make food?
- **F.** soil, air, water
- **G.** water, sunlight, oxygen
- **H.** water, sunlight, carbon dioxide
- **J.** soil, water, carbon dioxide

13. How does a cactus survive in a very dry environment?
- **A.** Its roots are deep.
- **B.** It has no stem or leaves.
- **C.** Its shallow roots quickly collect rainfall.
- **D.** It doesn't need water to make food.

14. Which is a way by which scientists group plants?
- **F.** by root length
- **G.** by plant height
- **H.** by color
- **J.** by leaf type

15. MAIN IDEA AND DETAILS Which of the following describes deciduous plants?
- **A.** They make food in the winter.
- **B.** They stay green all year long.
- **C.** They lose their leaves each fall.
- **D.** They have leaves shaped like needles.

16. What do grasses, shrubs, and many trees share?
- **F.** size and shape
- **G.** need for oxygen
- **H.** losing leaves
- **J.** flowers and seeds

Inquiry Skills

17. Compare a cherry tree and a spruce tree.

18. Asnerlie went hiking in the woods. A few days later, she had an itchy skin rash on her arm. Draw a conclusion about what might have caused the rash.

Critical Thinking

19. As Samantha helps her mother plant flower bulbs, she asks why she isn't planting seeds. Do plants that grow from bulbs make seeds? Explain your answer.

20. Zach's younger brother Jake will not eat vegetables. Zach jokes that Jake still eats green plants. How could this be true?

CHAPTER 3
Types of Animals

Animals can be classified by their traits.

Essential Questions

Lesson 1
What Do Animals Need to Live?

Lesson 2
What Are Vertebrates?

Lesson 3
What Are Invertebrates?

Student eBook
www.hspscience.com

What do you wonder?

A flamingo has an interesting way of eating. Its beak is upside down when it collects food from the water. How can you tell what kind of animal a flamingo is? How does this relate to the **Big Idea?**

Lesser flamingos

LESSON 1

Investigate to find out how animal homes are alike and different.

Read and Learn what animals need to survive.

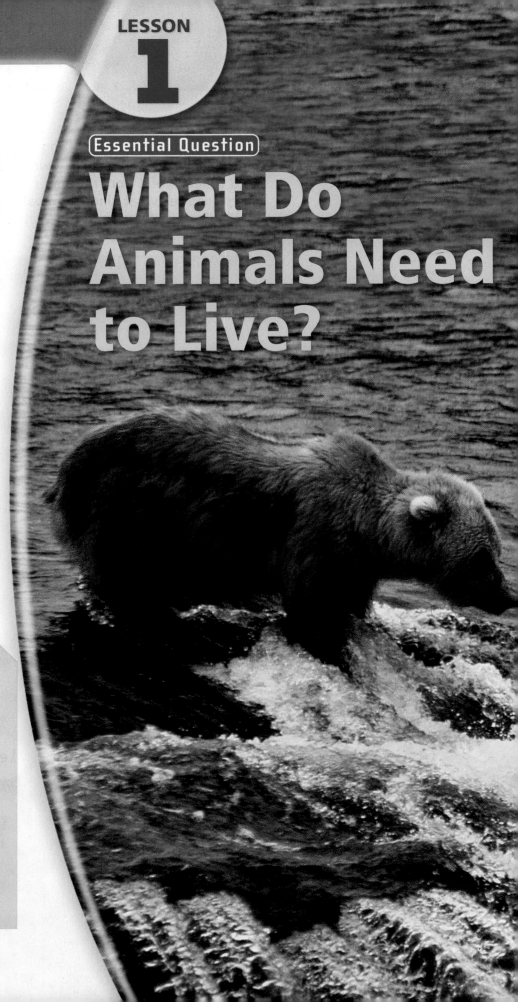

Essential Question
What Do Animals Need to Live?

Fast Fact

Big Bears
The Alaskan brown bear is one of the world's largest meat-eating animals. Bears eat so much that they can spend most of the winter in their homes, called dens. In the Investigate, you will compare some animals' homes.

Vocabulary Preview

oxygen [AHK•sih•juhn] A gas that all living things need and that plants give off into the air (p. 117)

Investigate

Animal Homes

Guided Inquiry

Start with Questions

People who keep animals as pets provide homes for them. Wild animals have to find their own homes.

- In what kinds of homes might wild animals live?
- How is the home for this dog different from a home for a wild animal?

Investigate to find out. Then read and learn to find out more.

Prepare to Investigate

Inquiry Skill Tip

When you compare objects, make a list of how they are alike. Make a separate list of how they are different. Then write a short summary of the similarities and differences.

Materials
- picture sorting cards
- index cards
- reference books
- markers

Make an Observation Chart

Animal	Home

Follow This Procedure

1. Look at the picture sorting cards your teacher gave you.
2. As you **observe** each picture, notice the animal's home.
3. With a partner, make a matching game. On an index card, write the name or draw a picture of an animal. Name or draw its home on another card. Use the picture sorting cards or reference books if you need help.
4. Play your matching game. As you match the animals to their homes, discuss the different types of homes. Talk about the ways the homes are alike and the ways they are different. Then **classify** the animals by the types of homes in which they live.

Draw Conclusions

1. Describe the homes of the foxes and the owl. How are the homes alike? How are they different?
2. **Inquiry Skill Compare** two of the animal homes you observed. Explain how each home protects the animal that lives there.

Independent Inquiry

Study the animal pictures again. **Draw conclusions** about why each animal uses the type of home it does.

Read and Learn

VOCABULARY
oxygen p. 117

SCIENCE CONCEPTS
▶ what animals need in order to live and grow

MAIN IDEA AND DETAILS
Look for details about what animals need in order to live.

```
        Main Idea
       /    |    \
   detail detail detail
```

Animals and Their Needs

Have you ever cared for a cat or a dog or had a hamster or a parakeet? These animals are good pets. They are fun to play with and to watch. Just like plants, pets have needs. They need food, water, and places to live. People give pets these things.

Wild animals have the same needs as pets. However, they must find their own food, water, and shelter.

MAIN IDEA AND DETAILS What do all animals need?

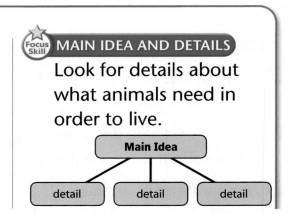

This elf owl finds shelter in a saguaro (suh•GWAR•oh) cactus. How is the owl different from a pet parakeet? ▼

◀ Pandas look cuddly, but they are not pets. They must find their own food, water, and shelter. They eat bamboo leaves.

116

Animals Need Oxygen

Animals and plants share many of the same needs. Both have structures to take in the air they need to live and grow. Animals need oxygen in the air. **Oxygen** is one of the gases that makes up air.

Many animals that live on land get oxygen by breathing air. Some animals get oxygen in other ways. Insects, for example, get oxygen through tiny holes in their bodies. Many water animals, such as fish, get oxygen from the water. Other water animals, like whales, rise to the surface of the water to breathe air.

▲ These hippos rise above the water's surface to breathe air.

 **MAIN IDEA AND DETAILS** Name two ways animals get oxygen.

Make a Model
Choose an animal shown in this lesson. Use clay, sticks, and other natural materials to make a model of that animal's home. What need does this home meet?

Animals Need Water

Just like plants, animals need water. Did you know that more than half of your body is water? Every day, water leaves your body in your breath, in your sweat, and in your urine. To be healthy, you must drink to replace lost water.

Look at the graph on this page. Just like people, elephants must replace water that leaves their bodies. It's hard to imagine drinking so much water. Most animals get the water they need by drinking it. Other animals get water mainly from the foods they eat.

 **MAIN IDEA AND DETAILS**

Name two ways animals get water.

Math in Science
Interpret Data

How many times as much water does an elephant need as a horse needs?

Daily Water Needs of Animals

(Bar graph showing Amount (gallons) on y-axis from 0 to 50, and Animal on x-axis: Elephant ≈ 50, Adult human ≈ 1, Horse ≈ 10)

▲ This dog gets the water it needs from the people who care for it.

On the plains of Africa, animals must find their own drinking water. ▼

Animals Need Food

Like plants, animals need food to live and grow. However, animals can't make their own food. They must get food by eating plants or other animals.

Animals have body parts that help them get food. For example, an elephant uses its trunk to pull leaves from trees. How do the body parts of the animals shown on these pages help them get food?

MAIN IDEA AND DETAILS
Why do animals need food?

The giraffe's long neck enables it to eat leaves that other animals can't reach.

▲ This pelican has a large bill that helps it catch fish.

Animals Need Shelter

Think about a stormy day. How do you stay warm and dry? Like other animals, you seek shelter—at home or in another safe place. Shelters help protect animals from the weather and from other animals.

There are many kinds of shelters for animals. Some animals use different parts of plants for shelter. Some birds build nests from twigs, grass, and mud. Beavers build shelters called lodges from mud and wood. Other animals, such as rabbits and moles, make burrows in the ground.

Why do animals need shelter?

These white tent bats find shelter in an odd place—under a leaf! ▼

◀ This robin's nest is high above the ground, where other animals can't bother the young birds.

Lesson Review

Essential Question

What do animals need to live?

In this lesson, you learned that all animals have needs that must be met. All animals need food, water, shelter, and oxygen to survive.

1. **Focus Skill — MAIN IDEA AND DETAILS**
   Draw and complete a graphic organizer for this main idea: Animals have many needs.

   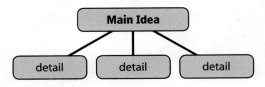

2. **SUMMARIZE** Write a three-sentence summary of this lesson. Tell what animals need to survive.

3. **DRAW CONCLUSIONS** After a walk, Jacob notices that his pet dog is panting. What should Jacob do? How do you know?

4. **VOCABULARY** Use your own words to describe how oxygen and air are related.

Test Prep

5. What kind of shelter do beavers build?
   A. burrow
   B. den
   C. lodge
   D. nest

Make Connections

Writing

Expository
Write an essay **comparing** your main body parts with the body parts of another animal.

Social Studies

Human Shelters
Humans build their shelters. Many different skills are needed to construct, or build, shelters. Research to find out what those skills are. Report your findings to your class.

121

LESSON 2

Essential Question
What Are Vertebrates?

Investigate to find out how animals keep warm.

Read and Learn about different types of vertebrates.

Fast Fact

Going Batty
These bats keep warm by wrapping their wings around their bodies. Bats can also keep warm by crowding together in groups, or colonies. In the Investigate, you will make a model of another way animals keep warm.

Fruit bats

Vocabulary Preview

vertebrate [VER•tuh•brit] An animal with a backbone (p. 127)

mammal [MAM•uhl] A type of vertebrate that has hair or fur and feeds its young with milk from the mother; most mammals give birth to live young (p. 128)

bird [BERD] A type of vertebrate that has feathers (p. 129)

reptile [REP•tyl] A type of vertebrate that has dry skin covered with scales (p. 130)

amphibian [am•FIB•ee•uhn] A type of vertebrate that has moist skin, begins its life in water with gills, and develops lungs as an adult to live on land (p. 131)

fish [FISH] A type of vertebrate that breathes through gills and spends its life in water (p. 132)

Investigate

Keeping Warm

Guided Inquiry

Start with Questions

Animals that live in cold places need ways of keeping warm.

- Can you observe anything about this polar bear that might help it stay warm?
- What other ways do you think it stays warm?

Investigate to find out. Then read and learn to find out more.

Prepare to Investigate

Inquiry Skill Tip

When you make a model, it is important that the model match as closely as possible the object it represents. In this Investigate, for example, shortening is a good model for blubber because both are kinds of fat.

Materials

- large plastic bowl
- large zip-top plastic bag
- disposable plastic gloves
- water
- ice
- solid vegetable shortening
- large spoon

Make an Observation Chart

Hand	Observations
Without Shortening	
With Shortening	

Follow This Procedure

1. Fill a bowl with water and ice. Use a spoon to half-fill a bag with vegetable shortening.

2. Put a glove on each of your hands.

3. Put one of your gloved hands in the zip-top bag. Work your hand into the vegetable shortening. Keep your other hand outside the bag. Use it to mold the shortening so that the shortening evenly covers the hand in the bag.

4. Take your hand out of the zip-top bag, and put both hands in the water. **Compare** the ways your hands feel. **Record** your **observations**.

Draw Conclusions

1. Which hand felt warmer in the ice water? Why?

2. **Inquiry Skill** Scientists **use models** to study things they can't easily observe. In this Investigate, you **made a model** of a mammal with a layer of blubber, or fat. Why was making a model easier than observing the animal?

Independent Inquiry

How does fur protect animals from cold? Write a **hypothesis**. Conduct a **simple investigation** to test it.

Read and Learn

VOCABULARY
vertebrate p. 127
mammal p. 128
bird p. 129
reptile p. 130
amphibian p. 131
fish p. 132

SCIENCE CONCEPTS
▶ that animals with a backbone are grouped by traits

MAIN IDEA AND DETAILS
Look for details about different kinds of vertebrates.

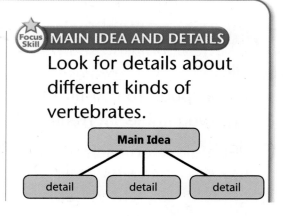

Vertebrates

You have probably seen animals like these in a zoo, in a pet store, or in books. They don't look very much alike, do they? However, the monkeys, bird, snake, frog, and fish all share an important trait. Each of these animals has a backbone. You have a backbone, too. You can feel it in the center of your back.

These monkeys are mammals. They have fur, and the mothers make milk for their young.

Toucans (TOO•kanz) are birds—vertebrates that have feathers and lay eggs.

Animals with a backbone are called **vertebrates** (VER•tuh•brits). Most vertebrates have large brains and sharp senses. These traits help the vertebrates survive. Vertebrates, like most plants and other animals, also have structures that enable them to reproduce.

Scientists classify vertebrates into groups based on their traits. The five major groups of vertebrates are mammals, birds, reptiles, amphibians (am•FIB•ee•uhnz), and fish. Compare and contrast the animals on these pages. What traits does the fish have that the monkeys don't have?

MAIN IDEA AND DETAILS
What trait do all vertebrates have?

The tree frog started its life as an egg in water. But this amphibian lives on land now.

The snake's long, slithery body may make it seem to have no backbone. But this reptile does have a backbone—a very flexible one.

This scaly fish is a vertebrate that spends its entire life in water.

Mammals

Mice, horses, and bats all have something in common with you. Just like you, these animals are mammals. **Mammals** are vertebrates that have hair or fur.

Mammals also use lungs to breathe. Even mammals that live in the water, such as whales, have lungs. Whales must rise to the top of the water to get oxygen from the air.

Young mammals drink milk made by their mothers. Most mother mammals give birth to live young.

MAIN IDEA AND DETAILS What are three traits that all kinds of mammals share?

▲ Pigs aren't furry, but they do have hair—one of the traits of mammals.

Like other mammals, this tiger has fur and makes milk for its young. ▼

▲ The kangaroo is one of just a few kinds of mammals that carry their young in a pouch.

These birds look very different from one another. But they all are vertebrates with two legs, feathers, wings, and beaks.

Birds

Birds share some traits, such as a backbone and lungs, with mammals. Yet their body covering is different. **Birds** are vertebrates that have feathers. Some feathers keep birds warm. Others help them fly. Some birds, such as penguins, can't fly, but they have feathers. Penguins have wings that are like flippers.

Unlike most mammals, birds don't give birth to live young. Instead, they lay eggs. The young birds hatch from the eggs. Mother birds don't feed milk to their young.

 MAIN IDEA AND DETAILS
What are some traits of birds?

Reptiles

When you think of a snake, do you think of something slimy? Many people do. However, snakes aren't slimy. They have dry skin covered with scales. They are reptiles.

Reptiles are vertebrates that have dry skin covered with scales. Like mammals and birds, reptiles breathe with lungs. Some reptiles, like crocodiles, spend a lot of time under water. These animals must go to the surface to get the oxygen they need. Most reptiles, like the ones shown here, hatch from eggs laid on land.

This iguana (ih•GWAH•nuh) hatched from an egg that was laid on land. Like other reptiles, it is dry and scaly.

MAIN IDEA AND DETAILS
What are three traits of reptiles?

The alligator breathes with lungs. It is often in water but must have air to get the oxygen it needs.

This sea turtle lives in the water, but it lays its eggs on land.

This frog hatched as a tadpole from an egg laid in water. It then grew into a frog. Tadpoles live under water. Even though the frog stays near water, it lives on land and breathes with lungs.

A salamander's speed and colors help it survive.

Amphibians

Have you ever watched a frog grow? Frogs are amphibians. Young frogs, called tadpoles, hatch from eggs that were laid in water. Tadpoles don't get oxygen from air. They have gills that get oxygen from water. As tadpoles grow, they develop lungs. Then they can live on land.

Amphibians are vertebrates that have moist skin as adults. Most adult amphibians have legs and usually stay close to water so they can keep their skin moist. Most of them also lay their eggs in water.

MAIN IDEA AND DETAILS

What are three traits of amphibians?

Classify Animals

Your teacher will give you picture sorting cards of different vertebrates. Look at the body coverings of the animals on the picture sorting cards. Classify the animals into different vertebrate groups.

Fish

Have you ever watched a goldfish swim? If so, you might have noticed a flap of skin moving on each side of the fish's head. These flaps cover the gills. Gills take oxygen from water. **Fish** are vertebrates that take in oxygen through gills and spend their whole lives in water.

Most fish lay eggs and have scales. Scales help protect the fish. They are made of a thin, strong material. It is much like the material of your fingernails. Fish don't have arms and legs. Instead, they have fins to help them swim.

MAIN IDEA AND DETAILS
What are some traits of fish?

The fish shown here may look very different, but they all have scales and gills, and they all live in water. ▶

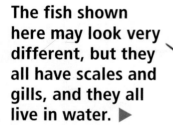

132

Lesson Review

Essential Question

What are vertebrates?

In this lesson, you learned that some animals are classified as vertebrates. Vertebrates are animals that have a backbone. Mammals, birds, reptiles, amphibians, and fish are all vertebrates.

1. **MAIN IDEA AND DETAILS** Draw and complete a graphic organizer for this main idea: Vertebrates are classified by their traits.

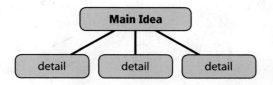

2. **SUMMARIZE** Write four sentences that summarize this lesson. Tell about the different types of vertebrates.

3. **DRAW CONCLUSIONS** Rosita finds an animal she has never seen before. She wonders if it is an amphibian or a reptile. How might she tell the difference?

4. **VOCABULARY** Use the lesson vocabulary to write a paragraph about vertebrates.

Test Prep

5. **Critical Thinking** Explain how the way a mammal takes in oxygen is different from the way a fish takes in oxygen.

Make Connections

 Writing

Narrative
Suppose you had no backbone. Write a funny **story** describing what might happen while you were doing your daily activities.

 Literature

Comparing Animals
Read *Stellaluna,* by Janell Cannon. As you read, make two lists. In one, list the ways in which the animals are alike. In the other, list the ways in which they are different.

LESSON 3

Essential Question
What Are Invertebrates?

Investigate to find out how worms interact with their environment.

Read and Learn about different types of invertebrates.

Fast Fact

Butterfly Travelers
In the fall, these butterflies travel long distances to warmer places. In the Investigate, you will draw conclusions about the travels of another animal.

Monarch butterflies

Vocabulary Preview

invertebrate
[in•VER•tuh•brit] An animal without a backbone (p.138)

Investigate

A Worm Farm

Guided Inquiry

Start with Questions

Farmers till the soil to mix and loosen it before planting crops.

- How is the soil in this picture being tilled?
- Is soil mixed in nature without machines?

Investigate to find out. Then read and learn to find out more.

Prepare to Investigate

Inquiry Skill Tip

When you draw a conclusion in an investigation, you should use logic to explain facts. After you write a conclusion, ask, "Does this conclusion make sense? Does it explain all the facts?"

Materials

- canning jar with lid
- soil
- uncooked oatmeal
- sand
- earthworms
- square of dark fabric

Make an Observation Chart

Time of Observation	Observations	Sketch of Jar
Start		
One week later		

Follow This Procedure

1. Put 2 cm of moist soil into the jar. Sprinkle a very thin layer of oatmeal over the soil. Add 2 cm of moist sand.

2. Repeat the layers of soil, oatmeal, and sand until the jar is almost full. About 5 cm from the top, add a last layer of soil. Do not sprinkle any oatmeal on top of the last layer. Put several earthworms on top of the soil.

3. Place the fabric square over the opening of the jar. Screw the lid onto the jar, or use a rubber band to hold the fabric in place. Put the jar in a dark place.

4. After one week, observe the jar. Compare the way it looks now with the way it looked when you set it up.

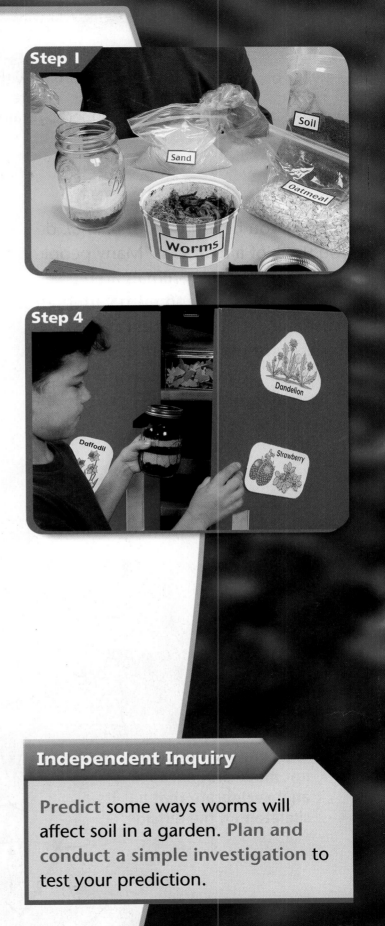

Draw Conclusions

1. What changes did you observe?

2. **Inquiry Skill** Draw a conclusion about why these changes occurred.

Independent Inquiry

Predict some ways worms will affect soil in a garden. Plan and conduct a simple investigation to test your prediction.

Read and Learn

VOCABULARY
invertebrate p. 138

SCIENCE CONCEPTS
▶ that animals without a backbone are grouped by traits

COMPARE AND CONTRAST
Look for ways the groups of invertebrates are alike.

| alike | different |

Invertebrates

When you hear the word *animal*, do you think of a mammal? Many people do. However, most animals aren't mammals. Most aren't even vertebrates. Most animals are **invertebrates** (in•VER•tuh•britz), or animals without a backbone. There are many more types of invertebrates than there are types of vertebrates.

All the animals shown here are invertebrates. They don't have a backbone.

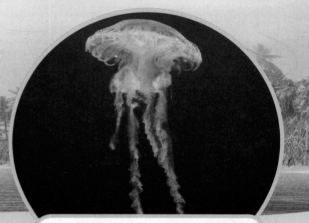

▲ Unlike the corals, the jellyfish moves in water.

▲ The corals in this colony do not move around. Each is an invertebrate that has its skeleton on the outside.

This scallop's shell has two halves. ▶

◀ Insects, such as this firefly, are invertebrates. Insects have six legs.

Spiders aren't insects. They are in a group of invertebrates that have eight legs. ▼

Like all other invertebrates, the horseshoe crab has no backbone. Unlike the firefly or the spider, it has 10 legs. ▼

There are more than a million kinds of invertebrates. Scientists group invertebrates by ways they are alike. Look at the pictures on these pages. Why might scientists place the firefly, the horseshoe crab, and the spider in different groups?

Invertebrates may have simple or complex bodies. They may have a lot of legs—or no legs at all! Even though there are many different invertebrates, they all have one common trait. None of them has a backbone.

COMPARE AND CONTRAST How are invertebrates like vertebrates? How are they different?

▲ The honeybee, luna moth, and dragonfly are part of the largest group of invertebrates—the insects.

Insects

There are more kinds of insects than of other invertebrates. In fact, there are more kinds of insects than of all other animals put together. Insects live everywhere on Earth. Even cold arctic areas have many insects during the summer.

Insects have three body parts and six legs. They have no backbone, but they have a hard outer covering. Many insects, such as those shown on this page, have wings.

 **COMPARE AND CONTRAST** How are the animals on this page alike?

Make a Model
The bodies of many invertebrate animals are symmetrical (sih•MEH•trih•kuhl). This means that they are the same on each side. Make a model of a ladybird beetle. How did you make it symmetrical from side to side?

Spiders and Ticks

Spiders and ticks look like insects, but they aren't insects. They belong to another group, because they have traits different from those of insects. Spiders and ticks have eight legs and only two body parts. Like insects, spiders and ticks have an outer body covering. The covering helps protect them.

The banana spider and the deer tick are grouped together. They both have two body parts, eight legs, and an outer body covering. ▶

COMPARE AND CONTRAST How are spiders and ticks like insects? How are they different from insects?

Science Up Close

For more links and animations, go to www.hspscience.com

Spider Webs

1. The orb spider's body makes silky thread from protein. The web begins with a sticky thread called the bridge.

2. The spider anchors the next thread to the ends of the bridge. This is the first frame thread, or support for the web.

3. From the first frame thread, the spider anchors more frame threads. The spider spins nonsticky radial threads from the center to the frames.

4. The spider walks on radial threads when walking on the web. Finally, the spider forms a spiral of sticky silk from the edges of the web to its middle.

Snails, Clams, and Squids

Snails, clams, and squids are in the same group. They all have soft bodies. Also, most animals in this group have a head.

Some of these animals, like the squid, have tentacles. Some, like the clam and the snail, have hard shells to protect their soft bodies. They also have a kind of "foot." The foot pokes out of the shell and helps the animal move.

The Atlantic long-finned squid has no shell. It swims freely, using its tentacles for movement and for getting food. ▶

 COMPARE AND CONTRAST How is the coquina clam like the squid? How is it different?

The sea slug has no shell. It swims or crawls along surfaces to get food. ▼

This land snail has a shell that protects its soft body. Its muscular "foot" helps it move. ▶

▲ The Florida coquina (koh•KEE•nuh) clam's shell has two halves that protect the animal's soft body. Like the snail, the clam has a muscular "foot" that helps it move.

The starfish is covered with spines that protect it. ▶

When the sea anemone (uh•NEM•uh•nee) opens its "mouth," its tentacles make it look like a beautiful flower. ▼

▲ Like insects and spiders, this blue crab has a hard outer covering.

Other Kinds of Invertebrates

All the animals shown on this page are invertebrates. The worms you observed in the Investigate are another kind of invertebrate. Worms have no shells, legs, or eyes.

Scientists have not yet found or grouped all of Earth's animals. When a new animal is found, the first thing scientists ask is likely to be, "Does it have a backbone?"

COMPARE AND CONTRAST How is the starfish like the blue crab?

Invertebrates Are Important

Invertebrates are important to other living things. Bees and butterflies, for example, move pollen from one flower to another. This helps the flowers make seeds. As you saw in the Investigate, earthworms improve the soil. Plants grow better in soil that has earthworms than in soil that doesn't.

Many invertebrates are important foods for other animals. People can eat clams and crabs. People also eat the honey and use the wax made by bees.

COMPARE AND CONTRAST
How are earthworms like bees?

▲ Honeybees provide more than just their famous treat, honey. They also make beeswax. One way people use beeswax is to make candles.

◀ The movement of earthworms mixes air into the soil.

Lesson Review

Essential Question

What are invertebrates?

In this lesson, you learned that an invertebrate is an animal that does not have a backbone. There are many kinds of invertebrates, including insects, spiders, snails, and jellyfish.

1. **COMPARE AND CONTRAST** Draw and complete a graphic organizer to compare and contrast two kinds of invertebrates.

2. **SUMMARIZE** Write a paragraph that summarizes this lesson. Tell about 6 different types of invertebrates.

3. **DRAW CONCLUSIONS** While looking for seashells at the beach, Alex finds an animal with no backbone and no shell. How can he tell if it belongs in the same group of invertebrates as the clams?

4. **VOCABULARY** Use the lesson vocabulary to write a sentence.

Test Prep

5. Which animal is **not** an invertebrate?
   A. clam
   B. earthworm
   C. goldfish
   D. snail

Make Connections

 Writing

Persuasive
Everyone has an **opinion** on what is the best. Choose one of your favorite invertebrates. Write at least two paragraphs explaining why this animal is the best invertebrate.

 Math

Make a Graph
With a classmate, brainstorm a list of the types of invertebrates you might find in a backyard, park, or garden. How many are insects? How many live in water? What other invertebrates are there? Make a bar graph to show your results.

Science Spin — From Weekly Reader
TECHNOLOGY

Better Than Nature Can Make?

Replacing Nature

For thousands of years, clothes have been made by using natural materials, such as cotton, wool, and silk. These materials are all found in nature. For example, cotton comes from a plant, wool is made from a sheep's fur, and silk is spun by caterpillars.

The first synthetic materials were produced less than 100 years ago. These materials, such as rayon, nylon, spandex, and polyester, were all first made in laboratories. Today, synthetic materials have replaced natural fabrics for many purposes.

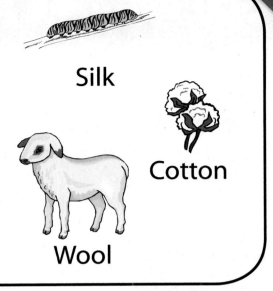

Silk

Cotton

Wool

Many times, scientists come up with new ideas by looking at nature. For example, swimsuit makers made a fabric for Olympic athletes after studying the skin of sharks.

Recently, scientists in Turkey, a country in Europe, also turned to nature for ideas. Scientists there came up with an idea for a new fabric by looking at water lilies. Water lilies are plants that grow in ponds and have big leaves that float on the surface.

Looking at Lilies

The scientists were inspired by how waterproof a water lily's leaves are. They wanted to make a material that shed water similar to the way the plant's leaves shed water.

The scientists then began working with different materials, such as plastic, to form a new kind of material. Like the plant, when this new material gets wet, the water does not soak in but stays on the surface and rolls off.

The new material is synthetic. *Synthetic* is another way of saying "made by people." The new waterproof material will be useful for many people, especially firefighters, who need to keep dry while they work.

Think and Write

1. Why did people start making synthetic materials?
2. Do you think synthetic materials should replace natural materials? Why or why not?

Find out more. Log on to
www.hspscience.com

Chapter 3 Review and Test Preparation

Vocabulary Review

Use the terms below to complete the sentences. The page numbers tell you where to look in the chapter if you need help.

oxygen p. 117

vertebrate p. 127

mammal p. 128

bird p. 129

reptile p. 130

amphibian p. 131

fish p. 132

invertebrate p. 138

1. An animal with hair or fur that feeds its young milk is a _____.

2. An animal without a backbone is an _____.

3. A dry, scaly animal that lays eggs and lives on land is a _____.

4. An animal with a backbone is a _____.

5. A vertebrate with two legs, wings, and feathers is a _____.

6. A gas, found in air or water, that animals need is _____.

7. A moist-skinned animal that lives near water is an _____.

8. A vertebrate that lives its whole life in water is a _____.

Check Understanding

Write the letter of the best choice.

9. **MAIN IDEA AND DETAILS** Which detail given below is **not** a basic need of animals?
   A. color
   B. food
   C. shelter
   D. water

10. Why do animals need shelter?
    F. for food
    G. for protection
    H. for quiet
    J. for water

11. Which animals are grouped together?
    A. snails and squids
    B. spiders and ants
    C. goldfish and whales
    D. frogs and snakes

12. How do mammals that live in water get oxygen?
 F. They take it in through gills.
 G. Their skin absorbs it.
 H. They go to the top of the water to breathe air.
 J. They don't need oxygen.

13. Which is **not** a trait of mammals?
 A. fur
 B. giving birth to live young
 C. feathers
 D. feeding milk to their young

14. COMPARE AND CONTRAST In what way are birds and mammals alike?
 F. They both have feathers.
 G. They both have fur.
 H. They both breathe with lungs.
 J. They both give birth to live young.

15. To which vertebrate group does the animal below belong?

 A. amphibians **C.** mammals
 B. birds **D.** reptiles

16. What is one important way scientists group animals?
 F. by their ages
 G. by their lengths
 H. by their colors
 J. by whether they have a backbone

Inquiry Skills

17. Compare amphibians with reptiles.

18. You see an animal in a pond. Then the animal comes out of the pond. What **conclusion** can you draw?

Critical Thinking

19. Why do many amphibians stay near water for their entire lives?

20. A bat is a mammal. Use what you know about mammals to identify some traits of bats.

Visual Summary

Tell how each picture shows the **Big Idea** for its chapter.

CHAPTER 1 Big Idea
Cells are the building blocks of all the many types of living things.

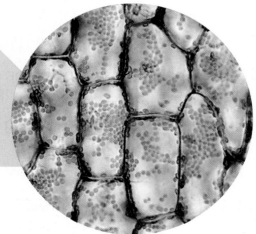

CHAPTER 2 Big Idea
Plants have parts that help them meet their needs.

CHAPTER 3 Big Idea
Animals can be classified by their traits.

Living Things Interact

UNIT B — LIFE SCIENCE

CHAPTER 4
Where Living Things Are Found . . 152

CHAPTER 5
Living Things Depend on
One Other 196

Unit Inquiry

Changing Environments

Environments can change over time. In some places, there are many lakes during the rainy season. During the dry season, the lakes evaporate. How are plants affected as a lake dries up? Plan and conduct an experiment to find out.

CHAPTER 4
Where Living Things Are Found

What's the Big Idea?
All the living, once-living, and nonliving things in an environment interact in an ecosystem.

Essential Questions

Lesson 1
What Are Ecosystems?

Lesson 2
What Are Some Types of Ecosystems?

Lesson 3
How Do Living Things Survive in Ecosystems?

Lesson 4
How Do Ecosystems Change?

Go online
Student eBook
www.hspscience.com

What do you wonder?

Everglades National Park, in southern Florida, is one of the largest wilderness areas in the United States. Alligators, herons, and Florida panthers are just a few of the animals that call the Everglades home. What traits do you think this Florida panther has that help it survive in its ecosystem? How does this relate to the **Big Idea?**

Florida panther

LESSON 1

Investigate to find out how environments can be similar and different.

Read and Learn about parts of an ecosystem.

Fast Fact

Using Ecosystems
Every day, the average adult horse needs about 1 gallon of water for each 100 pounds it weighs. Most horses get their water from the environment in which they live. In the Investigate, you will observe one type of environment.

Essential Question
What Are Ecosystems?

Horses in a field

Vocabulary Preview

environment
[en•vy•ruhn•muhnt] The things, both living and nonliving, that surround a living thing (p. 159)

ecosystem
[EE•koh•sis•tuhm] The living and nonliving things that interact in an environment (p. 160)

population
[pahp•yuh•LAY•shuhn] A group of organisms of the same kind that live in the same place (p. 160)

community
[kuh•MYOO•nuh•tee] All the populations of organisms that live in an ecosystem at the same time (p. 160)

habitat
[HAB•ih•tat] The place where an organism lives in an ecosystem (p. 161)

Investigate

Observe an Environment

Guided Inquiry

Start with Questions

A tidal marsh is one type of ecosystem.

- What kinds of plants and animals live here?
- How is this ecosystem different from other ecosystems where you live?

Investigate to find out. Then read and learn to find out more.

Prepare to Investigate

Inquiry Skill Tip

Before you infer, you should define any important terms. In this Investigate, for example, you must understand what *living* and *nonliving* mean before you can infer which things fit in each category.

Materials

- safety goggles
- wire coat hanger

Make a Data Table

Living Things	Number	Nonliving Things	Number

Follow This Procedure

CAUTION: Put on safety goggles.

1. Bend the coat hanger into a square. Ask your teacher for help, if necessary.

2. Go outside. Place the square on the ground. Closely **observe** the ground inside the square. This square of ground is an environment.

3. Copy the table. **Record** all the living things you **observe** and how many there are of each. Then record all the nonliving things you observe and how many there are of each.

4. Share your table with a classmate. **Compare** the environments you observed. How are they alike? How are they different?

Draw Conclusions

1. **Compare** the things you found in your environment with the things a classmate found. Why do you think you found different things?

2. **Inquiry Skill** How did you **infer** which things were living and which things were nonliving?

Step 1

Step 2

Independent Inquiry

Compare the environment you **observed** at school with an environment you observe at or near your home.

157

Read and Learn

VOCABULARY
environment p. 159
ecosystem p. 160
population p. 160
community p. 160
habitat p. 161

SCIENCE CONCEPTS
▶ what an environment is
▶ what an ecosystem is

MAIN IDEA AND DETAILS
Look for details about environments and ecosystems.

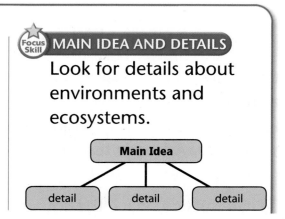

Where Things Live

Living things need places to live and grow. Fish live in water. Many birds live in trees and fly through the air. Plants grow where there is soil, water, and sun.

Living things can be found almost everywhere on Earth. Some fish can live in the deepest parts of the oceans. Some plants can live on the tops of high mountains. Scorpions can survive in dry deserts. Cattails can grow in swamps.

Some plants live between the cracks in pavement.

The living and nonliving things that surround a living thing make up its **environment**. Plants and animals use things from their environments to meet their needs. The things they use are called resources. The resources in an environment include food, water, air, space, and shelter.

Many living things may share an environment and its resources. If the environment has too little of any resource, the living things compete with one another to get what they need.

MAIN IDEA AND DETAILS What resources do living things use from their environments?

▲ Koalas in Australia live in eucalyptus trees. The leaves of the trees are their only food.

These prairie dogs can survive and grow in a grassy environment. ▼

This stork makes its nest on a chimney high above the ground. ▼

Parts of an Ecosystem

The organisms living in and around a pond interact with one another. A fish might eat an insect. A frog might sit on a lily pad. Together, the living things and the nonliving things they interact with make up an ecosystem. An **ecosystem** (EE•koh•sis•tuhm) is made up of all the living and nonliving things in an environment.

Different types of organisms live in the same ecosystem. A group of organisms of the same kind living in the same place is a **population** (pahp•yuh•LAY•shuhn). For example, the frogs in a pond make up one population. The waterlilies in the pond make up another population.

All of the populations that live in an ecosystem at the same time form a **community** (kuh•MYOO•nuh•tee).

Science Up Close

For more links and animations, go to **www.hspscience.com**

Pond Ecosystem

Ponds are rich ecosystems. They are filled with living and nonliving things that interact with one another.

Fish live in the water. They may reach up to eat insects just above the surface.

Freshwater turtles are faster in water than on land. They eat frogs, small fish, worms, and plants.

Each organism in this picture is a part of the pond's community.

All the members of a community live in the same ecosystem. However, they don't all live in the same part of the ecosystem. Fish swim in water, but birds build nests in trees. The place in its ecosystem where a population lives is its **habitat**. The habitat includes both living and nonliving things.

MAIN IDEA AND DETAILS What is the difference between a population and a community?

Ecosystems Around You
Find out about an ecosystem that is close to where you live. Research what types of plants and animals live there. Draw and color the ecosystem. Label the animals and plants.

Birds such as this kingfisher hunt for fish and frogs near the pond's edge.

Waterlilies float on the surface of the pond. Frogs often rest on their leaves.

Snails crawl on plants that live in shallow water near the pond's edge.

Organisms and Their Habitats

Some organisms can survive only in certain habitats. A polar bear, for example, could not find the water it needs in a desert. A rain forest would be too wet for a desert owl.

An organism's habitat gives it everything it needs to survive. For example, a pond has the water, food, and oxygen that a fish needs. The fish could not survive without these things. What does your habitat provide for you?

 **MAIN IDEA AND DETAILS** What does a living thing's habitat provide for it?

▲ The scarlet macaw lives high in the trees of the rain forest. It feeds on the fruits and large nuts that grow there.

◀ Mountain goats live on rocky cliffs, where they feed on plants growing among the rocks.

Lesson Review

Essential Question
What are ecosystems?

In this lesson, you learned that ecosystems are made up of all the living and nonliving things in an environment. Many organisms share the resources in an environment. Some animals can live only in certain ecosystems.

1. **MAIN IDEA AND DETAILS** Draw and complete a graphic organizer for this main idea: There are several parts to an ecosystem.

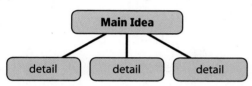

2. **SUMMARIZE** Write a three-sentence summary for the lesson. Tell about the living and nonliving parts of an ecosystem.

3. **DRAW CONCLUSIONS** Nonliving things are part of an ecosystem. Why is this important?

4. **VOCABULARY** Write two sentences that explain how an environment and an ecosystem are different.

Test Prep

5. Which is a nonliving part of an ecosystem?
   - A. bird
   - B. water
   - C. snail
   - D. cattail

Make Connections

 Writing

Descriptive
Research different types of wetland ecosystems from your state. Select one and write two paragraphs that **describe** the living and nonliving things in the ecosystem. Include how the living things depend on their environment.

 Math

Organize Data
In your state, there may be many kinds of endangered animals. Choose one of them, and research its population. How many are there now? How many have there been in each of the last 10 years? Organize your findings in a bar graph.

LESSON 2

Investigate to find out about different plant parts.

Read and Learn about different types of ecosystems.

Essential Question

What Are Some Types of Ecosystems?

Fast Fact

Ice-Cold Tundra
The tundra (TUHN•druh) is cold and snowy. Plants that grow in the tundra are different from plants that grow in other ecosystems. In the Investigate, you will observe one type of plant found in an ecosystem.

Arctic tundra

Vocabulary Preview

desert [DEZ•ert] An ecosystem that is very dry (p. 168)

grassland [GRAS•land] An area of land that is generally hot in the summer and cold in the winter. The main plants found in this ecosystem are grasses. (p. 169)

forest [FAHR•ist] An ecosystem in which many trees grow (p. 172)

Investigate

Grass Roots

Guided Inquiry

Start with Questions

This grass grows in a grassland ecosystem. It is the main kind of plant in this type of ecosystem.

- What kind of roots do you think the grass has?
- How are the roots of grasses different from the roots of other plants?

Investigate to find out. Then read and learn to find out more.

Prepare to Investigate

Inquiry Skill Tip

Collect and analyze as much data as you can before you infer. Look for patterns in the data that might help you infer about a new situation.

Materials

- plastic gloves
- hand lens
- grass plants
- ruler
- sheet of white paper

Make a Data Table

Height of Tallest Blade	Length of Longest Root

Follow This Procedure

1. Put on the plastic gloves. **Observe** the different types of grass plants that your teacher has for you.

2. The leaves of a grass plant are called blades. Carefully hold up one grass plant by its blades. Gently shake the plant. **Observe** what happens to the soil.

3. Very carefully remove the soil from around the roots.

4. Place the grass plant on a sheet of white paper. **Observe** it with the hand lens.

5. **Measure** and record the height of the tallest blade and the length of the longest root.

Draw Conclusions

1. **Compare** the height of the tallest blade with the length of the longest root.

2. **Inquiry Skill** Infer how the roots of a tree might be different from the roots of a grass plant.

Independent Inquiry

Carefully remove a plant from a pot or from the ground. **Compare** the roots of a grass plant with the roots of the plant you have chosen.

Read and Learn

VOCABULARY
desert p. 168
grassland p. 169
forest p. 172

SCIENCE CONCEPTS
▶ how ecosystems are different
▶ how ecosystems support plants and animals

MAIN IDEA AND DETAILS
Look for details about different ecosystems.

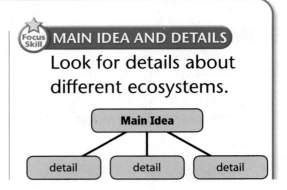

Desert Ecosystems

Deserts are very dry ecosystems. You might think that a desert is just sand. However, there is life in the desert. Desert plants and animals can survive with very little water.

Desert plants, such as cacti, have thick stems that store water. The roots of a cactus lie just below the soil and spread far from the plant. This lets them take in a lot of water quickly when it rains.

Temperatures in some deserts rise very high by day. Most animals find shady places to sleep and come out only after sunset. Some animals, such as kangaroo rats, drink hardly any water. They get all the water they need from the plants they eat. Toads keep moist by staying under the soil.

MAIN IDEA AND DETAILS

How do animals in a desert stay cool?

Arches National Park, in Utah, is located in a desert ecosystem. ▼

▲ Until the 1800s, as many as 60 million bison lived on the grasslands of North America.

Grassland Ecosystems

Another kind of ecosystem is the grassland. **Grasslands** are dry, often flat areas of land that are hot in the summer and cold in the winter. They get more rain and snow than deserts do but less than most other ecosystems. Food crops grow well there.

As you may have guessed, the main plant in a grassland ecosystem is grass. A few small bushes and wildflowers also grow. Since grasslands are dry, few trees are able to grow there except along rivers and streams.

A grassland ecosystem includes both large animals, such as bison and coyotes, and small animals, such as mice, rabbits, and snakes.

 MAIN IDEA AND DETAILS

Why is a grassland a good place to start a farm?

▲ The shape of a bottlenose dolphin's body helps it glide through the water.

Saltwater Ecosystems

If you have ever tasted ocean water, you know that it is salty. Oceans cover about three-fourths of Earth's surface, so there are more saltwater ecosystems than any other kind.

Sharks, sea turtles, corals, and octopods are all ocean animals. So are whales and seals. Plantlike organisms, such as kelp and seaweed, are also part of some saltwater ecosystems. Organisms that live in the oceans need salt water. They would not survive in a freshwater environment, such as a pond.

 MAIN IDEA AND DETAILS

About how much of Earth is NOT covered by oceans?

Freshwater Ecosystems

Rivers, ponds, and streams have fresh water. They make up freshwater ecosystems. Unlike ocean water, fresh water does not have much salt.

Ducks and some kinds of insects live in freshwater ecosystems. So do trout and turtles. Deer and foxes may visit there, too. Many kinds of plants live in freshwater ecosystems.

 **MAIN IDEA AND DETAILS** Where might you find a freshwater ecosystem?

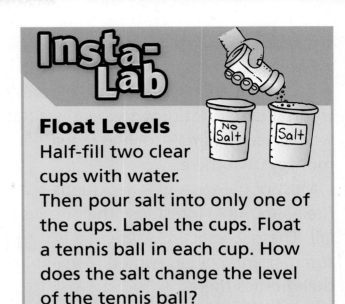

Float Levels
Half-fill two clear cups with water. Then pour salt into only one of the cups. Label the cups. Float a tennis ball in each cup. How does the salt change the level of the tennis ball?

A river is one type of freshwater ecosystem. ▼

Forest Ecosystems

Forests are ecosystems in which many trees grow. There are different kinds of forests. A tropical rain forest grows where it is hot and wet all year. Animals such as jaguars and monkeys live there.

Most of the trees in a deciduous (dee•SIJ•oo•uhs) forest lose their leaves in the fall. These forests grow where it is warm in summer and cold in winter. Many animals make their homes there.

Evergreen trees, such as pines and firs, grow in a coniferous (koh•NIF•er•uhs) forest. Most trees in coniferous forests have cones. These forests have cool summers and cold winters.

MAIN IDEA AND DETAILS What kind of forest ecosystem grows where it is hot and wet?

▲ A long-eared owl searches for food in the forest.

Squirrels eat nuts and fruits that grow in the forest. ▼

▼ Deciduous forests provide habitats for rabbits, bears, deer, and foxes.

Lesson Review

Essential Question

What are some types of ecosystems?

In this lesson, you learned that there are many types of ecosystems. Each ecosystem has characteristics that affect which organisms can live there.

1. **MAIN IDEA AND DETAILS** Draw and complete a graphic organizer for the main idea: *Different kinds of ecosystems are made up of different kinds of living things.*

   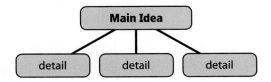

2. **SUMMARIZE** Write a summary of this lesson. Begin with this sentence: *Different environments have different ecosystems.*

3. **DRAW CONCLUSIONS** Suppose all the plants in one ecosystem died. What do you think would happen to that ecosystem? Why?

4. **VOCABULARY** Write a paragraph comparing the living things in a desert ecosystem to those in a grassland ecosystem.

Test Prep

5. **Critical Thinking** A friend asks you to help in identifying the ecosystem to which a plant belongs. The plant has wide, spreading roots and a thick stem that stores water. Which kind of ecosystem does this plant probably live in?

Make Connections

 Writing

Persuasive

Write an **opinion** article for the school newspaper. Explain how cutting down trees for lumber affects the forest ecosystem.

 **Math**

Collect Data

Research an ecosystem you learned about in this lesson. Use reference sources to find out the ecosystem's average monthly temperature and rainfall. Make two bar graphs to show the information you collected.

LESSON 3

Investigate to find out how insects hide.

Read and Learn about how living things survive.

Essential Question

How Do Living Things Survive in Ecosystems?

Fast Fact

Chew on This

Beavers use their sharp teeth to chew through the trunks and branches of small trees and shrubs. Their brown color helps them blend in with their surroundings. In the Investigate, you'll infer how color helps other living things hide.

Beaver

Vocabulary Preview

adaptation [ad•uhp•TAY•shuhn] Any trait that helps a plant or an animal survive (p. 178)

instinct [IN•stinkt] A behavior that an animal knows without being taught (p. 178)

hibernate [HY•ber•nayt] To go into a deep, sleeplike state for winter (p. 180)

migrate [MY•grayt] To travel from one place to another and back again (p. 181)

camouflage [KAM•uh•flahzh] Colors, patterns, and shapes that disguise an animal and help it hide (p. 182)

mimicry [MIM•ik•ree] The imitating of the look of another animal (p. 182)

175

Investigate

How Insects Hide

Guided Inquiry

Start with Questions

This insect looks like the area around it. Can you see it?

- How does being able to hide help an insect?
- What makes this insect hard to see?

Investigate to find out. Then read and learn to find out more.

Prepare to Investigate

Inquiry Skill Tip

The word infer comes from the Latin word *inferre*, which means "to bring in." Use this meaning to remind you to "bring in" as much information as possible before you infer.

Materials

- construction paper
- crayons
- scissors
- watch or clock
- chenille sticks
- tape
- ruler

Make a Data Table

Minute	Number of Insects Found	Description of Insect Found
1		
2		
3		
4		
5		

Follow This Procedure

1. Look around the classroom for a "habitat" for a model insect you will make. **Observe** the colors and shapes of things in the habitat.

2. Draw a construction-paper rectangle 5 cm long and 3 cm wide. This will be the size of your insect.

3. Color the body so your insect blends into its habitat. Make legs and wings.

4. Tape your insect in its habitat. Don't hide it behind anything.

5. Ask a classmate to be a "bird." Ask the bird to look for one minute for your insect and other classmates' insects. **Record** results in a table. Continue until the bird finds all the insects.

Draw Conclusions

1. Which insects did the bird find first? Why were they easy to find?
2. **Inquiry Skill** Infer why some insects were hard to find.

Independent Inquiry

Draw conclusions about why the fur of some animals, such as foxes and rabbits, changes color with the seasons.

Read and Learn

VOCABULARY
adaptation p. 178
instinct p. 178
hibernate p. 180
migrate p. 181
camouflage p. 182
mimicry p. 182

SCIENCE CONCEPTS
▶ how organisms adapt to their environments

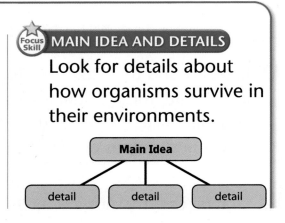

MAIN IDEA AND DETAILS
Look for details about how organisms survive in their environments.

How Living Things Survive

All living things have ways to survive. Any trait that helps an animal survive is an **adaptation**. An adaptation can be physical. For example, the arctic hare in the picture changes colors in summer and winter. An adaptation can also be a behavior. A snake hides in the shade when it is hot. Animals learn some behaviors. Other behaviors are instincts. An **instinct** is a behavior an animal knows without being taught.

In the winter, this hare has white fur. Its fur changes colors to blend in with the environment.

Plant Structures

The leaves of a bromeliad (broh•MEE•lee•ad) collect rainwater because of their shape. The stems, roots, and leaves of plants are adaptations that help the plants survive.

Roots with this special shape are called prop roots. These roots help support tall, thin plants, such as corn, and plants that live in swampy areas.

The stem of this vine forms tendrils that hold its leaves up to gather sunlight.

Plants also have adaptations that help them survive. Plant parts are physical adaptations. Remember that the stems of some desert plants store water. Some rain-forest plants have very large leaves. These leaves help them take in more sunlight for making food in the shady forest. Even roots have physical adaptations. Some roots grow deep into the ground to get water from far below the surface.

MAIN IDEA AND DETAILS What is an example of an adaptation?

Insta-Lab

Thumbs Down
Tuck your thumb into the palm of your hand. Without moving your thumb, try to pick up objects. Try to write your name without moving your thumb. Share your observations with a classmate. How is the thumb a useful adaptation for humans?

Hibernation

In the fall, when the weather gets cooler, some animals start eating more food than usual. Then, they **hibernate** (HY•ber•nayt), or go into a sleeplike state for the winter. A hibernating animal doesn't move. Its body temperature drops, and its heartbeat rate slows down. The animal might breathe only once in half an hour.

An animal loses more body heat during winter than at other times. If it needed to stay warm and be active, it would need to eat more—but food is harder to find in winter. Because a hibernating animal is not active, it needs less energy. It lives off the fat stored in its body, so it doesn't need to find food.

MAIN IDEA AND DETAILS What are two things that happen to an animal's body when it hibernates?

This chipmunk spends time hibernating during the winter. It curls into a ball to stay warm. ▼

Math in Science
Interpret Data

Find the difference between each animal's active and hibernating heartbeat rates.

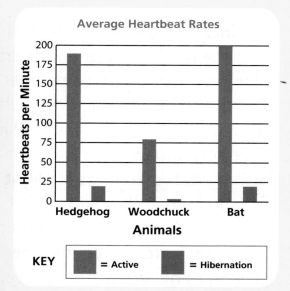

Average Heartbeat Rates

KEY ■ = Active   ■ = Hibernation

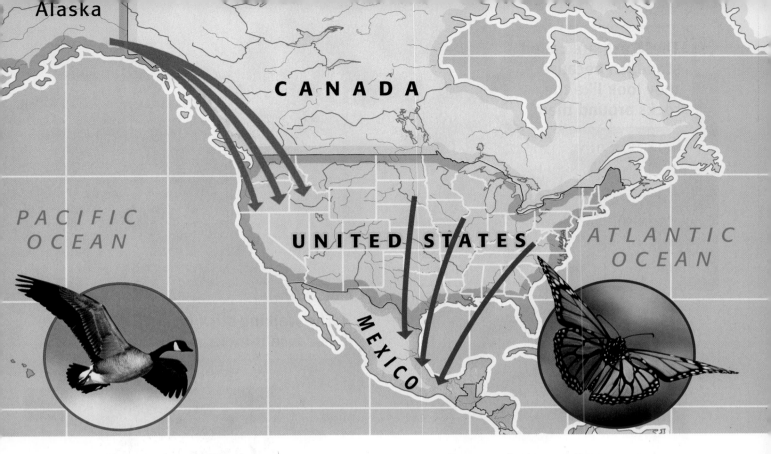

▲ In fall, Canada geese fly south from northern Canada and Alaska. Groups fly in two lines that form a V shape in the sky.

▲ Millions of monarch butterflies fly from the United States to Mexico each fall. When spring comes, they fly north again to lay their eggs.

Migration

Many animals **migrate** (MY•grayt), or travel from one place to another and back again. Most birds in the Northern Hemisphere fly south in the fall. Southern areas usually have warmer climates and more food. These birds return north in the spring to raise their young.

In the fall, gray whales migrate from the cold waters near Alaska. They spend the winter in the warm waters near California. Female whales stay there to give birth to their young. In the spring, they return to Alaska.

 **MAIN IDEA AND DETAILS** Why do some animals migrate?

These fish are called stonefish because they look like the rocks around them.

The walking stick insect is hard to see because it looks much like the twig.

A chameleon can change its color to look like what is next to it.

The "eyes" on the wings of this moth confuse a bird that tries to eat it. The bird thinks that another animal, much larger than the moth, is looking at it.

Hide and Seek

Some animals can hide without trying. These animals are hidden by their shapes, colors, or patterns. Such disguises are called **camouflage**. For example, many female birds are brown. This helps them blend in with their nests as they sit on their eggs.

Some animals look very much like other animals. Some snakes that are not harmful look just like a harmful kind of snake. Since animals don't know which is harmful, neither snake gets eaten. Imitating the look of another animal is called **mimicry**.

MAIN IDEA AND DETAILS What is camouflage?

Lesson Review

Essential Question

How do living things survive in ecosystems?

In this lesson, you learned that living things have adaptations that help them survive in their environments.

1. **MAIN IDEA AND DETAILS** Draw and complete a graphic organizer for the main idea: Living things have adaptations that help them survive.

   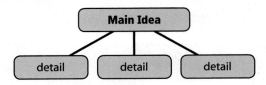

2. **SUMMARIZE** Write a paragraph to summarize how living things survive.

3. **DRAW CONCLUSIONS** How do camouflage and mimicry help animals survive in their environments?

4. **VOCABULARY** Write a sentence that explains how the terms *adaptation, hibernation,* and *migration* are related.

Test Prep

5. Which of the following is a physical adaptation?
   A. camouflage
   B. hibernation
   C. instinct
   D. migration

Make Connections

 Writing

Narrative
Choose an animal that hibernates. Write a **story** about how that animal gets ready for hibernation.

 Drama

Migration Play
Research an animal that migrates during the winter. With five other classmates, write a play that tells about the animal's migration path. Perform your play for the class.

LESSON 4

Essential Question
How Do Ecosystems Change?

Investigate to find out how ecosystems can change.

Read and Learn about factors that can change ecosystems and the effects of those changes.

Fast Fact

Hoover Dam
Hoover Dam, on the Colorado River, is on the border between Nevada and Arizona. The reservoir, or human-made lake, that formed behind it changed the land forever. In the Investigate, you will observe how an environment can change.

184

Hoover Dam

Vocabulary Preview

resource [REE•sawrs] A material that is found in nature and that is used by living things (p. 189)

185

Investigate

Changing the Environment

Guided Inquiry

Start with Questions

When too much rain falls too quickly, flooding can result. This street has been flooded.

- Why do you think this street flooded?
- How might the land change because of this flood?

Investigate to find out. Then read and learn to find out more.

Prepare to Investigate

Inquiry Skill Tip

A good prediction is based on past observations, measurements, and data. If you can explain why you think a prediction will be true, then it is a good prediction—whether it turns out to be correct or not.

Materials

- moist sand
- wooden block
- shallow cardboard box
- leaves and twigs
- water
- watering can
- small stones

Make an Observation Chart

Observation (Slow Water)	Observation (Fast Water)

Follow This Procedure

1. With a partner, pack the sand into the box. Use your fingers to make hills, valleys, and a streambed. Push the leaves and twigs into the sand to represent plants.

2. Carefully lift one end of the box off the table. Place the wooden block beneath that end.

3. Slowly pour a little water from the watering can into the streambed you made. **Observe** and record what happens to the sand, water, and plants.

4. Pour the water more quickly into the streambed. **Observe** and record again.

Draw Conclusions

1. What happens to the sand, water, and plants when only a little water is poured into the streambed?

2. **Inquiry Skill** Add several stones along the streambed. **Predict** what will happen if you pour a lot of water into the streambed. Try it. Was your prediction correct?

Independent Inquiry

Predict what will happen to the sand, plants, and water if you make a dam across your streambed. Try it.

Read and Learn

VOCABULARY
resource p. 189

SCIENCE CONCEPTS
▶ how ecosystems change over time

CAUSE AND EFFECT
Look for the causes of changes in ecosystems.

cause → effect

How Ecosystems Change

Ecosystems change over time. Sometimes, the changes are natural. Fires and floods cause changes. They may destroy habitats and kill many living things. Yet other living things survive. Seeds that survive grow into plants. Animals move back into the area to eat the plants.

Living things change ecosystems, too. When beavers cut down trees to build a dam, they change the forest. The dam changes the way the stream flows.

CAUSE AND EFFECT When beavers build a dam, what are the effects on the ecosystem?

The eruption caused a landslide that swept away the top of the mountain.

Mount Saint Helens, a volcano in Washington State, was 2,950 meters (9,680 ft) high before it erupted in May 1980.

▲ After the eruption, an area of 180 square kilometers (70 sq mi) was covered with ash. Many habitats were destroyed.

Notice how the area has changed after construction. There is now even a lake and a pond that weren't there before. ▼

▲ This is what the area looked like before construction.

How People Change Ecosystems

People change ecosystems when they use the materials, or **resources**, from them. They cut down trees to build houses. They dig up rocks and stones to make roads. They use water that animals and plants depend on. Car exhaust pollutes the air. Trash pollutes rivers and fields. These kinds of changes can harm the ecosystem.

People also can help ecosystems. Planting trees after a fire helps develop new habitats. Bringing water to dry areas helps more plants grow.

 **CAUSE AND EFFECT** How are ecosystems affected by people?

Trashy Items
Make a list of five things you have thrown into the trash today. Compare your list with those of two classmates. Together, decide how much of the trash that you threw away could have been recycled.

189

Effects of Changes in Ecosystems

When one part of an ecosystem changes, it can affect the whole ecosystem. Changes such as fires cause some animals to run away. Those that stay must compete for the resources that are left. If there are not enough resources, populations get smaller. Some members die, and others leave. If a change brings more resources, populations grow.

Some organisms are able to adapt to changes in an ecosystem. People build homes in the habitats of bears, deer, and other animals. The people and animals then have to learn to share the habitats.

CAUSE AND EFFECT What can happen to organisms when an ecosystem changes?

▲ People often see moose in Alaska.

▲ In many places, ducks live near people.

▲ On some roads in Florida, drivers must stop their cars when alligators are crossing.

Deer often dart out in front of cars. Drivers must watch for them. ▼

Lesson Review

Essential Question

How do ecosystems change?

In this lesson, you learned that natural events can change ecosystems. Living things can also change ecosystems. Changes in an ecosystem can affect everything in that ecosystem.

1. **CAUSE AND EFFECT** Draw and complete a graphic organizer that shows causes of changes to an ecosystem.

2. **SUMMARIZE** Write a summary of this lesson. Begin with this sentence: Changes in ecosystems have two main causes.

3. **DRAW CONCLUSIONS** What are some ways in which people can change an ecosystem for the better?

4. **VOCABULARY** Write a sentence that identifies three resources in an ecosystem.

Test Prep

5. **Critical Thinking** Explain how living things can change an ecosystem.

Make Connections

 Writing

Narrative

Suppose people are building a neighborhood in an area where a group of bears lives. Write a **story** about what the bears do. You may make your story funny or serious.

 Health

Make a Poster

Interview a nurse or doctor. Ask him or her how air pollution changes a person's health. Make a poster of your findings. Share your poster with your class.

Science Spin
From Weekly Reader
TECHNOLOGY

HIDING IN PLAIN SIGHT

Cell phone tower

Over the past few years, there has been a huge increase in cell phone use in the United States. With that increased use has come a not-so-pretty side effect: ugly cell phone towers. Some companies, however, are working to make cell phone towers blend in with their surroundings.

Faster, Better, Uglier?

Cell phone companies are working to make cell phone technology faster and better than ever. Cell phone companies divide parts of a state into cells, or areas. Each cell has a base station with a tower and other equipment. Base stations send calls from one cell phone to another.

The problem is that not many people like seeing the tall metal towers. A company in Arizona has one way to solve that problem, however, and the idea came from observing nature!

The company makes trees, plants, leaves, and bushes from special materials that camouflage, or hide, cell towers. Because the United States has many types of environments—forests, deserts, and wetlands, for example—these hidden towers must be adapted to each environment.

A Tree or Not a Tree?

In most cases the towers are so well hidden that most people don't see them as towers. For example, the company made a cactus from a special plastic. The cactus fits with its desert environment and hides a cell phone tower. The company has made pine trees for mountain environments and palm trees for warmer areas.

Think and Write

1. How would you camouflage a cell tower in the area where you live?
2. What good and bad effects do these towers have on the communities around them?

Find out more. Log on to www.hspscience.com

Chapter 4 Review and Test Preparation

Vocabulary Review

Use the terms below to complete the sentences. The page numbers tell you where to look in the chapter if you need help.

environment p. 159
ecosystem p. 160
population p. 160
habitat p. 161
adaptation p. 178
resource p. 189

1. A material from an ecosystem is a _____.

2. A group of living things of the same kind is a _____.

3. A living thing is surrounded by its _____.

4. A living thing tries to get everything it needs to survive from its _____.

5. The living and nonliving things in an environment that interact make up an _____.

6. Any physical trait or behavior that helps a living thing survive is an _____.

Check Understanding

Write the letter of the best choice.

7. Which is a nonliving part of an environment?
   A. an animal   C. a root
   B. an insect   D. oxygen

8. Which kind of adaptation does the animal in the picture have?
   F. camouflage   H. migration
   G. instinct   J. mimicry

9. In which ecosystem can only plants and animals that need very little water survive?
   A. desert   C. grassland
   B. forest   D. tundra

10. If a bird flies south each winter, what is it doing?
    F. adapting
    G. camouflaging
    H. hibernating
    J. migrating

194

11. Which of these plants would most likely grow well in a desert?

    A. cactus 　　C. spider plant

    B. pine tree 　　D. sunflower

12. **MAIN IDEA AND DETAILS** Which is an example of an instinct?

    F. a dog sitting on command
    G. a raccoon getting food from a garbage can
    H. a leopard's spots
    J. a chipmunk eating extra food before it hibernates

13. What happens to an animal during hibernation?

    A. It eats a lot.
    B. It leaves its ecosystem often.
    C. It is inactive.
    D. It moves from one habitat to another.

14. Which is an example of mimicry?

    F. a white arctic hare in the snow
    G. a fly with stripes like a bee's
    H. a frog burrowing deep into the mud for winter
    J. ducks traveling to find food

15. Which of these includes all the other answer choices?

    A. an environment
    B. an ecosystem
    C. a population
    D. a community

16. **CAUSE AND EFFECT** Which can cause changes in the environment?

    F. animals and plants only
    G. nature only
    H. living things and nature
    J. people only

Inquiry Skills

17. You're walking in the forest and you find a small animal. It is not moving, but you see that it is breathing very slowly. What can you *infer* about this animal?

18. You see a volcano erupt on television. *Predict* what might happen next.

Critical Thinking

19. How might a blizzard change the habitats of plants and animals living in a forest ecosystem?

20. People build a new house near a tree in which a raccoon lives. How might the raccoon adapt its behavior to get food in its changed environment?

The Big Idea

195

CHAPTER 5
Living Things Depend on One Another

What's the Big Idea? All living things need energy to survive and grow.

Essential Questions

Lesson 1
How Do Plants and Animals Interact?

Lesson 2
What Are Food Chains?

Lesson 3
What Are Food Webs?

 Student eBook
www.hspscience.com

Snowy-bellied hummingbird

What do you wonder?

Hummingbirds use their long, thin beaks to get nectar from flowers. What are some other ways that animals get their food? How do this hummingbird and plant show the flow of energy within an ecosystem? How does this relate to the Big Idea?

LESSON 1

Investigate to find out how animals' teeth relate to the foods they eat.

Read and Learn about ways organisms get energy.

Essential Question

How Do Plants and Animals Interact?

Fast Fact

Going Ape
Orangutans use their teeth to eat. Their diet is usually fruit. In the Investigate, you will observe some animal teeth.

Bornean orangutan

Vocabulary Preview

producer [pruh•DOOS•er] A living thing that makes its own food (p. 203)

consumer [kuhn•SOOM•er] A living thing that gets its energy by eating other living things as food (p. 203)

decomposer [dee•kuhm•POHZ•er] A living thing that breaks down dead organisms for food (p. 203)

herbivore [HER•buh•vawr] An animal that eats only plants (p. 204)

carnivore [KAHR•nuh•vawr] An animal that eats other animals (p. 205)

omnivore [AHM•nih•vawr] A consumer that eats both plants and animals (p. 206)

199

Investigate

Checking Teeth

Guided Inquiry

Start with Questions

There are many kinds of horses, but most have teeth like the ones below.

- What do the horse's teeth look like?
- Why do horses have this kind of teeth?

Investigate to find out. Then read and learn to find out more.

Prepare to Investigate

Inquiry Skill Tip

To help you infer, start by making a list of all your observations.

Materials

- paper and pencil
- small mirror
- picture sorting cards

Make an Observation Chart

Kind of Animal	Drawing of Teeth	Description of Teeth	Kind of Food

Follow This Procedure

1. Copy the table onto paper.
2. **Observe** the picture sorting cards that your teacher has provided.
3. On your table, **record** the name of an animal. Draw the shape of its teeth. It might have teeth of different shapes.
4. **Record** words that describe the teeth.
5. Read the back of the picture sorting card. Record the foods the animal eats.
6. Repeat Steps 2–5 for four other animals.
7. Use the mirror to **observe** your own teeth. Add yourself to the table.

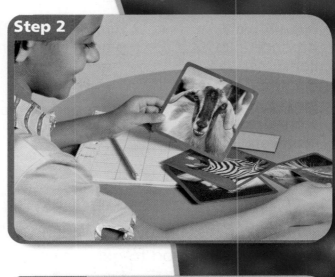
Step 2

Step 3

Draw Conclusions

1. Which of the animals in your table catch other animals for food? Which of the animals eat plants?
2. **Inquiry Skill** Scientists learn by **observing**. Then they use what they **observe** to **infer** the reasons for something. What can you **infer** about the shape of an animal's teeth and the kind of food it eats?

Independent Inquiry

Choose another animal. Find out what it eats, and **predict** what kind of teeth it has. Then find out if you are correct.

201

Read and Learn

VOCABULARY
producer p. 203
consumer p. 203
decomposer p. 203
herbivore p. 204
carnivore p. 205
omnivore p. 206

SCIENCE CONCEPTS
▶ how living things get energy
▶ how animals depend on plants

 **COMPARE AND CONTRAST**
Compare how different animals get energy.

alike —— different

Making and Getting Food

All living things need food. When you're hungry, your body needs food to get energy. You can make a sandwich, but can you really "make" your own food? Can you make the peanuts in your peanut butter?

Unlike you, plants are able to make their own food by using sunlight, air, and water. Plants store in their parts the extra food that they make. The food is full of energy.

Animals cannot make their own food. Some animals, such as rabbits, eat plants to get energy. The energy in the plants is taken into the rabbits' bodies. Other animals get energy by eating the rabbits.

COMPARE AND CONTRAST
How do plants and animals get food differently?

◀ The energy on Earth's surface comes from the sun. Without the sun, plants would have no energy. They could not make food and grow. Without plants, animals could not get energy and could not grow.

Producers, Consumers, and Decomposers

Plants are producers. A **producer** (pruh•DOOS•er) is a living thing that makes, or produces, its own food. Producers use this food to grow. Grass is a producer. So are trees and bushes.

Animals are consumers. A **consumer** (kuhn•SOOM•er) is a living thing that gets energy by eating other things as food. Consumers cannot make their own food. Deer, eagles, frogs, and even people are consumers.

Some kinds of living things are decomposers. A **decomposer** (dee•kuhm•POHZ•er) breaks down dead things for food. Earthworms, bacteria, and mushrooms are decomposers. Most of the decomposers are very small. You can see them only with a microscope.

COMPARE AND CONTRAST How are producers, consumers, and decomposers alike and different?

Sunflowers are producers. They store food in their seeds. The bird is a consumer. It gets energy by eating the seeds.

Earthworms eat dead plant parts and break them into tiny bits. Worms help put nutrients from the dead plants back into the soil. The nutrients will be taken in by the roots of next year's sunflowers.

Herbivores

There are three kinds of consumers: herbivores, carnivores, and omnivores. A **herbivore** (HER•buh•vawr) is a consumer that eats only plants. Tiny bees are herbivores, and so are giant pandas. People who eat only fruits and vegetables are herbivores, too.

Each herbivore has body parts that help it eat plants. A hummingbird has a long beak and tongue that help it reach nectar, a sweet liquid made by flowers. The nectar, found deep inside flowers, is the bird's food. A cow has flat teeth for chewing grass.

 COMPARE AND CONTRAST

How is a herbivore different from a producer?

▲ Caterpillars are herbivores because they eat only leaves.

▲ Horses are also herbivores. When they chew grass, their lower jaws move sideways instead of up and down.

This Galápagos (guh•LAH•puh•gohs) tortoise eats only plants. It is a herbivore. Loggerhead sea turtles eat fish and crabs, so they are not herbivores. Other turtles eat both plants and animals.

▲ Wolves are carnivores. They hunt in packs and eat only meat.

Carnivores

A **carnivore** (KAHR•nuh•vawr) is a consumer that gets its food by eating other animals. Carnivores have body parts that help them hunt and eat their food.

A red-tailed hawk, for example, has sharp eyesight. It can see a rabbit a mile away! The hawk has claws to help it catch the rabbit. Its beak tears meat apart easily.

A leopard's spots hide it as it sneaks up on its next meal. Its sharp teeth and claws help it catch, kill, and eat its food.

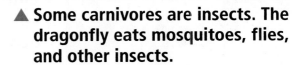

▲ Some carnivores are insects. The dragonfly eats mosquitoes, flies, and other insects.

Jobs for Teeth
Try eating a pretzel with just your front teeth. What happens? What does this tell you about the purposes of your front teeth and your back teeth?

 **COMPARE AND CONTRAST**

What is the same about a herbivore and a carnivore?

Omnivores

An **omnivore** (AHM•nih•vawr) is a consumer that eats both plants and animals. Do you eat hamburgers? If you do, then you are an omnivore. The meat is from a cow. The bun is from wheat, a plant.

The chimpanzee is an omnivore. It eats mostly fruit, but it also eats ants and other insects. Sometimes, it hunts and eats small mammals.

Most omnivores have teeth that help them eat both plants and animals. Sharp teeth in the front of your mouth help you tear meat. Flat teeth in the back of your mouth help you grind plants.

> The grizzly bear is the biggest omnivore in North America. It has a good sense of smell. As the bear looks for food, it often stops to sniff the air.

COMPARE AND CONTRAST
How are herbivores, carnivores, and omnivores different?

◀ Birds have beaks. Chickens use beaks to eat seeds as well as insects they catch.

Raccoons eat nearly everything— duck eggs, mice, frogs, sweet corn, insects, and fruit. ▶

Lesson Review

Essential Question

How do plants and animals interact?

In this lesson, you learned how organisms get energy. Organisms that make their own food are producers. Consumers get their energy by eating other living things. Decomposers break down dead things to get energy.

1. **COMPARE AND CONTRAST** Draw and complete a graphic organizer that compares and contrasts different types of consumers and what they eat.

2. **SUMMARIZE** Write a summary of this lesson that tells about herbivores, carnivores, and omnivores.

3. **DRAW CONCLUSIONS** In what way does a wolf depend on plants for energy?

4. **VOCABULARY** Make a crossword puzzle from the vocabulary words.

Test Prep

5. **Critical Thinking** Which one would a carnivore eat?
   A. a carrot
   B. a leaf
   C. a mosquito
   D. a seed

Make Connections

Writing

Expository
You have discovered a new kind of animal. Now you need to figure out if the animal is a herbivore, a carnivore, or an omnivore. Write an **explanation** of how you will do that.

Art

Consumer Mobile
Make a mobile showing the three types of consumers. Include at least two examples of each kind of consumer.

207

LESSON 2

Essential Question
What Are Food Chains?

Investigate to find out what a food chain is.

Read and Learn about how energy is passed from one organism to another.

Fast Fact

Lunch Time!
An osprey plunges down to the water with its talons out to catch a fish. Talons are claws on the osprey's toes that keep the fish from getting away. In the Investigate, you will find out what some other animals eat.

Osprey catching food

Vocabulary Preview

food chain [FOOD CHAYN] The path of food from one living thing to another (p. 212)

energy pyramid [EN•er•jee PIR•uh•mid] A diagram that shows how energy gets used in a food chain (p. 214)

predator [PRED•uh•ter] An animal that hunts another animal for food (p. 216)

prey [PRAY] An animal that is hunted by a predator (p. 216)

Investigate

Making a Food Chain

Guided Inquiry

Start with Questions

This squirrel is eating a seed to get energy.

- What other items are good food for squirrels?
- Are squirrels food for anything?

Investigate to find out. Then read and learn to find out more.

Prepare to Investigate

Inquiry Skill Tip

Some models have many parts. When you make a model, make sure that all the parts work together to get your idea across.

Materials

- 5 index cards
- tape
- marker
- 4 pieces of string or yarn
- tape

Make an Observation Chart

Organism	What It Eats	What Eats It

Follow This Procedure

1. Number five index cards 1 through 5 in the bottom right-hand corner.

2. On Card 1, draw and label grass. Draw and label a cricket on Card 2. On Card 3, draw and label a frog. Draw and label a snake on Card 4. On Card 5, draw and label a hawk.

3. Put the cards in order by number. Use the yarn and the tape to connect them.

4. Lay the connected cards on a table. You have **made a model** of a food chain!

Draw Conclusions

1. Which part of your food chain is a producer? Which parts are consumers?

2. **Inquiry Skill** Scientists **make a model** to study and understand a process. How does the model you made help you understand food chains?

Independent Inquiry

Find out what some other kinds of organisms eat. **Make a model** that shows the relationship between a producer and your consumers.

21

Read and Learn

VOCABULARY
food chain p. 212
energy pyramid p. 214
predator p. 216
prey p. 216

SCIENCE CONCEPTS
▶ how energy from food is passed to living things in a food chain

 **SEQUENCE**
Look for ways energy passes through a food chain.

☐ → ☐ → ☐

Food Chains

You are at the end of many food chains. A **food chain** shows the path of food from one living thing to another.

For example, one food chain begins with an apple tree that uses the energy in sunlight to help make its food. The tree stores some energy in its apples. When you eat an apple, you get the energy that was stored in the apple. That energy first came from sunlight.

A plant is the producer in this food chain. The plant uses sunlight to make its own food.

A grasshopper eats the tree's leaves. It gets energy that was stored in the leaves.

Another food chain begins with corn plants, which make their own food. Next, a chicken eats corn from the plant. Then, the chicken lays an egg, and you eat the egg. The energy in the egg that you eat started with energy from sunlight.

A food chain always begins with a producer. Some food chains are long, and others are short. In a food chain, smaller animals are usually eaten by larger animals. When an animal or a plant dies, decomposers become part of the food chain. Decomposers break down dead plants and animals, which become part of the soil again.

What happens to the energy in a seed after a bird eats it?

A shrew eats the grasshopper. It uses the energy from the insect to live and grow.

The owl is the last consumer in this food chain. It must eat many small animals to get the energy it needs.

Energy Pyramids

A mouse gets energy by eating plants. The mouse uses some of the energy to live and grow. If a fox eats the mouse, the fox gets only the energy the mouse did not use.

An **energy pyramid** is a diagram that shows how energy is used in a food chain. The producers are the biggest group in the pyramid. Energy from the producers is passed to the herbivores that eat them. The herbivores use some of that energy. The rest of the energy is then passed to the animals that eat the herbivores.

 SEQUENCE

What two kinds of consumers get energy after the herbivores?

A bobcat must catch and eat many small animals to get enough energy to live and grow. ▼

The View from the Top

Using Picture Sorting Cards 34–37, draw an energy pyramid on a separate sheet of paper. There are fewer living things at the top of the pyramid. Draw an appropriate number of living things at each level.

Science Up Close

For more links and animations, go to www.hspscience.com

Energy Pyramid

Each level in an energy pyramid has fewer living things than the level below it. The reason is that every level has less energy than the level below it.

A hawk is the top consumer in this energy pyramid.

The weasels are carnivores. They get their energy from eating the birds.

These birds are carnivores that get their energy by eating grasshoppers. Most of the energy is used to live. The rest of the energy is stored.

Grasshoppers are herbivores that get their energy by eating grass. Most of the energy is used to live. The rest of the energy is stored.

Grass, a producer, uses the energy of sunlight to make food.

Predator and Prey

You've learned that some animals eat other animals. An animal that hunts another animal for food is a **predator** (PRED•uh•ter). A wolf is a predator. So is an anteater.

An animal that is hunted for food is **prey** (PRAY). Rabbits and mice are often prey for the wolf. Ants are the prey of the anteater.

Some animals are both predators and prey. A small bird that eats insects is a predator. If a hawk hunts the small bird, then the bird becomes the prey and the hawk is the predator.

Focus Skill SEQUENCE

A medium-size fish is a predator. What could happen next to make that fish become prey?

▲ The puffin can carry up to 10 fish at one time to feed its chicks.

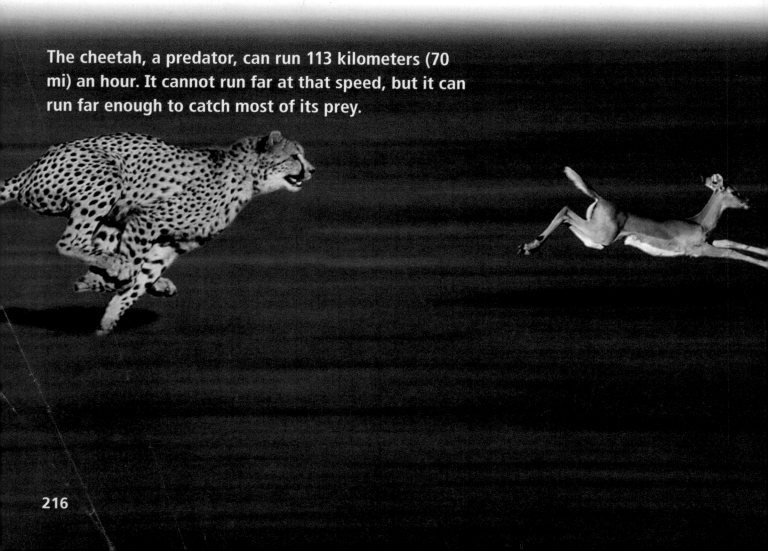

The cheetah, a predator, can run 113 kilometers (70 mi) an hour. It cannot run far at that speed, but it can run far enough to catch most of its prey.

Lesson Review

Essential Question

What are food chains?

In this lesson, you learned that food chains show the connections between what animals eat and what eats them. An energy pyramid shows how much energy is lost at each level of a food chain.

1. **SEQUENCE** Draw and complete a graphic organizer to show a food chain sequence.

2. **SUMMARIZE** Write a paragraph to summarize what producers, consumers, and decomposers do in a food chain.

3. **DRAW CONCLUSIONS** What effect would less sunlight have on an energy pyramid?

4. **VOCABULARY** Make up a matching quiz with this lesson's vocabulary words.

Test Prep

5. **Critical Thinking** Every night, you observe an owl catching mice. Explain how the owl gets energy that came from the sun, even though the owl hides in the dark shadows during the day.

Make Connections

 Writing

Expository
In 1972, a law stopped people from using DDT to kill insects. Find out how DDT affected eagles and food chains. Write a brief **report** about it.

 Math

Name a Fraction
A food chain is made up of a producer, an herbivore, an omnivore, and a carnivore. What fraction of living things in this food chain eat plants?

LESSON 3

Essential Question
What Are Food Webs?

Investigate to find out how food chains overlap.

Read and Learn about food webs and about how animals defend themselves.

Fast Fact

Tongue Tale
An anteater catches ants by using a very long, sticky tongue! Yet ants are not its only food. An anteater eats termites, too. In the Investigate, you will learn about other animals that are part of more than one food chain.

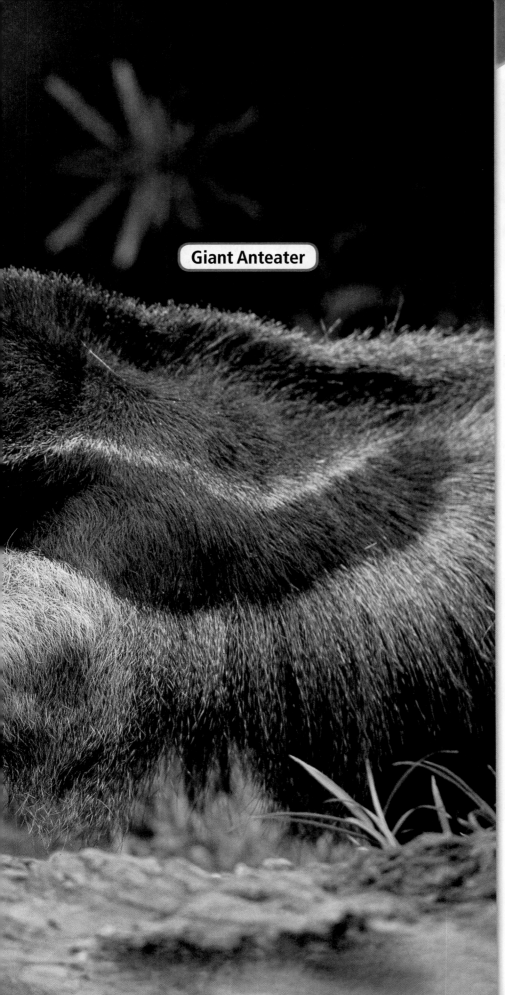

Giant Anteater

Vocabulary Preview

food web [FOOD WEB]
Food chains that overlap
(p. 222)

Investigate

Making a Food Web

Guided Inquiry

Start with Questions

This girl is eating lunch. The ingredients in her sandwich come from many sources.

- What types of food is the girl eating?
- Could you draw a food chain to show her eating three different types of food?

Investigate to find out. Then read and learn to find out more.

Prepare to Investigate

Inquiry Skill Tip

You have many ways to communicate. When you have an idea you want to share, decide whether writing, speaking, or using pictures or diagrams will work best.

Materials

- index cards, cut into fourths
- poster board
- tape or glue
- crayons

Make a Data Table

Living Thing	How Many It Eats	How Many Eat It
clover		
grasshopper		
frog		
snake		
owl		
mouse		

Follow This Procedure

1. Write on its own card the name of each living thing from the table.
2. Tape or glue the cards in a circle on the poster board.
3. Use the table to make two different food chains. **Record** them on paper.
4. Use a crayon to draw arrows between the parts of one food chain. Use a different color to draw arrows between the parts of the other food chain.
5. **Observe** where the food chains overlap. You have **made a model** of a food web.

| Food Web Table ||
Living Thing	What It Eats
clover	uses the sun to make its own food
grasshopper	clover
frog	grasshopper
snake	frog, mouse
owl	snake, mouse
mouse	clover

Step 4

Draw Conclusions

1. Why should both food chains start with clover?
2. **Inquiry Skill** Scientists often use models, graphs, and drawings to **communicate** ideas. How does your model help **communicate** what a food web is?

Independent Inquiry

Choose an animal you like. Find out what it eats and what eats it. Then **make a model** of a food web that includes this animal.

Read and Learn

VOCABULARY
food web p. 222

SCIENCE CONCEPTS
- what food webs are and how they can change
- how animals stay safe

MAIN IDEA AND DETAILS
Look for details about food webs.

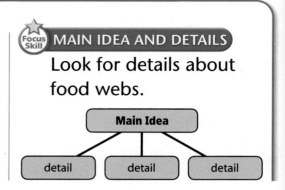

Food Web

Most animals eat many different things. You do, too. This means that you and most animals are part of many food chains. Overlapping food chains are called a **food web**.

MAIN IDEA AND DETAILS What is a food web?

This heron is a predator. It eats fish, lizards, frogs, mice, and insects. A young heron can also be an alligator's prey. ▶

▲ Every food chain and food web begins with producers. These plants provide food for animals. Some also provide shelter.

Some Ways Animals Defend Themselves

Mice are prey of hawks, owls, and other animals. A mouse might seem weak and helpless. Yet even small animals such as mice have ways to defend themselves.

For example, the mouse's color helps it blend in with its surroundings. Mice go out mostly at night. The darkness helps hide them. Most mice can hear, see, and smell very well. Their senses help them know if a predator is nearby.

When in danger, an opossum lies on its side with its eyes open, playing dead. Predators do not often attack dead animals. ▶

Some snakes defend themselves by biting. The poison released from their fangs may kill their prey—or their predators.

▲ Skunks spray a smelly liquid at their predators. If you see a skunk raise its tail, run away fast!

▲ Wildebeests stay in herds to be safe. A wildebeest that strays from the herd may become prey to predators.

▲ When an octopus is in danger, it squirts ink into the water. The cloud of ink hides the octopus so it can escape.

Insects can be prey, too. To help them hide, many insects have colors that blend in with their surroundings. Others have shapes like leaves or sticks that fool predators.

Some animals, like the gazelle, can outrun predators. A gazelle can run 80 kilometers (50 mi) an hour. Yet its main predator, the cheetah, runs even faster at 113 kilometers (70 mi) per hour. However, the gazelle runs in a zigzag pattern. That makes it harder to catch. If the cheetah does not catch the gazelle quickly, it gives up.

▲ The anemone stings other fish but not the clown fish. This helps the clown fish stay safe from predators.

 MAIN IDEA AND DETAILS

What are three ways that animals defend themselves?

Hiding from Predators

Put 10 peas and 10 kernels of corn on a sheet of yellow paper. Close your eyes for a minute. Then open them, and pick up the first 5 seeds you see. Compare results with classmates. What do you notice?

225

Changes in Food Webs

Many things can change a food web. An increase in plants often causes an increase in herbivores. More herbivores can lead to more carnivores and omnivores. On the other hand, fewer plants mean fewer animals. A change to one part of a food web can change many other parts of the food web.

Adding a new plant or animal can also change a food web. A new plant may crowd out other plants. A new animal may eat many other animals or plants.

 **MAIN IDEA AND DETAILS**

Name two things that can change a food web.

Math in Science
Interpret Data

The cane toad was introduced in Australia to get rid of the cane beetle. It was not known at the time that the toad would reproduce quickly and interrupt other food webs. If the population increase of the cane toad isn't stopped, what will happen in the next step of this graph?

Lesson Review

Essential Question

What are food webs?

In this lesson, you learned that food chains that overlap form food webs. Food webs can be changed in different ways. Many animals have ways to protect themselves from being eaten.

1. **MAIN IDEA AND DETAILS** Draw and complete a graphic organizer for this main idea: There are three parts of a food web.

   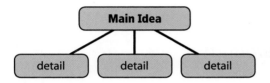

2. **SUMMARIZE** Write a paragraph that tells the most important information in this lesson.

3. **DRAW CONCLUSIONS** What is the difference between a food web and an energy pyramid?

4. **VOCABULARY** Write a sentence to explain to a younger student in your school what a food web is.

Test Prep

5. Which of these is **least** likely to get eaten in a food web?
   A. herbivore
   B. predator
   C. prey
   D. producer

Make Connections

Writing

Persuasive

To get rid of mosquitoes in your town, some people want to bring in a bird from Brazil that eats mosquitoes. Write a **letter** to the editor of your town's newspaper telling your opinion.

Art

Paper Food Web

Make a model of a food web out of paper-chain links. Draw a plant or an animal on each link. You might connect one link, such as grass, to several other links.

People in Science

Margaret Morse Nice

Perhaps, like Margaret Morse Nice, you have watched birds in your back yard. By the age of 12, she was recording her observations about birds. What began as fun became her lifework.

After college, Morse married Leonard Nice, whom she had met there. Each time her family moved, Nice studied the birds around her home.

Nice was one of the first to use colored bands to identify birds. To follow birds over time, she would first capture a bird. Then she would place a colored band of plastic on the bird's leg and let the bird go. In this way, Nice studied how birds mated, built nests, and raised their young.

Nice and her husband worked together to write a book on the birds of Oklahoma. Nice later published a two-part study of song sparrows. She also translated into English many articles that people from other countries had written about birds.

▶ **MARGARET MORSE NICE**
▶ Ornithologist, a person who studies birds

Think and Write

1. Why is it helpful to have someone translate articles written by scientists in other countries?
2. What birds can you observe from your home?

Rodolfo Dirzo

► RODOLFO DIRZO
► Tropical Ecologist

When he was growing up in Mexico, Rodolfo Dirzo used to watch bugs. That led him to study snails and slugs far away in Wales. After completing his education, he returned to Mexico.

Dirzo studies two different forests in Mexico. In one, the forest has not been disturbed. In the other, some of the forest has been cut down. As a result, many animals that would be expected to live there are no longer there. Dirzo compares the two places. Without animals to help spread seeds and to trample down vegetation, the forest plants may change. Dirzo expects that there will be fewer kinds of trees without the animals to help provide places for different plants to grow.

 Think and Write
1. What might cause animals to leave a forest?
2. Why does Dirzo compare two different forests?

Career Environmental Scientist

The U.S. government has laws that protect our air, soil, and water. Environmental scientists help make sure those laws are obeyed. These scientists are on the lookout for pollution. If they find it, they study the pollution to try to figure out ways to stop it.

Chapter 5 Review and Test Preparation

Vocabulary Review

Use the terms below to complete the sentences. The page numbers tell you where to look in the chapter if you need help.

producers   p. 203
consumers   p. 203
decomposers   p. 203
herbivore   p. 204
carnivore   p. 205
food chain   p. 212
predator   p. 216
food web   p. 222

1. An animal that eats only meat is a _____.
2. Dead plants and animals are broken down by _____.
3. Food moves from one living thing to another in a path called a _____.
4. Carnivores and omnivores are both _____.
5. The overlapping paths of food are shown by a _____.
6. A cow is both a consumer and a _____.
7. A cheetah is a consumer, a carnivore, and a _____.
8. The only living things that can make their own food are _____.

Check Understanding

Write the letter of the best choice.

9. What is the source of energy for plants?
   A. food webs   C. producers
   B. prey   D. sunlight

10. **SEQUENCE** How can energy pass from a wheat plant to a wolf?
    F. through a carnivore
    G. through a herbivore
    H. through a predator
    J. through a producer

11. What can cause an increase in herbivores?
    A. an increase in predators
    B. a decrease in prey
    C. an increase in producers
    D. an increase in carnivores

12. To which group does this mushroom belong?
    F. carnivore   H. herbivore
    G. decomposer   J. prey

230

13. Which of these can cause a change in a food web?
  A. animals that protect themselves by running fast
  B. animals that blend in with their surroundings
  C. animals that are new to the ecosystem
  D. animals that live in herds

14. MAIN IDEA AND DETAILS What is the main purpose of this diagram?

  F. to show what eats what
  G. to outline an ecosystem
  H. to show that producers make their own food
  J. to show that the amount of energy that passes from one level to the next becomes less

15. What are most people?
  A. carnivores
  B. herbivores
  C. omnivores
  D. producers

16. Which of these does **not** happen?
  F. Plants become food.
  G. A predator becomes prey.
  H. An omnivore eats meat.
  J. A carnivore eats plants.

Inquiry Skills

17. How could you use index cards to make a model of a food chain?

18. A turtle is chewing on weeds growing beside a pond, and it snaps up a dragonfly that gets too close. To which group of consumers can you infer that the turtle belongs?

Critical Thinking

19. What would happen to food webs if thick clouds from a volcanic eruption hid the sun for a year?

20. How would your life change if you were part of only one food chain?

Visual Summary

Tell how each picture shows the **Big Idea** for its chapter.

 Big Idea

All the living, once-living, and nonliving things in an environment interact in an ecosystem.

 Big Idea

All living things need energy to survive and grow.

EARTH SCIENCE

ILLINOIS

Illinois Excursions and Projects
Illinois Climate 234
Herbert Trackman Planetarium 236
Apple River Canyon State Park 238
Stories in Rocks 240

UNIT C
EARTH'S LAND 241

CHAPTER 6
Minerals and Rocks 242

CHAPTER 7
Forces That Shape the Land 280

CHAPTER 8
Conserving Resources 316

UNIT D
WEATHER AND SPACE 363

CHAPTER 9
The Water Cycle 364

CHAPTER 10
Earth's Place in the Solar System 402

Illinois Excursions

Illinois Climate

Illinois has four seasons. Each year the weather patterns follow the same cycle. What is your favorite season? Each season has things to enjoy.

Crops are ready to be harvested in the fall.

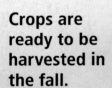

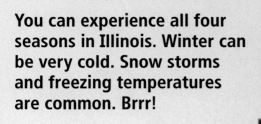
You can experience all four seasons in Illinois. Winter can be very cold. Snow storms and freezing temperatures are common. Brrr!

Four Seasons

Have you ever jumped into a pile of leaves? The red, purple, and yellow leaves on the trees tell you that fall is here. The shorter days and cooler temperatures also remind you that fall has arrived. Bundle up! Winter comes after fall. Winter in Illinois is usually cold! It can bring snow and very cold temperatures.

Spring follows the winter. Spring brings milder temperatures, but also brings severe weather. Many plants begin to grow again. Summer is hot in Illinois. It's a good time to play outside, but don't get too hot.

The seasons are caused by Earth's tilt on its axis. The tilt causes different parts of Earth to face the sun more directly at some times of year and less directly at other times. It is easy to tell what season of the year it is by looking at your environment.

Think And Write

1. **Scientific Thinking** Which season is your favorite? Compare it to the other seasons.
2. **Scientific Inquiry** The weather in Illinois follows a predictable cycle. Identify and explain something else in nature that follows a cycle.

Illinois Excursions

Joliet

Herbert Trackman Planetarium

Are you a stargazer? At the Herbert Trackman Planetarium, in Joliet, you can learn a lot about the universe and the night sky. The planetarium is a special kind of theater. It is shaped like a dome. Images of the night sky are shown on the inside of the dome.

Patterns in the Night Sky

You have probably seen the Big Dipper at night. It is a part of the Great Bear, one of the many constellations, or star patterns, that you can see in the night sky. Maybe you have noticed the Little Bear before. In its pattern, you see Polaris, also called the North Star. In the Northern Hemisphere, you can see Polaris in the same position each night.

Did you know that Earth is always moving? Have you noticed how the moon appears to change each night? Did you know that some of the bright things in the sky are planets? The planets, like the stars, move in patterns. The planetarium is a good place to learn about patterns in the night sky. It is out of this world!

Star charts like this one can help you find stars in the night sky.

Think And Write

1. **Science and Technology** Use the Internet to research some constellations. Write a paragraph naming the constellations you researched and describing their patterns.

2. **Scientific Thinking** Polaris is located directly above the North Pole. Explain why Polaris is useful for travelers.

Illinois Excursions

Apple River Canyon State Park

APPLE RIVER CANYON STATE PARK

Do you like to camp? If so, you will enjoy Apple River Canyon State Park. The park is in the hills of northwestern Illinois. At the park, you can see many kinds of wildlife. The river also has a variety of fish for people who enjoy fishing. You may even spot an eagle high up in a tree while you are there.

The Apple River looks quiet and peaceful here. Over time, the river has made big changes in the landscape.

Fishing at the park

Water has weathered some of this rock wall.

Weathering and Erosion

While at the river, notice the canyon's steep walls. The moving waters of the Apple River formed the canyon. The river slowly wore away pieces of limestone and shale. This process is called weathering. As the pieces of rock were broken into smaller pieces, the river carried them away to new locations. This process is called erosion.

If you explore the state park, you can see other changes weathering and erosion have made. Who knows? The rock you sit on after a long hike may have come from a place far away!

Think And Write

1. **Scientific Thinking** How are weathering and erosion alike? How are they different?

2. **Scientific Inquiry** Research to find out how erosion and weathering can change the course of a river. Write a paragraph to describe how the flow of a river can be affected by these factors.

Illinois Projects and Activities

Project: Stories in Rocks

Materials
- collection of rocks
- hand lens
- Internet connection

Procedure
1. Use the Internet to find images of igneous, metamorphic, and sedimentary rocks.
2. Examine each of your rocks. Compare them with the images you researched.
3. Sort the rocks into groups based on their types. Use the images you researched to help you sort them.

Draw Conclusions
1. How did you decide whether the rocks were metamorphic, igneous, or sedimentary?
2. What events might have contributed to the formation of your rocks? What do the rocks tell you about the land from which they came?

Earth's Land

UNIT C EARTH SCIENCE

CHAPTER 6
Minerals and Rocks 242

CHAPTER 7
Forces That Shape the Land 280

CHAPTER 8
Conserving Resources 316

Unit Inquiry

Hurricane Damage

Hurricanes can be dangerous storms if you are not prepared. Scientists and engineers have designed buildings to withstand the strong winds of a hurricane. What types of structures are the most stable? Plan and conduct an experiment to find out.

CHAPTER 6
Minerals and Rocks

What's the Big Idea? Rocks are made up of different minerals, and they go through changes during the rock cycle.

Essential Questions

Lesson 1
What Are Minerals and Rocks?

Lesson 2
What Are the Types of Rocks?

Lesson 3
What Are Fossils?

 Student eBook
www.hspscience.com

What do you wonder?

A starfish fossil like this one gives clues that tell about both the starfish and the rock around it. What could scientists learn about life in the past from this fossil? How does this relate to the **Big Idea?**

Fossil of starfish

LESSON 1

Essential Question

What Are Minerals and Rocks?

Investigate to find one way to help identify minerals.

Read and Learn about the different types of minerals and how to identify them.

Fast Fact

Hard Minerals
Not all minerals are the same. One of the ways minerals differ is in their hardness. Some minerals are hard enough to cut steel. In the Investigate, you will learn more about the hardness of minerals.

Vocabulary Preview

mineral [MIN•er•uhl] A solid object found in nature that has never been alive (p. 248)

rock [RAHK] A naturally formed solid made of one or more minerals (p. 252)

Rock and mineral samples

Investigate

Testing Minerals

Guided Inquiry

Start with Questions

There are many types of minerals. They all have characteristics that make them different. One characteristic that can be tested is hardness.

- How could you test how hard a mineral is?
- How could you identify the types of minerals in the photo?

Investigate to find out. Then read and learn to find out more.

Prepare to Investigate

Inquiry Skill Tip

To put a series of objects in order, begin by placing two objects in order. Then fit another object into the sequence. Continue, adding one object at a time.

Materials

- safety goggles
- minerals labeled A through G

Make an Observation Chart

Mineral to Test	Minerals It Scratches	Minerals That Scratch It
Sample A		
Sample B		
Sample C		
Sample D		
Sample E		
Sample F		
Sample G		

Follow This Procedure

1. Make a table like the one shown.

2. **CAUTION: Put on safety goggles.** A harder mineral scratches a softer mineral. Try to scratch each mineral with Sample A. **Record** which minerals Sample A scratches.

3. A softer mineral can be scratched by a harder mineral. Try to scratch Sample A with each of the other minerals. **Record** the minerals that scratch Sample A.

4. Repeat Steps 2 and 3 for each mineral.

5. Using the information in your table, **order** the minerals from softest to hardest.

Draw Conclusions

1. Which mineral is hardest? Which is softest? How do you know?

2. **Inquiry Skill** Scientists often put objects in **order**. How did you decide how to **order** the minerals? How can putting minerals in **order** of hardness help you identify them?

Step 1

Step 2

Independent Inquiry

Test the hardness of each mineral again. This time, use a penny and your fingernail. **Classify** the minerals by what scratches them.

Read and Learn

VOCABULARY
mineral p. 248
rock p. 252

SCIENCE CONCEPTS
▶ what minerals and rocks are

MAIN IDEA AND DETAILS
Look for main ideas about minerals and rocks.

```
        Main Idea
       /    |    \
  detail  detail  detail
```

Minerals

You might know that minerals are found in some of the foods you eat. You may have even seen someone wearing minerals. If an object is solid, formed in nature, and has never been alive, it is likely to be a **mineral**. There are many different minerals, and no two kinds are exactly alike. Gold, for example, is shiny. Graphite (GRAF•yt) is dull and dark, and it is so soft that you can write with it. Yet diamonds are so hard that people use them to cut steel.

Only about 100 minerals are common. One of the most common minerals is quartz.

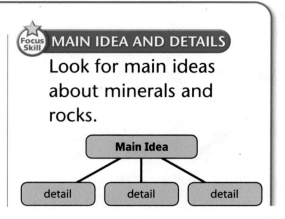
Mica

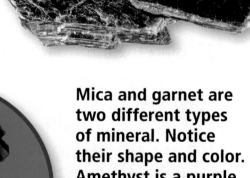
Garnet

Mica and garnet are two different types of mineral. Notice their shape and color. Amethyst is a purple form of quartz.

Amethyst

Quartz

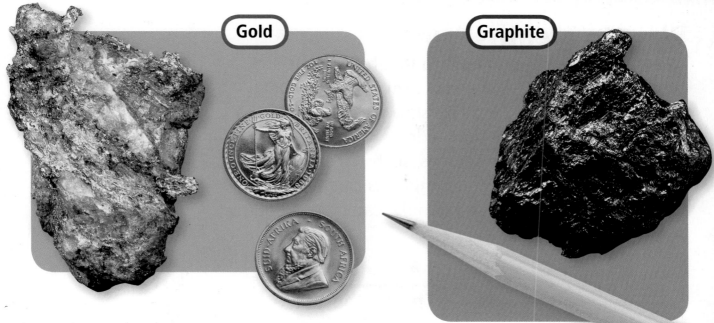

You use some of these minerals every day.

You use minerals throughout the day. If you put salt on your food, you are using a mineral. Minerals are also in many of the things around you. For example, quartz is used to make the glass in a window. Iron that is used in buildings comes from minerals, such as hematite. Minerals such as gold and diamond are often used to make jewelry. If you have a penny dated before 1983, it is made mostly of copper. Copper comes from minerals.

What are three ways minerals are used?

Ways to Identify Minerals

In the Investigate, you learned that hardness is one property of minerals. How hard a mineral is can help identify that mineral. The scale shown on this page is a tool scientists use to tell the hardness of a mineral.

Minerals have other properties, such as color, that can also help identify them. Some minerals can have more than one color. Quartz, for example, can be pink, purple, white, clear, or even black.

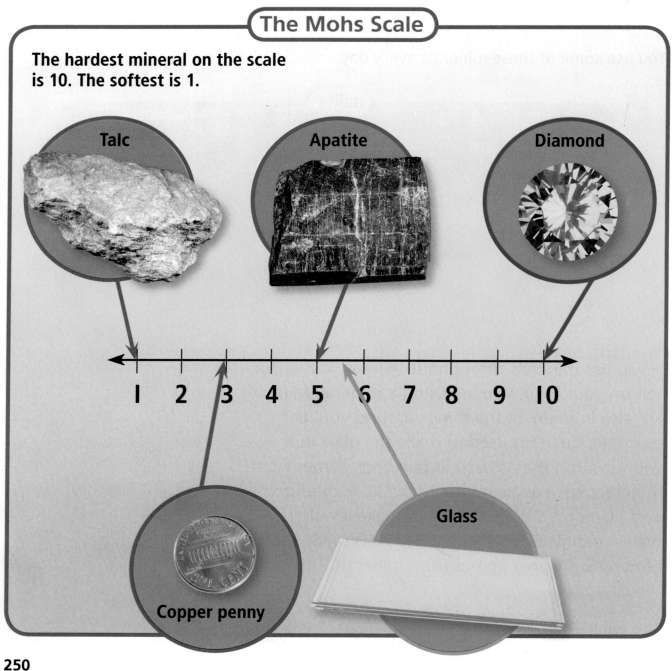

The Mohs Scale

The hardest mineral on the scale is 10. The softest is 1.

A streak test is one way to identify a mineral. ▼

Streak is another way to identify minerals. Streak is the color of the powder left behind by a mineral when it is rubbed against a rough white tile. The streak is usually the same color as the mineral. However, the streak can be a different color than the mineral's outside color.

 MAIN IDEA AND DETAILS

What are two ways to identify minerals?

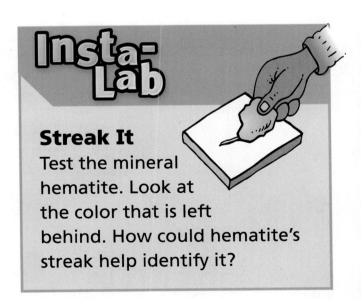

Streak It
Test the mineral hematite. Look at the color that is left behind. How could hematite's streak help identify it?

Rocks

Earth is made mostly of rock. A **rock** is a naturally formed solid made of one or more minerals. Look at the rock below.

Granite is made of several different minerals. Feldspar, quartz, and different types of mica can be found in this granite. Some granite may also have traces of other minerals.

Some rocks are made of only one type of mineral. Limestone is made up of many tiny grains of calcite. Even though there are many grains, calcite is the only mineral in limestone.

MAIN IDEA AND DETAILS
What is rock made of?

The Texas State House

▲ Builders use granite because it is beautiful and very strong.

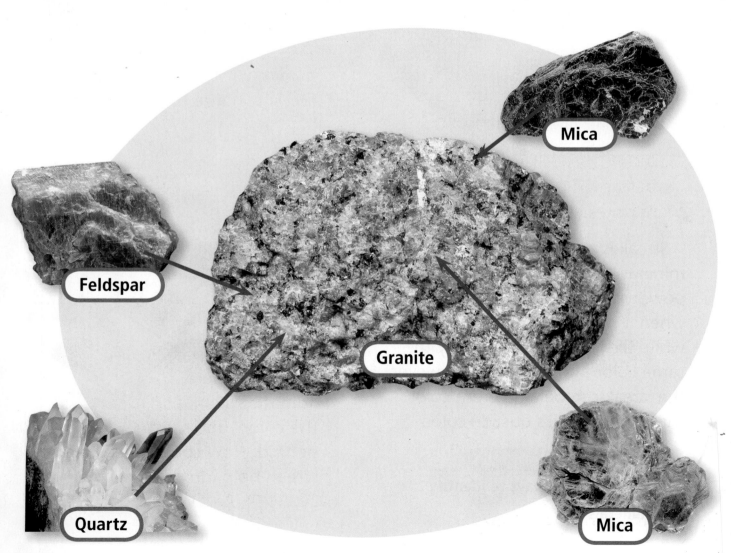

252

Lesson Review

Essential Question

What are minerals and rocks?

In this lesson, you learned that there are many types of minerals. They can be identified by how hard they are and by their streak. Rocks are made up of one or more minerals.

1. **MAIN IDEA AND DETAILS** Draw and complete a graphic organizer for the main idea: Minerals and rocks are closely related.

   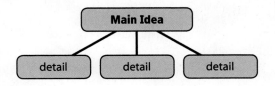

2. **SUMMARIZE** Write a summary of this lesson. Tell about some ways to identify minerals.

3. **DRAW CONCLUSIONS** Why is it better to identify a mineral by its hardness than by its color?

4. **VOCABULARY** Write at least two sentences comparing and contrasting a rock and a mineral.

Test Prep

5. Which is true of all minerals?
   - A. They are all hard.
   - B. They were never alive.
   - C. They are made of rocks.
   - D. They were once animals.

Make Connections

 Writing

Expository
Look at the rocks where you live. Choose one that looks interesting to you. Write a **description** of the rock. Tell about the color of the rock, how it feels, and what it looks like.

 Social Studies

The Changing Penny
The United States penny has changed through the years. Research the different materials the penny has been made of. Make a poster to share your findings.

LESSON 2

Investigate to find out how one group of rocks is formed.

Read and Learn about how the different groups of rocks are formed and how people use them.

Essential Question

What Are the Types of Rocks?

Fast Fact

Rock House
These cliff houses were built more than 1,000 years ago. In the Investigate, you will model a way these rocks might have formed.

Cliff houses

Vocabulary Preview

igneous rock [IG•nee•uhs RAHK] Rock that was once melted and then cooled and hardened (p. 258)

sedimentary rock [sed•uh•MEN•ter•ee RAHK] Rock made when materials settle into layers and get squeezed until they harden into rock (p. 259)

metamorphic rock [met•uh•MAWR•fik RAHK] Rock that has been changed by heat and pressure (p. 259)

Investigate

Make a Model Rock

Guided Inquiry

Start with Questions

This truck is pouring cement. Certain types of rock are used in making cement.

- How is cement used in your neighborhood?
- How is this cement made?

Investigate to find out. Then read and learn to find out more.

Prepare to Investigate

Inquiry Skill Tip

When you can use a model to make observations, make a list of ways you can compare the model and the actual object.

Materials

- newspaper
- wax paper
- paper or plastic cup
- plastic spoon
- sand
- gravel
- white glue
- water
- hand lens

Make an Observation Chart

Property	Observation
Color	
Texture	
Hardness	
Drawing of Rock	

Follow This Procedure

1. Spread newspaper over your work area. Place a smaller sheet of wax paper on the newspaper.
2. Place 1 spoonful of sand in the cup.
3. Add 1 spoonful of gravel to the cup. Stir the sand and gravel.
4. Add 1 spoonful of glue to the cup.
5. Stir the mixture until it forms a lump. You may need to add a little water.
6. Pour the mixture onto the wax paper, and let it dry. You have made a model of a rock.

Draw Conclusions

1. Use the hand lens to observe the dried mixture you made. What does the mixture look like?
2. **Inquiry Skill** Scientists often use models to understand processes they can't easily observe. Rocks can form when sand and gravel are somehow cemented or glued together. How is the model you made like a rock?

Independent Inquiry

Make models of rocks by using different materials. Explore how changing the materials changes the rock.

257

Read and Learn

VOCABULARY
igneous rock p. 258
sedimentary rock p. 259
metamorphic rock
  p. 259

SCIENCE CONCEPTS
▶ what the three groups of rocks are
▶ how rocks form

COMPARE AND CONTRAST
Look for the words *alike* and *different* when reading about types of rocks.

alike —— different

Types of Rocks

Suppose you want to start a rock collection. How would you group the rocks? By color, by grain size, by whether they have layers? Rocks can look very different, but there are just three main groups of rocks. Rocks are grouped by how they form.

Igneous (IG•nee•uhs) **rock** is rock that was once melted and then cooled and hardened. Some igneous rocks cool quickly and look like glass. Other igneous rocks cool more slowly and have large grains.

Igneous Rock

Obsidian (uhb•SID•ee•uhn)

Granite

Rock that forms from material that has settled into layers is called **sedimentary** (sed•uh•MEN•ter•ee) **rock**. The layers are squeezed together until they form rock.

The third group of rocks is called metamorphic rock. **Metamorphic** (met•uh•MAWR•fik) **rock** is rock that has been changed by temperature and pressure.

 COMPARE AND CONTRAST

How is metamorphic rock different from igneous rock?

Fizzy Rock
Calcite (KAL•syt) bubbles when it comes in contact with vinegar. Use a dropper to place several drops of vinegar on limestone and on sandstone. Which rock has calcite?

How Rocks Form

The three groups of rock form differently. Melted rock can reach Earth's surface through a volcano. Then it cools and becomes igneous rock.

Sedimentary rock, by contrast, begins with the breaking of rocks. Wind and water break rock into bits. Then the wind and water carry away the bits of rock and soil. The bits settle into layers. After a long time, the layers harden into rock.

Metamorphic rock forms deep in Earth's crust in yet another way. The pressure and temperature there change rock into metamorphic rock.

COMPARE AND CONTRAST

Contrast the ways the three types of rock form.

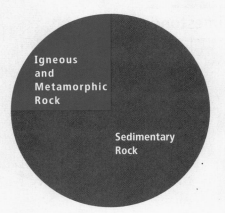

Math in Science
Interpret Data

Rocks That Make Up Earth's Surface

What fraction of Earth's surface is covered by sedimentary rock?

Igneous and Metamorphic Rock

Sedimentary Rock

Vents in volcanoes move melted rock to the surface, where it cools. The cooled rock is igneous rock. ▶

▲ Wind and water break rock into small bits. Then wind and water carry away the bits of rock. Layers of these bits of rock form and, under pressure, become sedimentary rock.

◀ Rock inside Earth softens from Earth's high temperature. Pressure in Earth's crust then squeezes the rock. The rock changes into metamorphic rock.

The Rock Cycle

Over time, one kind of rock can become any other kind. The process of rocks changing from one kind of rock to another kind of rock is called the rock cycle. The diagram on the right-hand page shows this cycle.

Wind and water break down all kinds of rocks to form sedimentary rocks. Any kind of rock that melts and cools can become igneous rock. Any rock can end up in Earth's crust and be pressed and heated. Then that rock can become metamorphic rock.

How are all rocks alike and different?

This Colorado mountain range has all three kinds of rocks. ▼

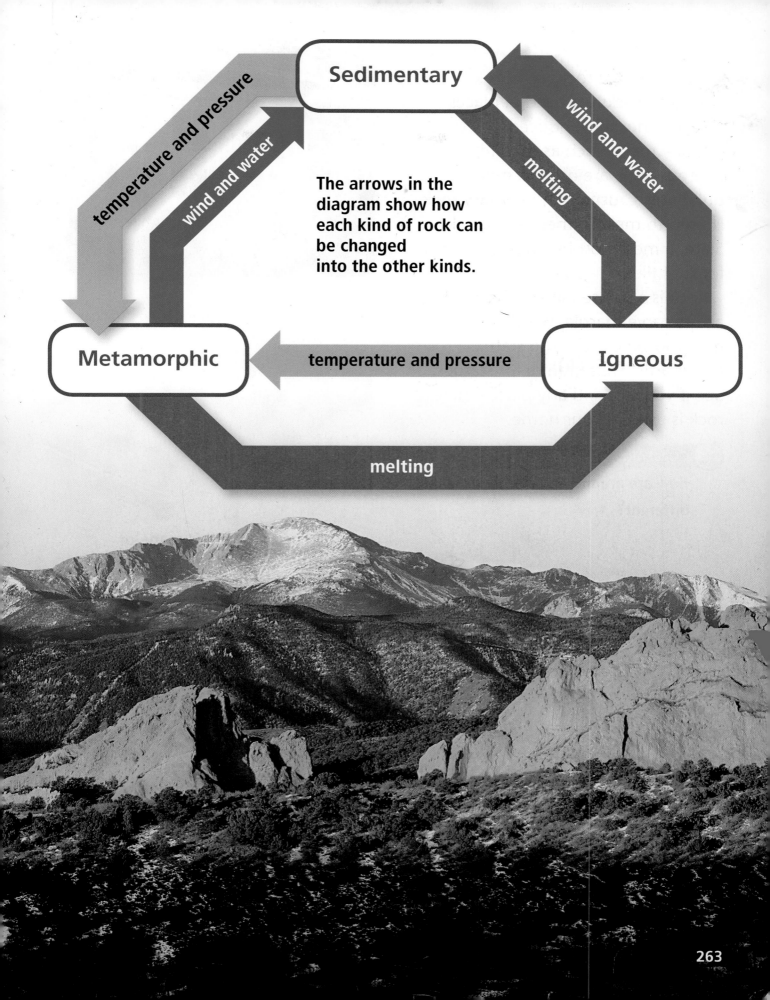

How People Use Rocks

You often see rocks outside. However, rocks are used in places you may not have noticed.

Many rocks, such as granite, an igneous rock, are used to make buildings. Crushed sedimentary rock is used to make cement and bricks. Slate, a metamorphic rock, is used for roof tiles.

Rock is also used in art. Some artists carve marble, which is a metamorphic rock, into statues. Other artists use granite for statues. See if you can find any place where rock is used in your home.

COMPARE AND CONTRAST
How are granite and marble alike and different?

The cheese board is made of marble, a metamorphic rock. Mount Rushmore is made up of mostly igneous and metamorphic rock. Stones for houses come from all three types of rock.

Lesson Review

Essential Question

What are the types of rocks?

In this lesson, you learned that the three groups of rocks are igneous, sedimentary, and metamorphic. The rock cycle is the process by which one kind of rock becomes another.

1. **COMPARE AND CONTRAST** Copy and complete a graphic organizer that compares and contrasts the different kinds of rocks.

2. **SUMMARIZE** Write a three-sentence summary for this lesson. Tell about the rock cycle.

3. **DRAW CONCLUSIONS** Why are few sedimentary rocks found deep inside Earth?

4. **VOCABULARY** Write a definition for *metamorphic rock.*

Test Prep

5. **CRITICAL THINKING** Suppose you're walking near a river. You see a dark gray rock that has layers like thick pages in a book. You pick up the rock, and some of the "pages" start to separate in your hands. What type of rock is it? Explain.

Make Connections

Writing

Expository
Interview someone who works with rocks or minerals. Find out how he or she uses them. Write a **report** about what you learn, and share your report with the class.

Drama

Rock Cycle Play
With four other classmates, write a play about the rock cycle. Your play can be funny or serious. Perform the play for the class.

265

LESSON 3

Essential Question
What Are Fossils?

Investigate to find out how some fossils form.

Read and Learn about the different types of fossils and how they form.

Fast Fact

Learning from the Past
Animals like this one no longer live on Earth, but fossils show that they looked similar to some animals alive today. In the Investigate, you will find out about how some fossils form.

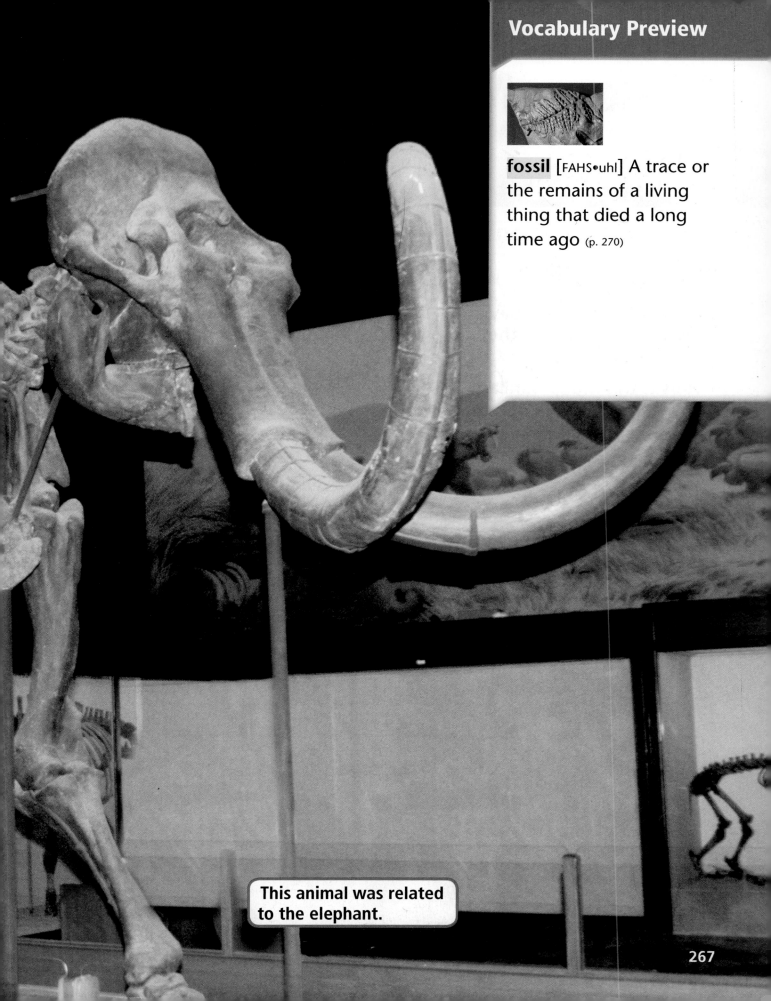

Vocabulary Preview

fossil [FAHS•uhl] A trace or the remains of a living thing that died a long time ago (p. 270)

This animal was related to the elephant.

267

Investigate

Make a Model Fossil

Guided Inquiry

Start with Questions

No dinosaurs are alive today. This model shows what some dinosaurs may have looked like when they were alive.

- How is this model similar to living animals that you've seen before?
- How do scientists know what this dinosaur may have looked like?

Investigate to find out. Then read and learn to find out more.

Prepare to Investigate

Inquiry Skill Tip

When you use a model, you can make many kinds of observations. You can observe how the model looks and how the model was made. Compare these observations to the real object's appearance and the way it formed.

Materials

- seashell
- petroleum jelly
- modeling clay
- small bowl or paper plate
- white glue

Make an Observation Chart

Kind of Fossil	Description
Modeling Clay Fossil	
Glue Fossil	

Follow This Procedure

1. Coat the outside of the seashell with a thin layer of petroleum jelly.
2. Press the seashell into the clay to **make a model** of a fossil.
3. Remove the seashell carefully from the clay.
4. Place the clay with the seashell's shape in the plastic bowl.
5. Drizzle white glue into the imprint. Fill it completely. This also makes a **model** of a fossil.
6. Let the glue harden for about a day. When it is hard, separate the hardened glue from the clay.

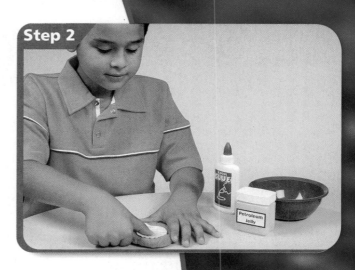

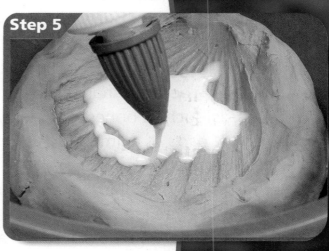

Draw Conclusions

1. You made two **models** of fossils. How do the fossils compare?
2. **Inquiry Skill** Scientists **use models** to better understand how things happen. How does pressing the seashell into the clay model how a fossil forms?

Independent Inquiry

Use at least four other once-living materials, such as fallen leaves, to **make models** of fossils. Which materials make the best fossils?

Read and Learn

VOCABULARY
fossil p. 270

SCIENCE CONCEPTS
▶ what fossils are
▶ how fossils form

MAIN IDEA AND DETAILS
Look for information about different kinds of fossils.

Fossils

Think about the "fossils" you made in the Investigate. Did they look like the shell you used? How were they different from the shell? A **fossil** is a trace or the remains of a living thing that died a long time ago. There are many kinds of fossils.

Some fossils, such as bones and teeth, look like the actual parts of animals. Slowly, minerals replaced the bones.

Other fossils, such as dinosaur tracks in mud, are only marks that were left behind. These marks are called trace fossils. The dinosaur made a footprint in mud. The mud hardened, and in time, it changed to rock. Amber fossils formed when insects became trapped in tree sap, which hardens over time.

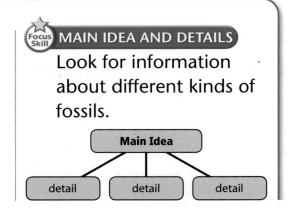
amber

cast   mold

A mold is the shape of a once-living thing left in sediment when the rock formed. The living thing that made the mold dissolved, leaving only a cavity shaped like the living thing.

A cast forms when mud or minerals later fill a mold. The cast has the shape of the living thing. In the Investigate, you made a model of a fossil mold and a fossil cast.

Plant fossils are not as common as animal fossils. That's because the soft parts of plants are easily destroyed as rocks form.

MAIN IDEA AND DETAILS
What are three kinds of fossils?

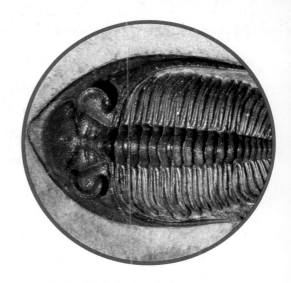

▲ This fossil shows a trilobite (TRY•luh•byt), a kind of animal that died out more than 200 million years ago.

◀ In petrified (PEH•truh•fyd) wood like this, minerals replaced the once-living plant.

Dinosaur tracks give clues to the animal's size and shape.

How Fossils Form

Places that have a lot of sedimentary rocks are better for fossil hunting than other places. Why? It's because what's left of a once-living thing is sometimes buried in the particles that form sedimentary rock. Fossils often form in limestone and shale.

Few fossils are preserved in metamorphic and igneous rock. The pressure and temperature that form these rocks often destroy plant and animal parts before they can become fossils.

Science Up Close

How a Fossil Forms

1. The soft parts of an animal decompose (dee•kuhm•POHZ), or rot away.

2. Hard parts of the animal, such as shells or bones, are buried under layers of sediment.

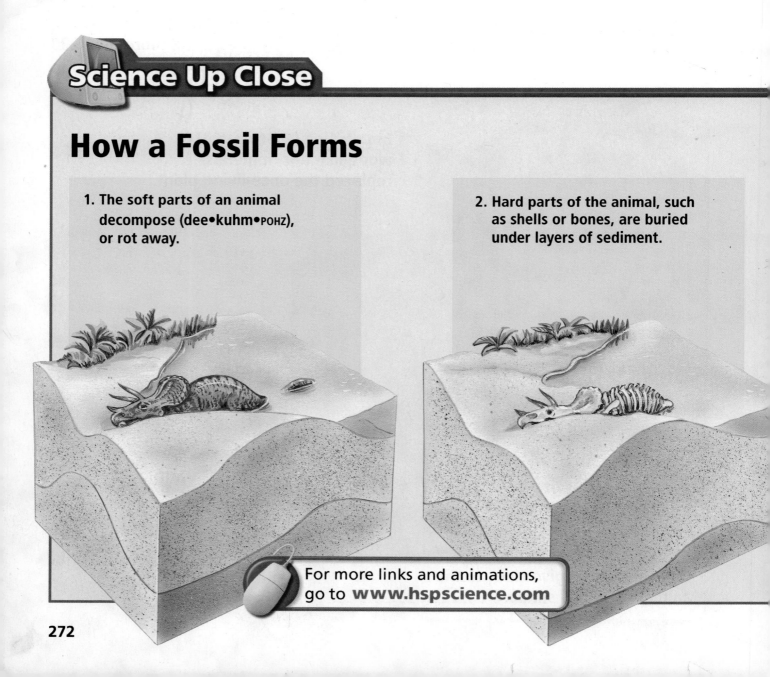

For more links and animations, go to www.hspscience.com

The Science Up Close shows how a fossil might have formed. After dying, the animal was quickly covered by layers of sediment. If the animal had not been covered quickly, another animal might have eaten it. After millions of years, the layers of sediment became sedimentary rock. What was left of the animal is now a fossil.

 **MAIN IDEA AND DETAILS**
Why are more fossils found in sedimentary rocks than in other rocks?

The fossil below shows what the triceratops (try•SAIR•uh•tahps) might have looked like. ▼

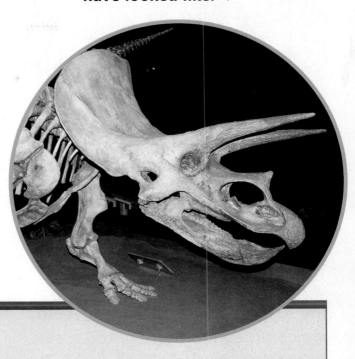

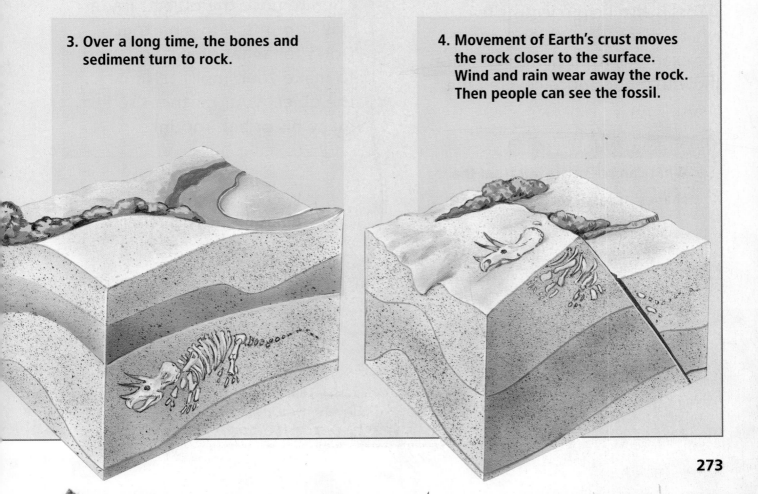

3. Over a long time, the bones and sediment turn to rock.

4. Movement of Earth's crust moves the rock closer to the surface. Wind and rain wear away the rock. Then people can see the fossil.

▲ What can a scientist tell about this fossil?

Learning from Fossils

Scientists today use fossils to learn about animals and plants that no longer live on Earth. For example, scientists learn what kind of foods animals ate by looking at the shapes of fossil teeth. The teeth are compared to those of today's animals.

Fossils also help scientists tell what a place was like long ago. For example, fossil clams in a place show that a sea once covered the area.

Focus Skill MAIN IDEA AND DETAILS

What can scientists tell from the shapes of fossil teeth?

Insta-Lab

Fossil Find
Press a hard object, such as a paper clip, into a ball of clay. Then remove the object. Trade clay imprints with a partner. Don't tell what object you used to make your imprint. Name the object your partner used to make his or her imprint.

▼ Why do scientists dig up dinosaurs very carefully?

Lesson Review

Essential Question
What are fossils?

In this lesson, you learned that fossils are traces or remains of long-dead living things. Cast, mold, and amber fossils are some types of fossils. Scientists study fossils to learn about living things of long ago.

1. **MAIN IDEA AND DETAILS** Draw and complete a graphic organizer about fossils:

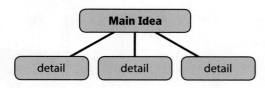

2. **SUMMARIZE** Write a short summary for this lesson. Tell how fossils form and why they're important.

3. **DRAW CONCLUSIONS** After walking through mud with his tennis shoes, Kyle notices an imprint of his track. What kind of fossil is formed this way?

4. **VOCABULARY** Write a sentence that tells the meaning of *fossil*.

Test Prep

5. Which is one thing scientists learn from fossils?
   A. how rocks form
   B. what sediments are
   C. what animals are like today
   D. about once-living things that no longer exist

Make Connections

 Writing

Narrative
Compare fossils, plants, and animals from similar environments in different places around the world. Write a **story** about spending a day in each place. Be sure to tell how the organisms are alike and different.

 Math

Solve a Problem
Write a word problem using information from this table. Have a classmate solve it.

Name of Dinosaur	Length
Stegosaurus	9 m (30 ft)
Brachiosaurus	25 m (82 ft)

275

Science Spin
From Weekly Reader
TECHNOLOGY

STUCK IN THE MUCK

Nearly 30,000 years ago, a thirsty woolly mammoth walked into a small pool to get a drink of water. The mammoth's feet sank into the dark pool, which was really a tar pit. Unable to get out of the sticky tar, the mammoth starved to death. Over time, the mammoth was covered by the tar.

Today, those tar pits are known as La Brea (BRAY•ah) site in Los Angeles. The tar oozes up from beneath the Earth's surface. Each summer, scientists spend two months digging through La Brea's Pit 91 looking for fossils.

Digging Into the Past

Scientists have unearthed thousands of fossils from Pit 91. "The information we captured will tell us more about what life was like in this area 35,000 years ago," scientist Chris Shaw told Weekly Reader.

Digging in the pit is a slow and careful process. The bottom of the pit is divided into a grid so that scientists can pinpoint the exact spot of each fossil. After scientists take measurements and make

drawings and pictures, the fossils are collected by using small hand tools. Those tools include dental picks, trowels, small chisels, and brushes.

The Search Continues

Shaw can't wait until next summer to start looking for more fossils. "The coolest thing about doing fieldwork," he said, "is realizing that the fossil you're touching has not seen the light of day for 35,000 years."

Think and Write
1. Why would scientists divide the bottom of Pit 91 into a grid?
2. Why do scientists have to be careful when digging up fossils?

Brrrrrr!

About 30,000 years ago, Earth was in an ice age. The weather was much cooler then, and a mile-thick layer of ice, or glacier, covered much of what is now the United States. By digging in the pit, scientists have learned that many cold-weather animals lived near the tar pits. Woolly mammoths, saber-toothed cats, camels, llamas, and even lions once roamed the area.

Find out more. Log on to
www.hspscience.com

Chapter 6 Review and Test Preparation

Vocabulary Review

Match the terms to the definitions below. The page numbers tell you where to look in the chapter if you need help.

mineral p. 248

rock p. 252

igneous rock p. 258

sedimentary rock p. 259

metamorphic rock p. 259

fossil p. 270

1. Trace or remains of a living thing that died long ago
2. Rock that was once melted and then cooled and hardened
3. An object that is solid, has never been alive, and was formed in nature
4. A rock formed from material that settled in layers
5. A solid made of one or more minerals
6. A rock changed by temperature and pressure

Check Understanding

Write the letter of the best choice.

7. **MAIN IDEA AND DETAILS** Which tells a way most minerals are identified?
   - A. beauty
   - B. heat
   - C. smell
   - D. streak

8. **COMPARE AND CONTRAST** How are casts different from molds?
   - F. Casts are only found in the mantle.
   - G. Casts are plant fossils. Molds are animal fossils.
   - H. Casts are materials that fill up a mold.
   - J. Casts are fossils. Molds are not.

9. Which do the drawings show?

   - A. how a fossil forms
   - B. how a mineral forms
   - C. how igneous rocks form
   - D. how to classify kinds of rocks

10. Which is most important in forming igneous rocks?
   F. fossil models
   G. melting
   H. cool temperatures
   J. dead plants and animals

11. Which is an imprint fossil?
   A. a petrified tree
   B. a dinosaur footprint
   C. a dinosaur bone
   D. an insect trapped in sap

12. Which process is shown in the drawing?

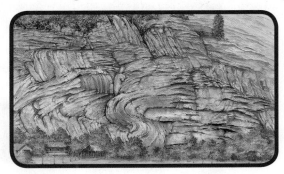

   F. how sediments build up
   G. how imprint fossils form
   H. how metamorphic rocks form
   J. how sandstone forms

13. Which would be needed to form sedimentary rocks?
   A. a volcano
   B. casts and molds
   C. pressure and temperature
   D. wind and water

14. Which is **not** a way to identify a mineral?
   F. by its outside color
   G. by its hardness
   H. by doing a streak test
   J. by trying to break it

15. What is the hardest mineral on the Mohs scale?
   A. apatite   C. diamond
   B. copper   D. talc

16. Which could a scientist infer about a place that has fish fossils?
   F. The place was once land.
   G. That place is a sea now.
   H. That place was once a sea.
   J. The fish did not live in water.

Inquiry Skills

17. How could you make a model to show the way sedimentary rocks form?

18. What could you use to order minerals by hardness?

Critical Thinking

19. A scientist finds a fossil. In what kind of rock was the scientist most likely digging? Explain.

The Big Idea

20. A scientist finds a dinosaur tooth that is flat, like a cow's tooth. What kind of food can the scientist infer the dinosaur ate? Why?

279

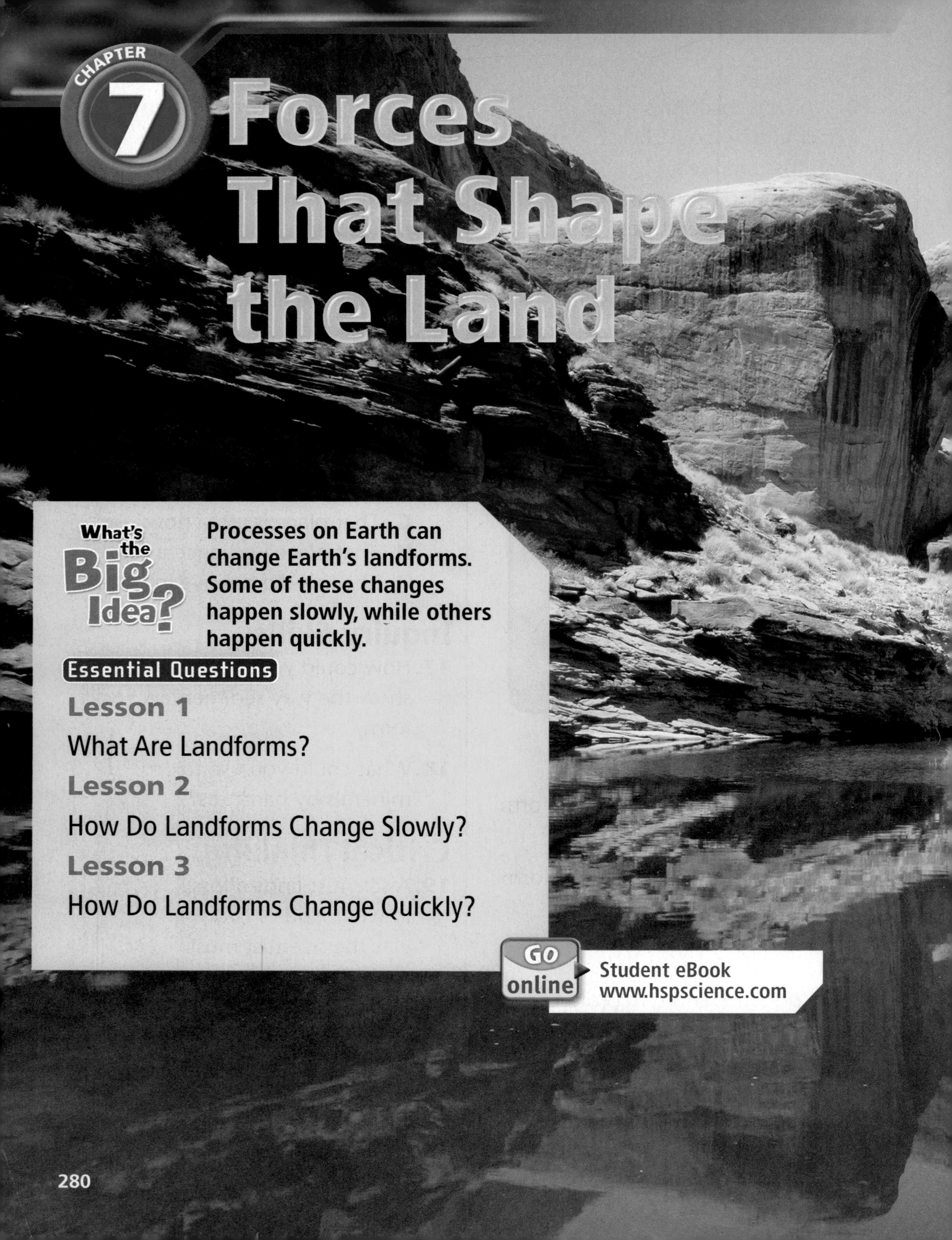

CHAPTER 7
Forces That Shape the Land

What's the Big Idea?

Processes on Earth can change Earth's landforms. Some of these changes happen slowly, while others happen quickly.

Essential Questions

Lesson 1
What Are Landforms?

Lesson 2
How Do Landforms Change Slowly?

Lesson 3
How Do Landforms Change Quickly?

Student eBook
www.hspscience.com

What do you wonder?

Rainbow Bridge in Utah is the largest natural stone bridge in the world. Its top is 88 meters (288 ft) above the river below. It is almost as high as the Statue of Liberty. Did this natural bridge form slowly or quickly? How does this relate to the **Big Idea?**

Rainbow Bridge

LESSON 1

Essential Question

What Are Landforms?

Investigate to find out how some mountains formed.

Read and Learn about different types of landforms on Earth.

Fast Fact

Folded Mountains
Many high mountains formed as rock was pressed and folded over millions of years. In the Investigate, you will model how mountains form as Earth's surface folds due to pressure.

Folded layers of rock on a mountain side

Vocabulary Preview

landform [LAND•fawrm] A natural shape on Earth's surface (p. 286)

mountain [MOUNT•uhn] A place on Earth's surface that is much higher than the land around it (p. 287)

valley [VAL•ee] A low area between higher land such as mountains (p. 288)

canyon [KAN•yuhn] A valley with steep sides (p. 288)

plain [PLAYN] A wide, flat area on Earth's surface (p. 289)

plateau [PLA•toh] A flat area higher than the land around it (p. 290)

283

Investigate

Folds in Earth's Crust

Guided Inquiry

Start with Questions

Mt. Fuji is the tallest mountain in Japan. It was formed by volcanic activity.

- How could volcanic activity cause a mountain to form?
- How else might mountains form?

Investigate to find out. Then read and learn to find out more.

Prepare to Investigate

Inquiry Skill Tip

When you use a model, identify what each part of the model represents. In this Investigate, the paper towels represent layers of rocks. The water helps the paper towels fold the way real rock layers do.

Materials

- 4 paper towels
- water in a plastic cup

Make an Observation Chart

	Observations	Sketch of Towels
Before pushing		
After pushing		

Follow This Procedure

1. Stack the paper towels on a table. Fold the stack in half.
2. Sprinkle water on both sides of the towels. They should be damp but not wet.
3. Place your hands on the ends of the damp towels.
4. Slowly push the ends toward the center. **Record** your **observations**.

Draw Conclusions

1. What happened as you pushed the ends of the towels together?
2. How did the height of the towels change as you pushed the ends?
3. **Inquiry Skill** Scientists use models to understand how things happen. How does this model help you understand how some mountains form?

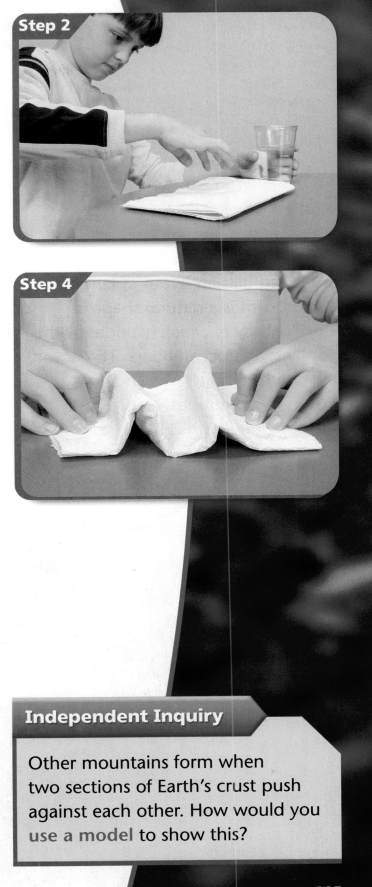

Independent Inquiry

Other mountains form when two sections of Earth's crust push against each other. How would you use a model to show this?

285

Read and Learn

VOCABULARY
landform p. 286
mountain p. 287
valley p. 288
canyon p. 288
plain p. 289
plateau p. 290

SCIENCE CONCEPTS
▶ what layers Earth has
▶ what landforms are

COMPARE AND CONTRAST
Look for ways landforms are alike and different.

alike — different

Surface Features

Maybe you live in a flat area. Perhaps you live in a hilly place. Flat areas and hills are types of landforms. A **landform** is a natural shape on Earth's surface, which includes both dry land and ocean floors. The ocean bottom has mountains and valleys, just as the land does.

The planet Earth is made up of three layers that are quite different. The outer layer, called the crust, includes both land and water features.

COMPARE AND CONTRAST

How are the layers of Earth different from one another?

Earth has three layers. The outside layer is the rocky crust. The mantle in the middle is very hot rock. The outer part of the core is liquid metal and the center is solid metal.

Mountains

Earth's highest landforms are mountains. A **mountain** is a place on Earth's surface that is much higher than the land around it. A mountain is at least 600 meters (about 2,000 ft) high. High landforms that are smaller than mountains are hills. In the Investigate, you saw how Earth's crust can fold to form mountains.

There are different kinds of mountains. Some are rocky and pointed, while others are rounded. Some mountains are low. Others are very high and have snow on their tops all year long. The longest mountain range on Earth is in the ocean.

 COMPARE AND CONTRAST

How are some mountains different from others?

This island is the top of a mountain. The rest of the mountain is under the water.

The Grand Tetons are mountains in Wyoming. They formed when forces pushed up a block of Earth's crust. ▼

Valleys

A **valley** is a low area between higher landforms, such as mountains. Some valleys are deep and narrow. They are V-shaped. Some valleys are wide and have flat bottoms. They are U-shaped. Their walls are far apart. Oceans have valleys, too.

A **canyon** is a steep-sided valley that has been carved by forces of nature. The biggest canyon in the United States is the Grand Canyon. It is 1.6 kilometers (1 mi) deep. The Colorado River wore away rock to form it. Canyons can even be found in the ocean.

Rivers can wear away rock and soil to form valleys. ▼

How is a valley different from a mountain?

It took millions of years for water to carve this canyon. ▼

The Great Plains of North America stretch for thousands of kilometers in the United States and Canada. Short grasses grow on much of the Great Plains.

Plains

A **plain** is a wide, flat area on Earth's surface. Most plains are lower than the land around them. The plants that grow on dry plains are mostly grasses. Plains that get more rainfall may have many trees.

A huge plain lies in the middle of the United States. This region is called the Great Plains. It is a dry area, but it has rich soil. Farmers grow crops such as wheat, corn, and oats there.

Plains are also found along some oceans, lakes, and rivers. The plains near ocean coasts are called coastal plains. Coastal plains in the eastern United States are very wide. They have flat, sandy beaches.

How is a coastal plain the same as the Great Plains?

Plateaus

A **plateau** (PLA•toh) is a flat area higher than the land around it. Most plateaus are wide. In some places, plateaus have very steep sides.

Plateaus can wear away and get smaller. A small plateau is a mesa (MAY•suh). *Mesa* means "table" in Spanish. Mesas look like tables made of rock. A mesa can wear away until it is just a flat-topped rock column called a butte (BYOOT).

Streams and rivers can cut deep valleys in plateaus. These valleys become canyons. The Grand Canyon is in the Colorado Plateau.

 **COMPARE AND CONTRAST**

How are a plateau and a mesa alike? How are they different?

Make a Landform
Use clay to make a model of a landform in this lesson. What features make this landform different from others?

Water and wind wear away plateaus. They may become mesas, like the one you see here. ▼

Lesson Review

Essential Question
What are landforms?

In this lesson, you learned that a landform is a natural shape on Earth's surface. Mountains, valleys, plains, and plateaus are all types of landforms.

1.  **COMPARE AND CONTRAST** Draw and complete a graphic organizer to compare and contrast different landforms.

   alike — different

2. **SUMMARIZE** Write a three-sentence summary about this lesson. Tell about the types of landforms.

3. **DRAW CONCLUSIONS** How are plateaus and mountains alike? How are they different?

4. **VOCABULARY** Make a crossword puzzle using the vocabulary terms from this lesson.

Test Prep

5. Which type of landform is in the middle of North America?
   A. a mountain
   B. a plain
   C. a plateau
   D. a valley

Make Connections

 Writing

Expository
Use the map on the back cover of your book to locate some landforms and bodies of water. Research some of the places you found. Then write a paragraph **describing** where they are and what they look like.

 Math

Construct a Bar Graph
Research the heights of four mountains. Make a bar graph showing their heights.

291

LESSON 2

Essential Question
How Do Landforms Change Slowly?

Investigate to find out how water can change bricks.

Read and Learn how weathering and erosion affect land.

Fast Fact

Rock Shapes
Moving water and wind can wear down rock into unusual shapes. What may happen to the top part of this rock in time? In the Investigate, you will model how water breaks down rock.

Weathered rock

Vocabulary Preview

weathering [WETH•er•ing]
The breaking down of rocks into smaller pieces (p. 296)

erosion [ee•ROH•zhuhn]
The movement of weathered rock and soil (p. 298)

glacier [GLAY•sher] A huge sheet or block of moving ice (p. 300)

Investigate

Water at Work

Guided Inquiry

Start with Questions

When it rains, or when snow melts, the water has to go somewhere. This water is running downhill over rocks.

- Have you ever seen water running in a ditch after a heavy rain?
- How does this waterfall affect the land around it?

Investigate to find out. Then read and learn to find out more.

Prepare to Investigate

Inquiry Skill Tip

When you interpret data, you explain the results you collected in an investigation. To do this, think about your data and observations. Then, give a reason for what happened.

Materials

- balance
- water
- small pieces of brick
- extra masses for the balance
- clear jar with lid

Make a Data Table

Time	Mass of Brick Pieces
Day 1	
1 week later	

Follow This Procedure

1. **Measure** the mass of the brick pieces. **Record** your results.
2. Fill the jar three-fourths full of water.
3. Put the brick pieces into the jar of water. Put the lid on the jar.
4. Take turns with a partner to shake the jar for 10 minutes. Do this three times a day for one week. Then **measure** and **record** the mass of the brick pieces.

Draw Conclusions

1. Do the brick pieces look different after a week of shaking them? If so, how are they different?
2. **Compare** the mass of the brick pieces before and after the shaking. Did the mass of the brick pieces change? If so, how did it change?
3. **Inquiry Skill** Scientists **interpret data** to understand how things work. Use your data to tell what happened to the brick pieces.

Independent Inquiry

Do large pieces of rock weather faster than small ones? **Plan and conduct an investigation** to test your prediction.

Read and Learn

VOCABULARY
weathering p. 296
erosion p. 298
glacier p. 300

SCIENCE CONCEPTS
▶ what weathering does
▶ what erosion does

 **SEQUENCE**
Look for the steps that lead to erosion.

How Rocks Are Broken Down

Earth's surface is rock. Rock wears down and breaks apart. When rocks are broken down into smaller pieces it is called **weathering** (WETH•er•ing).

Water causes weathering. Wind also breaks down rock. It picks up sand and smashes it against rocks. This slowly chips the rocks. Ice breaks down rock in a different way. You can see on the next page how it does this. Plant roots can grow in cracks in rock and widen them, too.

Weathering happens slowly. It can take thousands of years for rocks to break down into sand and soil.

 **SEQUENCE**
How does wind break down rocks?

Water flows through cracks in rock. It slowly wears down the rock edges into rounded shapes.

Weathering has cracked this rock. The layers on the surface have broken off. ▼

296

Science Up Close

How Ice Cracks Rock
Water in the form of ice can break down rock and concrete.

1. Water goes into a crack.

2. The water freezes and becomes ice. Frozen water takes up more space than liquid water. It pushes against the sides of the crack.

3. The crack becomes wider and breaks the concrete or rock.

 For more links and animations, go to www.hspscience.com

Making Rocks Move

After rocks are broken down, erosion moves the pieces. **Erosion** (ee•ROH•zhuhn) is the movement of weathered rock. Creep is a very slow type of erosion that moves rocks and soil. It can bend fences and walls over time.

Moving water is the cause of much erosion. Rainfall loosens sand and rock. It carries the sand and weathered rocks into rivers. The rivers carry the materials downstream.

When the water's flow slows down, the materials drop. Rivers drop sand along their banks. This is one way the shape of a river can change.

Plants help hold soil in place. They keep water from washing the soil away from farmlands.

Rivers drop sand and rock at their mouths. The sand and rock build up to form a landform called a delta. ▼

Rivers erode soil along their banks. This muddy river is carrying a lot of soil. ▼

Wind erosion blows sand into huge mounds called dunes. Dunes form in deserts and along sandy coasts.

Creep is very slow movement of rocks and soil. It happens too slowly to be seen. But it can bend fences and walls. Creep has moved these tombstones out of place.

Wind causes erosion, too. The stronger the wind, the more soil it can carry. When the wind slows down, the soil drops and makes landforms.

In deserts, wind forms big piles of sand called dunes. Sand dunes can be taller than a 20-story building.

Erosion is often slow. It can take hundreds or even thousands of years to change the land.

 **SEQUENCE**

What must happen before rock can be eroded?

Growing Ice

Fill $\frac{1}{2}$ of a small plastic cup with water. Mark the water line, and measure the mass of the water. Put the cup in a freezer. When the water is frozen, check the mark. What has changed? Measure the mass of the ice and how much it "grew." How does this relate to weathering?

How Glaciers Change the Land

Glaciers (GLAY•sherz) also shape the land through erosion. A **glacier** is a huge block of moving ice. Glaciers form where it is so cold that snow never melts. The snow piles up and turns to ice.

As glaciers move, they pick up rocks and soil. When they melt, they drop the rocks and soil, which form hills and ridges. Glaciers scrape the surface as they move. They make valleys wider.

SEQUENCE
How does a glacier form a hill?

Glaciers cover $\frac{1}{10}$ of Earth's surface. They form high in mountains and in areas near the poles. ▼

Math in Science
Interpret Data

Which glacier is the longest? Where is it?

Glaciers can carve deep grooves into rock. They can even form valleys.

Lesson Review

Essential Question

How do landforms change slowly?

In this lesson, you learned that weathering and erosion can change landforms. This process can take thousands of years. Weathering is the breaking down of rock into smaller pieces. Erosion is the moving of those pieces of rock from place to place. Wind, water, and glaciers all cause weathering and erosion.

1. **SEQUENCE** Draw and complete a graphic organizer to show the sequence of erosion and weathering.

2. **SUMMARIZE** Summarize how water causes erosion. Begin with rain falling on the side of a mountain.

3. **DRAW CONCLUSIONS** Would a tree root pushing up part of a sidewalk be an example of weathering or erosion? Explain.

4. **VOCABULARY** Draw a picture of a glacier. Then write a caption that explains what a glacier is.

Test Prep

5. **CRITICAL THINKING** Jamie visits her aunt every summer. She sees that her aunt's fence has moved from where it was. Explain how this may have happened.

Make Connections

 Writing

Narrative

Glaciers covered much of the land during the last Ice Age. Research the Ice Age. Then write a **story** about what a day might have been like for people living at that time.

 Math

Solve Problems

If a delta grows by 3 meters each year, how much will it grow in 5 years? How much will it grow in 10 years?

301

LESSON 3

Investigate to discover how volcanic eruptions can change landforms.

Read and Learn about what can cause landforms to change quickly.

Essential Question

How Do Landforms Change Quickly?

Fast Fact

A Tall Volcano
Mount Etna, shown in the picture, is the tallest active volcano in Europe. It is 3,350 meters (10,990 ft) high. In the Investigate, you will make a model of a volcano.

Mount Etna erupting

Vocabulary Preview

earthquake [ERTH•kwayk] The shaking of Earth's surface caused by movement in Earth's crust (p. 306)

volcano [vahl•KAY•noh] An opening in Earth's surface from which lava flows (p. 308)

flood [FLUHD] A large amount of water that covers normally dry land (p. 310)

Investigate

A Model Volcano

Guided Inquiry

Start with Questions

This island wasn't always here. It was formed many years ago.

- Have you ever been to an island?
- How did this island form?

Investigate to find out. Then read and learn to find out more.

Prepare to Investigate

Inquiry Skill Tip

It is a good idea to use a model when actual events occur too slowly or too quickly to see. You can also use a model when events are too dangerous to observe in nature.

Materials

- safety goggles
- lab apron
- 2 plastic jars
- measuring spoons
- water
- soil
- flour
- measuring cup
- large tray
- baking soda
- dropper
- white vinegar
- wax paper
- red and green food coloring

Make an Observation Chart

Color	Observations
Red	
Green	

Follow This Procedure

1. **CAUTION: Put on safety goggles and a lab apron.** Cover the tray with wax paper. Put a jar in the middle.

2. Mix $\frac{1}{2}$ tsp flour and 1 tsp baking soda in the jar. Add 10 drops of red food coloring.

3. Dampen the soil a little, and pack it around the jar in a cone shape. Make the top of the soil even with the top of the jar.

4. Slowly pour $\frac{1}{4}$ cup vinegar into the jar. **Observe**. Remove the jar carefully. Wait 15 minutes.

5. Repeat Steps 2 through 4 with green food coloring and the other jar. **Record** your **observations**.

Step 3

Step 4

Draw Conclusions

1. What happened when you poured the vinegar into the jar? What does the mixture represent?

2. **Inquiry Skill** Scientists use models to help them understand how things happen in nature. How did your model help you learn how volcanoes change the land?

Independent Inquiry

Learn about different types of volcanic eruptions. Choose one type. **Plan and conduct an investigation** to model it.

Read and Learn

VOCABULARY
earthquake p. 306
volcano p. 308
flood p. 310

SCIENCE CONCEPTS
▶ how earthquakes and volcanoes change the land
▶ how floods change the land

CAUSE AND EFFECT
Look for the effects of earthquakes.

cause → effect

Earthquakes

An **earthquake** is the shaking of Earth's surface. It is caused by movement in Earth's crust. Earth's crust is made up of rock. In some places, the rock has large cracks in it called faults.

When Earth's crust suddenly moves along these faults, the area above the fault can shake. The shaking of Earth's crust during an earthquake can cause many things to happen at Earth's surface.

Earthquakes can change landforms. They make big cracks in Earth's surface. They create uneven areas of ground.

Earthquakes can cause great damage. These homes in San Francisco tilted when the ground moved under them. ▼

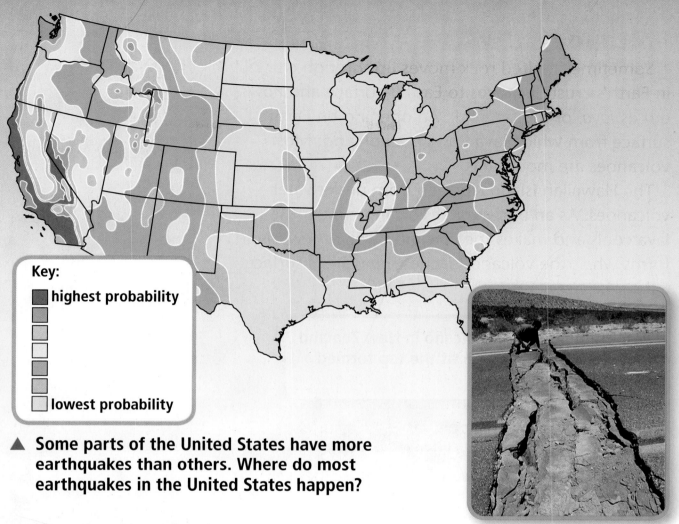

Key:
- highest probability
- lowest probability

▲ Some parts of the United States have more earthquakes than others. Where do most earthquakes in the United States happen?

▲ The movement of the ground can split roads and damage bridges.

Earthquakes can cause quick erosion. Near the coast, they can destroy beaches. In the mountains, they can cause large amounts of rock or soil to slide downhill. These movements are called landslides.

Earthquakes destroy buildings and other kinds of property. The landslides they cause can cover entire streets of houses. Even in undamaged homes, people may be without water or electricity for days after an earthquake.

 **CAUSE AND EFFECT**

What causes earthquakes?

Where's the Energy?
Hold one end of a ruler down on a desk. Pull up the other end. What will happen if you let the free end go? Let it go. Was your prediction correct? Infer why this happened.

307

Volcanoes

Sometimes melted rock moves up through cracks in Earth's crust. It moves to Earth's surface and flows out as *lava,* or molten rock. An opening on Earth's surface from which lava flows is a **volcano**. Most volcanoes are mountains.

The Hawaiian Islands formed from underwater volcanoes. As an underwater volcano erupts, the lava cools and makes the volcano taller. A new island forms when the volcano reaches the water's surface.

Heat from inside this volcano in New Zealand causes steam. The crater at the top formed during a past eruption.

When volcanoes erupt, they change Earth's surface quickly. The lava, ash, and rock they send out cover the land around them. Volcanoes also send out clouds of gas and ash that roll down the volcano's sides. The hot gas and ash burn down buildings and trees, affecting both human communities and animal habitats.

A volcano is a type of mountain. Lava and rock build up on its sides each time it erupts. The volcano mountain slowly becomes taller.

What effects does an erupting volcano have on land?

A river of lava flows down the sides of the volcano.

These houses are almost buried in lava and ash.

Floods

You know that water can change the land slowly. It can also change the land quickly. Sometimes, so much rain falls that rivers and streams cannot hold all the water. The water flows over their banks, causing a flood. A **flood** is a large amount of water that covers normally dry land.

Floods sweep over the land. They destroy buildings and wash away bridges, roads, and crops. Floods also carry soil. When the floodwaters go down, they leave the soil along banks of rivers. Farmers can grow crops in this fertile soil.

What causes a flood?

▲ Droughts happen when an area does not get enough rain. The dryness and heat cause cracks in the soil.

A river has flowed over its banks. It has flooded houses on a nearby plain. ▼

Lesson Review

Essential Question

How do landforms change quickly?

In this lesson, you learned that some natural events can change landforms very quickly. Earthquakes can make cracks in Earth's surface and cause areas of ground to become uneven. Volcanoes can form new land when the lava from them cools to become rock.

1. **CAUSE AND EFFECT** Draw and complete a graphic organizer that shows the effects of the movement of rock at a fault in Earth's crust.

2. **SUMMARIZE** Write a three-sentence summary of this lesson. Begin with the sentence: Earth's surface can change quickly.

3. **DRAW CONCLUSIONS** Suppose you lived in an area where earthquakes happen. What would you need to do to keep safe during an earthquake?

4. **VOCABULARY** Use a vocabulary term correctly in a sentence.

Test Prep

5. Which of these is an effect of earthquakes?
   - **A.** burned forest
   - **B.** large cracks in Earth
   - **C.** gas clouds
   - **D.** crops washed away

Make Connections

 Writing

Expository

Imagine that you are a TV reporter watching a volcano erupt. Write a **report** describing what you see. Make sure your information is correct.

 Social Studies

The Big Ones

Find out where the five biggest earthquakes have happened in the United States. Use an outline map. Put a dot where each earthquake happened. Write the names of the places beside the dots.

311

People in Science

Marisa Quinones

▶ **MARISA QUINONES**
▶ Geologist

When you think of a geologist, you may think only of rocks. But geologists do many things. They study the history of Earth, find water in dry climates, and search for precious rocks and minerals, such as gold and diamonds. For Marisa Quinones, geology involves looking for oil underground. Oil companies look for oil by setting off an explosion in the ground or by firing an air gun into the ocean floor. The rocks underground respond to the sound waves that result. Their responses are recorded on a computer. Quinones studies the patterns of rock responses. By studying them, she can decide whether oil is likely to be found in a certain spot.

Once drilling begins, Quinones looks at the cuttings, or rock chips, from the hole. The cuttings help her predict whether there could be any danger as the drilling is being done. They also let her know when the area that holds the oil has been reached.

Think and Write

1. What dangers might occur when people are drilling for oil?
2. Why do oil companies search for oil under the sea?

Charles Richter

▶ **CHARLES RICHTER**
▶ Inventor of the Richter Scale

Every so often, there are news reports about an earthquake. During the reports, you will probably hear the words "Richter scale." That's not a giant scale to weigh people. The Richter scale measures the overall strength of an earthquake.

That scale was created by Charles Richter (1900–1985). Richter was a *seismologist,* a scientist who studies earthquakes and the structure of the Earth. Seismologists try to determine where and when an earthquake may occur next.

Think and Write

1. Why would scientists want to know where an earthquake may occur next?
2. How is having a way to measure the strength of an earthquake helpful?

Career Volcanologist

Volcanologists are scientists who study volcanoes. They watch volcanic sites and gather information about the Earth. Volcanologists help keep people who live near volcanoes safe. They warn them of possible eruptions so they can leave the area.

Chapter 7 Review and Test Preparation

Vocabulary Review

Use the terms below to complete the sentences. The page numbers tell you where to look in the chapter if you need help.

canyon p. 288
plateau p. 290
weathering p. 296
erosion p. 298
glacier p. 300
earthquake p. 306
volcano p. 308
flood p. 310

1. The movement of weathered rock and soil is _____.
2. An opening in Earth's surface from which lava flows is a _____.
3. A valley with steep sides is a _____.
4. A huge block of moving ice is a _____.
5. The shaking of Earth's surface is an _____.
6. A flat area higher than the land around it is a _____.
7. The breaking down of rocks into smaller pieces is _____.
8. A large amount of water that covers the land is a _____.

Check Understanding

Write the letter of the best choice.

9. Which of these is a low area?
   A. mesa
   B. mountain
   C. plateau
   D. valley

10. Which of these words describes a plain?
    F. deep
    G. flat
    H. hilly
    J. steep

11. **CAUSE AND EFFECT** Which of these is an effect of weathering?
    A. mountain building
    B. water freezing
    C. volcano erupting
    D. rock splitting

12. Which landform does this illustration show?
    F. mountain
    G. plain
    H. plateau
    J. valley

314

13. Which of these causes the land to change slowly?
    A. flood   C. glacier
    B. earthquake   D. volcano

14. Where do most earthquakes happen?
    F. on buttes   H. on plains
    G. at faults   J. in valleys

15. **COMPARE AND CONTRAST** Which type of landform is a kind of mountain?
    A. volcano   C. plateau
    B. plain   D. valley

16. How does a flood change the land?
    F. It cracks rock.
    G. It drops soil.
    H. It lifts mountains.
    J. It sends out lava.

Inquiry Skills

17. Why would it be useful to make a model to study erosion caused by glaciers?

18. How could you use models to show the effects of weathering and the effects of erosion?

Critical Thinking

19. Which of these rocks has undergone more weathering? Explain your answer.

20. In which of these three areas would flooding most likely happen? Explain your answer.

A.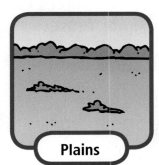
Plains

B.
Mountains

C.
Valleys

CHAPTER 8
Conserving Resources

What's the Big Idea?

Living things use Earth's resources to meet their needs. Some of these resources can be recycled or reused.

Essential Questions

Lesson 1
What Are Some Types of Resources?

Lesson 2
What Are Some Types of Soil?

Lesson 3
How Do People Use and Impact the Environment?

Lesson 4
How Can Resources Be Used Wisely?

Student eBook
www.hspscience.com

What do you wonder?

Resources are found almost everywhere. What resources do you see in this picture? How do people use these resources? Which are reusable? How does this relate to the **Big Idea?**

Niagara Falls

LESSON 1

Essential Question

What Are Some Types of Resources?

Investigate to find out how mining affects land.

Read and Learn about types of resources.

Fast Fact

Wind Power
These wind generators use wind to produce electricity. Energy from wind is grouped into seven classes. In the Investigate, you will model one way we get some of Earth's resources.

Wind farm

Vocabulary Preview

resource [REE•sawrs] A material that is found in nature and that is used by living things (p. 322)

renewable resource [ri•NOO•uh•buhl REE•sawrs] A resource that can be replaced within a human lifetime (p. 324)

reusable resource [ree•YOOZ•uh•buhl REE•sawrs] A resource that can be used again and again (p. 325)

nonrenewable resource [nahn•rih•NOO•uh•buhl REE•sawrs] A resource that, when it is used up, will not exist again during a human lifetime (p. 326)

Investigate

Mining Resources

Guided Inquiry

Start with Questions

These carts are filled with coal. Coal is burned by some companies to produce electricity. When you turn on a light, the electricity it uses may come from the energy released by burning coal.

- What other things do you use that need electricity?
- How might mining this coal have affected the land around the mine?

Investigate to find out. Then read and learn to find out more.

Prepare to Investigate

Inquiry Skill Tip

To help you infer, start by making a list of all your observations. Use that list to help you draw a conclusion.

Materials

- oatmeal-raisin cookie
- paper plate
- dropper
- water
- toothpick

Make a Data Table

Raisins Observed in Step 1	Total Number of Raisins Counted

Follow This Procedure

1. **Observe** your cookie. **Record** the number of raisins you see.

2. Put a few drops of water around each raisin. The cookie should become moist but not wet.

3. Use the toothpick to "mine" all the raisins from the cookie. If they are hard to get out, put a few more drops of water around them. **Record** the number you removed.

Draw Conclusions

1. Were there any raisins that you didn't see in the cookie the first time? Why didn't you see them?

2. How did the water help you dig out the raisins? How did the digging affect the cookie?

3. How is mining raisins from a cookie like mining resources from Earth?

4. **Inquiry Skill** Scientists use their observations to infer how similar things work. Use your observations to infer how mining could affect the land around the mine.

Step 2

Step 3

Independent Inquiry

Can the cookie be completely mined without tearing it up? **Hypothesize** how this might be done. Test your **hypothesis**.

321

Read and Learn

VOCABULARY
resource p. 322
renewable resource p. 324
reusable resource p. 325
nonrenewable resource p. 326

SCIENCE CONCEPTS
▶ what resources are
▶ which resources will not run out and which may run out

MAIN IDEA AND DETAILS
Look for details about resources.

```
        Main Idea
       /    |    \
  detail  detail  detail
```

Resources

What materials have you used so far today? You ate food for breakfast. Then you used water to brush your teeth. Now you are using this book. The food, the water, and even this book all came from resources. A **resource** is a material found in nature that is used by living things.

▼ People use fish as a resource.

People use many kinds of animals as resources. For example, cows are used for both meat and milk.▼

322

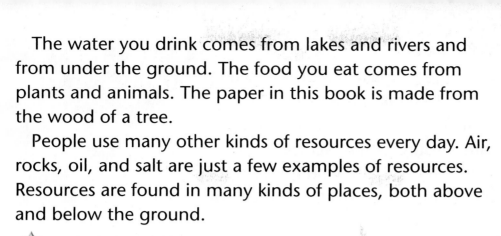

The water you drink comes from lakes and rivers and from under the ground. The food you eat comes from plants and animals. The paper in this book is made from the wood of a tree.

People use many other kinds of resources every day. Air, rocks, oil, and salt are just a few examples of resources. Resources are found in many kinds of places, both above and below the ground.

MAIN IDEA AND DETAILS
What is one resource that is found above the ground? Below the ground?

Crops are grown on this farm. People use some kinds of crops for food. Other kinds of crops, such as cotton, are used to make clothing, sheets, and towels. ▼

This drilling platform is used to drill for oil. Oil is used to make many products, including plastic. ▶

Renewable Resources

After people cut down a tree to make paper, they can plant another tree. The new tree takes the place of the tree that was cut down. Trees are renewable resources. **Renewable resources** are resources that can be replaced during a human lifetime. Plants and animals are renewable resources. Some kinds of energy are renewable resources, too. For example, energy from the sun is a renewable resource.

 MAIN IDEA AND DETAILS

What are some kinds of renewable resources?

The trees in this forest are renewable resources. They can be used to make houses, furniture, pencils, and paper. New trees can be planted to take the place of the ones that are cut down. ▼

▲ Air is a resource you need to breathe. Cars and factories make the air dirty. Plants, wind, and rain help clean the air to make it safe to breathe again.

Reusable Resources

Some kinds of resources can be used again and again. These kinds of resources are **reusable resources**. Air and water are two kinds of reusable resources. After you take a bath, the water is dirty. When you drain the tub, the water goes to a water treatment plant. There the water is cleaned so it can be reused.

MAIN IDEA AND DETAILS

What are some kinds of reusable resources?

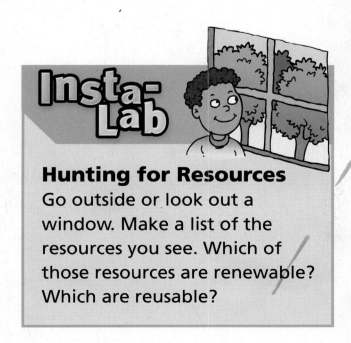

Hunting for Resources

Go outside or look out a window. Make a list of the resources you see. Which of those resources are renewable? Which are reusable?

Nonrenewable Resources

You probably have watched someone fill up a car's tank with gasoline. Cars burn gasoline for energy. When the gasoline is used up, the tank needs to be filled again.

The crude oil that is used to make gasoline is not renewable or reusable. One day, it will all be gone. Crude oil is a nonrenewable resource. **Nonrenewable resources** are resources that cannot be replaced in a human lifetime. When these resources are used up, there will be no more. Some other nonrenewable resources are coal, soil, and metals.

MAIN IDEA AND DETAILS
What are some nonrenewable resources?

Miners dig coal out of mines deep underground. Coal is burned to produce electricity. Coal takes thousands of years to form. ▼

There are only certain amounts of minerals on Earth. No more can be made. Some of Earth's mineral resources, such as copper ore and iron ore, are dug up at strip mines like this one. ▼

Lesson Review

Essential Question
What are some types of resources?

In this lesson, you learned that there are different types of resources. Renewable resources can be replaced during a human lifetime. Reusable resources can be used again and again. Nonrenewable resources cannot be replaced in a human lifetime.

1. **MAIN IDEA AND DETAILS** Draw and complete a graphic organizer for the main idea: There are different kinds of resources.

   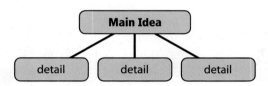

2. **SUMMARIZE** Write a summary of this lesson. Begin with this sentence: There are three types of resources.

3. **DRAW CONCLUSIONS** How is the sun a renewable resource?

4. **VOCABULARY** Write a short paragraph that describes the differences between renewable resources, reusable resources, and nonrenewable resources.

Test Prep

5. Which of the following resources is nonrenewable?
   - **A.** air
   - **B.** animals
   - **C.** coal
   - **D.** water

Make Connections

 Writing

Narrative
Choose a nonrenewable resource that you have used today. Research where it is found or how it is made. Write a *story* that describes how your life would be different without that resource.

 Math

Solve a Problem
Suppose your town uses 5 tons of coal a year and has enough coal to produce energy for 20 years. If your town used only 4 tons each year, how long would the coal last?

LESSON 2

Essential Question

What Are Some Types of Soil?

Investigate to find out how soils can be different.

Read and Learn about the layers of soil and types of soil.

Fast Fact

Slow-Forming Soil
It takes between 3,000 and 12,000 years to form soil that is good for farming. In the Investigate, you'll find out about two kinds of soil.

Farmer spraying crops

Vocabulary Preview

humus [HYOO•muhs] The part of some soil that is made up of broken-down parts of dead plants and animals (p. 332)

sand [SAND] Grains of rock that you can see with your eyes alone (p. 335)

silt [SILT] Grains of rock that are too small to see with your eyes alone (p. 335)

clay [KLAY] Particles of rock so small you need a microscope to see them (p. 335)

loam [LOHM] Soil that is a mixture of humus, sand, silt, and clay (p. 336)

Investigate

Observing Soil

Guided Inquiry

Start with Questions

This plant is a sundew. Instead of being eaten by insects, it eats them! Sundews grow in wet, swampy areas.

- Why might plants need to eat insects?
- How could different types of soil affect the types of plants that grow in them?

Investigate to find out. Then read and learn to find out more.

Prepare to Investigate

Inquiry Skill Tip

The observations you make during an investigation aren't just the things you can see. The senses of smell, sight, touch, and hearing often give important information.

Materials

- 2 soil samples
- microscope or hand lens
- small paper plates
- toothpick

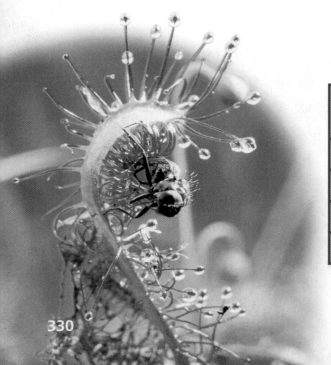

Make an Observation Chart

Characteristic	Sample 1	Sample 2
color		
shape		
size		
feel		

Follow This Procedure

1. Get a soil sample from your teacher. Place some of the soil on a paper plate.

2. Using a microscope or hand lens, observe the soil. Use a toothpick to move the soil grains around. Notice the colors, shapes, and sizes of the grains. Record what you observe by drawing the soil grains.

3. Pick up some soil from the plate. Rub it between your fingers. How does it feel? Record what you observe.

4. Repeat Steps 1 through 3 with the other soil sample.

Draw Conclusions

1. What senses did you use to observe the soil?
2. Describe your observations.
3. **Inquiry Skill** Scientists observe things so they can compare them. How were the soil samples alike? How were the soil samples different?

Step 2

Step 3

Independent Inquiry

Which soil holds more water—potting soil or sandy soil? Write a hypothesis. Then plan and conduct an investigation to find out.

Read and Learn

VOCABULARY
humus p. 332
sand p. 335
silt p. 335
clay p. 335
loam p. 336

SCIENCE CONCEPTS
▶ how soil is layered
▶ how soils are different

 **COMPARE AND CONTRAST**
Compare and contrast layers and types of soil.

alike — different

Layers of Soil

You might not think of soil as a resource, but it is a very important one. Without soil, plants couldn't grow. Many animals wouldn't have places to live.

Soil is a mixture of many different things. Soil is made up of water, air, humus, and tiny pieces of rock. **Humus** is the part of soil that is made up of parts of dead plants and animals. For example, when a tree loses a leaf, the leaf falls to the ground. As the leaf decays, it becomes humus.

This plant's roots grow in soil. ▼

Soil close to the surface may have a lot of humus. Soil that is deeper down has less humus and more small pieces of rock. If you cut into the soil, you would see different layers at different depths.

COMPARE AND CONTRAST

How is soil close to Earth's surface different from the soil deeper below the surface?

Science Up Close

Soil Layers

Topsoil
Soil forms in layers. The top layer is *topsoil*. Topsoil has a lot of humus. Many small animals, such as ants and earthworms, live in topsoil.

Subsoil
The layer beneath topsoil is *subsoil*. Subsoil does not have a lot of humus, but it does have small pieces of rock in it.

Bedrock
If you dig deep enough into soil, you will reach solid rock. This solid rock is called *bedrock*. Most of the small pieces of rock in soil come from the bedrock under it.

For more links and animations, go to www.hspscience.com

Different Types of Soil

Not all soil is the same. In the United States alone, there are more than 70,000 kinds of soil. Soils can have different colors. Some soils can hold more water than other soils.

Another difference among soils is the size of the rock grains found in them. The grain sizes make the soils feel different.

This soil is made up mostly of sand. The sand grains are big enough to see without a hand lens. ▼

This clay soil is made up of very tiny grains of rock. You would need a microscope to see the grains. ▼

Soils that have tiny grains of rock that you can see with your eyes alone are mostly **sand**. Soils with grains of rock that are difficult to see with only your eyes are **silt**. Soils with even tinier grains of rock are **clay** soils.

The main difference between sand, silt, and clay soils is their grain size. These three types of soils can also be made up of different minerals.

 COMPARE AND CONTRAST

Compare sand, silt, and clay soils.

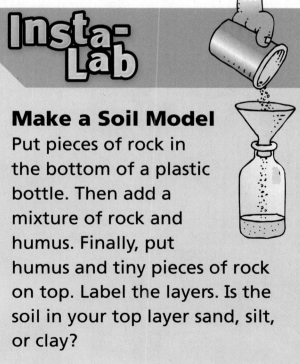

Make a Soil Model

Put pieces of rock in the bottom of a plastic bottle. Then add a mixture of rock and humus. Finally, put humus and tiny pieces of rock on top. Label the layers. Is the soil in your top layer sand, silt, or clay?

335

The Importance of Soil

What would the world be like without soil? People couldn't make bricks, pottery, or other items that are made from clay soil. Many animals would have nowhere to live, and plants would not be able to grow.

Soil is very important to people for growing fruits and vegetables. The best kind of soil to use to grow fruits and vegetables is loam. **Loam** is a mixture of humus, clay, silt, and sand. Most soil found on farms is loam.

Contrast the ways people use different kinds of soil.

This woman grows flowers and vegetables in loam. Clay soil would stay too wet. Sandy soil would dry out too quickly. ▼

Lesson Review

Essential Question

What are some types of soil?

In this lesson, you learned that soil may be made up of several parts, including humus, sand, silt, and clay. Different types of soil contain different amounts of each part. Loam is a mixture of all four soil parts.

1. **COMPARE AND CONTRAST** Draw and complete a graphic organizer that tells how soils are alike and different.

2. **SUMMARIZE** Write a three-sentence summary of this lesson. Tell about the types of soil.

3. **DRAW CONCLUSIONS** Why do most plants need soil to grow?

4. **VOCABULARY** Draw a graphic organizer that shows how humus, sand, silt, and clay are related to loam.

Test Prep

5. **Critical Thinking** Robin wants to start a vegetable garden in her backyard. The soil in her backyard is mostly clay. What should Robin do before she plants her vegetable seeds?

Make Connections

Writing

Persuasive

Suppose you are selling a type of soil. Write an **advertisement** that will persuade adults to buy your soil. Your advertisement should explain the best uses for the soil.

Physical Education

Take a Hike

Take a walk in your town with an adult family member. Collect small amounts of soil from three places. Compare the soils. Discuss other resources you saw on your walk.

Lesson 3

Investigate to find out how pollution can affect plants.

Read and Learn about how people use land and about the different types of pollution.

Essential Question

How Do People Use and Impact the Environment?

Fast Fact

Florida Orange Groves

Orange trees need rich soil, clean water, and warm weather. They get all these things in Florida, the leading state for orange growing. In the Investigate, you'll find out about the importance of clean water to plants.

Vocabulary Preview

pollution
[puh•LOO•shuhn] Any harmful material in the environment (p. 344)

Orange grove

Investigate

Pollution and Plants

Guided Inquiry

Start with Questions

These trees have been affected by acid rain. Acid rain is a form of pollution.

- Has the acid rain helped or harmed these trees?
- Would other forms of pollution be helpful or harmful to these trees?

Investigate to find out. Then read and learn to find out more.

Prepare to Investigate

Inquiry Skill Tip

When you compare things, think about the characteristics that are important for the investigation. Sometimes, you need to consider the way things look. Other times, you need to consider the way things behave.

Materials

- 3 clear plastic cups
- potting soil
- grass seeds
- measuring cup
- clean water
- salt water
- oily water

Make an Observation Chart

Day	Clean Water	Salty Water	Oily Water
1			
2			
3			
4			
5			
6			
7			

Follow This Procedure

1. Your teacher will provide you with three containers of water. One will have clean water, one will have water polluted with salt, and one will have water polluted with oil.

2. Fill three plastic cups with potting soil. Plant three seeds in each cup. **Measure** 10 mL of clean water. Water the seeds in the first cup with the clean water, and label the cup. Repeat with the other two cups and containers of water.

3. Place the cups in a sunny window. Every day, water each cup with the water it was first watered with.

4. **Observe** the cups for 10 days. Each day, **record** your observations.

Draw Conclusions

1. What did you **observe**?
2. Which plants grew best?
3. **Inquiry Skill** Scientists **compare** things to see how they are alike and how they are different. How are the plants grown with the three kinds of water alike and different?

Step 2

Step 3

Independent Inquiry

Would watering plants with water containing vinegar or dish detergent affect their growth? **Predict** what would happen. Then try it! Make sure to wear goggles.

Read and Learn

VOCABULARY
pollution p. 344

SCIENCE CONCEPTS
- how land is used
- what the different types of pollution are

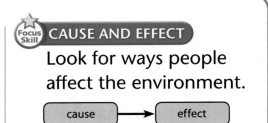

CAUSE AND EFFECT
Look for ways people affect the environment.

cause → effect

Uses of the Land

The land people live on is a very important resource. People use land in many different ways. People build on land. They use resources from the land to make buildings. Rocks and wood are building materials that come from the land.

Many other resources that are useful to people come from the land. Metals, gemstones, and coal are dug up, or mined, from the land. Land is also used for farming. Farmers use the soil to grow plants that are used for food, medicine, cloth, and more.

Some big cities spread out for many miles. Many uses of land are shown in this picture.

Whenever people use land, they change it in some way. Building, mining, and farming all change the land. Sometimes the effects are good. For example, planting trees helps keep the soil in place. But sometimes the effects are bad. For example, mining can tear up the land. This harms the plants and animals that live in the area.

What effects can people's actions have on land?

It is important for city planners to include open spaces such as parks for people to enjoy. ▶

Land Pollution

People can also change land by polluting it. **Pollution** (puh•LOO•shuhn) is any harmful material in the environment. There are many types of harmful materials, or pollutants. Solid waste, chemicals, noise, and even light can be pollutants.

One kind of pollution is land pollution. Land pollution can happen when people throw trash in the wrong places. Some trash can harm plants and animals. It can also pollute the water under the ground. Even if pollution does not harm living things, it can make the land ugly.

CAUSE AND EFFECT What are some effects of land pollution?

Math in Science
Interpret Data

A T-shirt can take 6 months to rot away. A rope can take a year. How many years does it take for a tin can to rot away? A glass bottle?

How Long Materials Take To Rot Away

Material	Months
Newspaper	1
Cotton T-shirt	6
Rope	12
Tin can	1,200
Glass bottle	6,000

Every year in the United States, more than 150 million tons of trash are put into landfills. ▼

▲ After a windy or rainy day, the pollution over the city is cleared away.

▲ If the air over the city does not move, the pollution stays in place. This makes the air unhealthful to breathe.

Air Pollution

Pollution affects more than just the land. Air can be polluted, too. Smoke, mostly from trucks, cars, and factories, is one cause of air pollution.

Air pollution can make it hard for people to breathe. Air pollution can also change the weather. For example, air polluted by car and factory gases traps heat from sunlight. This makes Earth warmer. Scientists call this *global warming*.

Focus Skill CAUSE AND EFFECT

What effect can air pollution have on people?

Insta-Lab

Seeing Air Pollution

Smear a circle of petroleum jelly in the center of a white paper plate. Leave the plate outdoors for a day or two. Observe the plate. How has it changed? What caused the change?

345

Water Pollution

Pollution in the air and on the land can get into water when there is rain. Rain washes pollutants from the air and land into the water. People who dump trash, oil, and other pollutants into the water also cause water pollution.

Drinking polluted water can make people sick. Animals that live in polluted water can also get sick. Some kinds of water pollution can be cleaned up. Water treatment plants can take pollutants out of water to make it clean again.

CAUSE AND EFFECT
What can cause water to become polluted?

Oil is being cleaned from this bird's feathers. Without this help, the bird would probably die. ▼

Sometimes, oil enters the oceans as a result of an accident. It takes many people to clean up an oil spill.

Lesson Review

Essential Question

How do people use and impact the environment?

In this lesson, you learned that people use the land in many ways. People live on the land and mine resources from the land. Sometimes, people harm the environment by polluting it. Pollution can be on the land, in the air, or in the water.

1. **CAUSE AND EFFECT** Draw and complete a graphic organizer to show the causes and effects related to how we use land.

2. **SUMMARIZE** Write a summary of this lesson by explaining the causes and effects of pollution.

3. **DRAW CONCLUSIONS** Can things other than people cause pollution? Explain.

4. **VOCABULARY** Describe what pollution is, and list three examples.

Test Prep

5. Which of the following resources comes from land?
   A. air
   B. fish
   C. trees
   D. water

Make Connections

Writing

Expository
Suppose nobody tried to clean up pollution now. Write a **description** of what Earth might look like 100 years from now.

Health

Technology and Health!
Find out how technology has affected health and farming and the way people get rid of trash. Make a poster to show what you learned.

347

LESSON 4

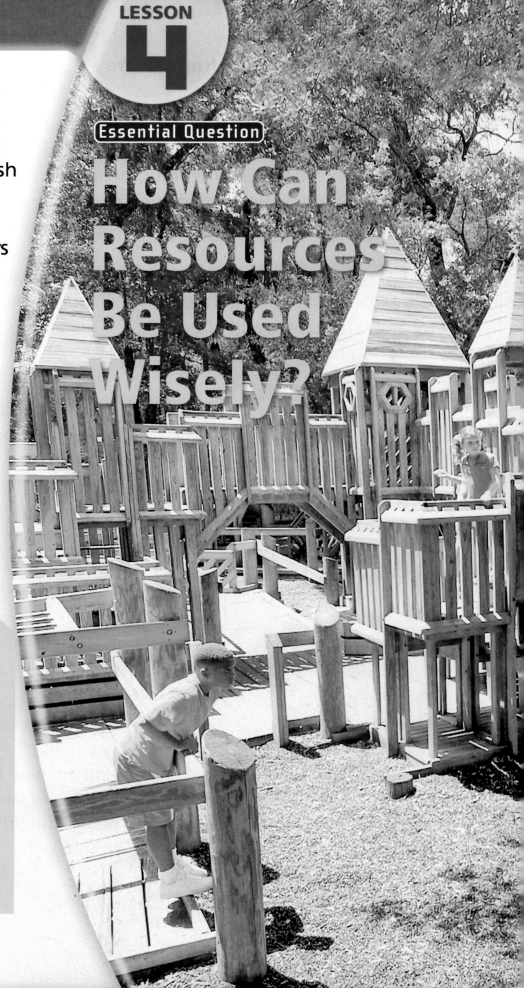

Investigate to find out how quickly trash is made.

Read and Learn about different ways to help protect the environment.

Essential Question

How Can Resources Be Used Wisely?

Fast Fact

From Trash to Treasure

Don't think of old plastic bottles, aluminum cans, and tires as trash. All of them can be recycled. In fact, parts of this playground were made from recycled materials! In the Investigate, you will try saving one kind of resource.

Wooden playground equipment

Vocabulary Preview

conservation [kahn•ser•VAY•shuhn] The saving of resources by using them wisely (p. 352)

reduce [ree•DOOS] To use less of a resource (p. 354)

reuse [ree•YOOZ] To use a resource again and again (p. 355)

recycle [ree•SY•kuhl] To reuse a resource by breaking it down and using it to make a new product (p. 356)

349

Investigate

Taking a Look at Trash

Guided Inquiry

Start with Questions

This is a landfill. When garbage is collected, most of it ends up in a landfill.

- How does garbage affect the environment?
- How quickly does the garbage in this landfill pile up?

Investigate to find out. Then read and learn to find out more.

Prepare to Investigate

Inquiry Skill Tip

When you use numbers to record data, you should be careful to record the correct numbers. If you record a number wrong, it could make the rest of your data wrong.

Materials

- large plastic trash bags
- bathroom scale
- calculator

Make a Data Table

Day	Weight of Trash (lb)
1	
2	
3	
4	
5	
Total Weight:	

Follow This Procedure

1. With the rest of your class, save all the paper you would normally throw away for one week.

2. At the end of each day, weigh the paper. **Record** the weight.

3. Use your data to make a line graph showing the weight of each day's collection.

4. Add up all the weights shown on your graph. The sum tells how many pounds of paper the class collected in one week.

Draw Conclusions

1. Suppose that 1 pound of paper takes up 2 cubic feet of space. How much landfill space would your class save by recycling the paper you saved this week?

2. **Inquiry Skill** There are many ways you can communicate an idea. In this investigation, you **used numbers** to describe the weight of the paper collected. How does using numbers help you tell people what you found out?

Step 2

Step 3

Independent Inquiry

Predict how much paper your class could save by using both sides of each sheet. **Plan and conduct an investigation** to find out.

Read and Learn

VOCABULARY
conservation p. 352
reduce p. 354
reuse p. 355
recycle p. 356

SCIENCE CONCEPTS
▶ how conservation saves resources
▶ how to reduce, reuse, and recycle

MAIN IDEA AND DETAILS
Look for details about how to protect the environment.

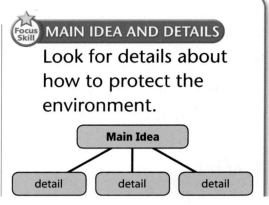

Ways to Protect the Environment

People use a lot of resources. You have read that many resources are nonrenewable. This means that if we use up those resources, we will have none of them left. To protect Earth's resources, it is important to practice conservation. **Conservation** (kahn•ser•VAY•shuhn) is saving resources by using them wisely. Making sure resources do not get used up is one way to protect the environment.

Manatees are protected by law.

Using certain resources can cause the environment to become polluted. You have read that pollution can harm people, animals, and plants. Because of this, it is important to keep pollution from getting into the environment. It is also important to clean up any pollution that has already gotten into it. In this lesson, you will learn how to conserve resources and cut down on pollution.

MAIN IDEA AND DETAILS
What are two ways to protect the environment?

When people protect the environment, they make it safer and more enjoyable.

Reduce

One of the ways you can protect the environment is to **reduce** the amount of resources you use. To reduce your use of resources means to use less of them. For example, you can use less water by taking showers instead of baths. You can use less electricity if you turn the TV off when you are not watching it. If people use fewer resources, there will be more resources to use in the future.

What are two ways you can reduce your use of resources?

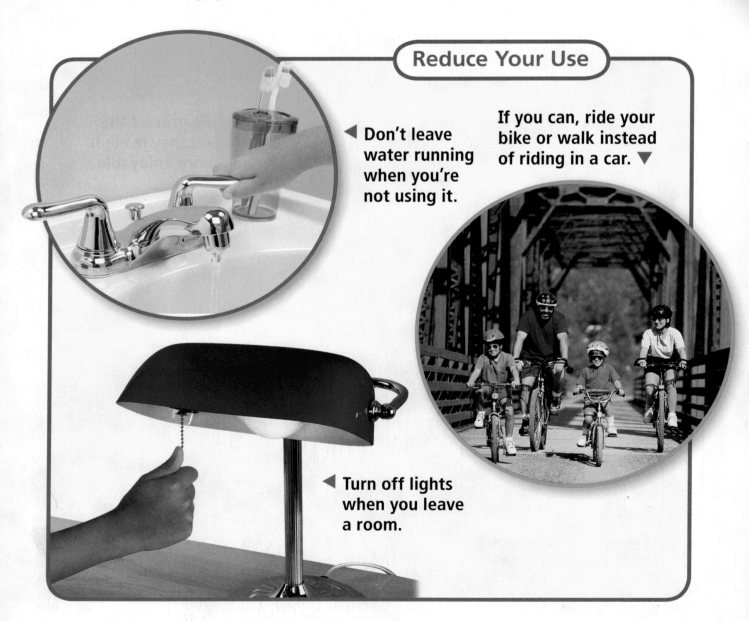

Reduce Your Use

◀ Don't leave water running when you're not using it.

If you can, ride your bike or walk instead of riding in a car. ▼

◀ Turn off lights when you leave a room.

▲ These students are reusing paper bags to make gift wrap.

Reuse

Do you wear clothes that no longer fit your older brother or sister? Do you fix toys when they break, instead of buying new ones? If so, you are reusing resources. When you **reuse** a resource, you use it again and again. Reusing helps reduce the amount of resources that would be needed to make new things. Reusing resources also saves people money.

 MAIN IDEA AND DETAILS

Why is reusing resources important?

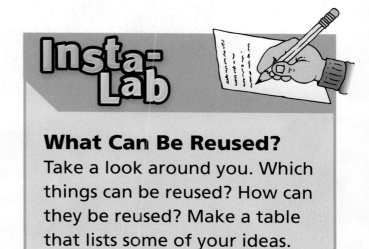

What Can Be Reused?
Take a look around you. Which things can be reused? How can they be reused? Make a table that lists some of your ideas.

Recycle

Another way to save resources is to recycle them. To **recycle** means to break a product down or change it in some way and then use the material to make something new. Many materials, such as paper, glass, the aluminum from cans, and some plastics can be recycled.

Recycled materials can be used to make many new products. Paper that is recycled can be used to make cards, paper towels, and newspaper. Plastic that is recycled can be used to make park benches, doormats, and much more.

MAIN IDEA AND DETAILS
What are some materials that can be recycled?

▲ This symbol means that a product can be recycled.

When you recycle, you help conserve resources by reusing the same resource. ▼

Lesson Review

Essential Question

How can resources be used wisely?

In this lesson, you learned that conservation is saving resources. Three types of conservation are reducing, reusing, and recycling.

1. **MAIN IDEA AND DETAILS** Draw and complete a graphic organizer for this main idea: There are three ways to protect the environment and conserve resources.

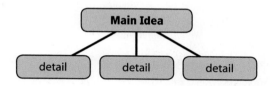

2. **SUMMARIZE** Write a summary for this lesson that tells the most important ideas about resources.

3. **DRAW CONCLUSIONS** List all the resources that you save when you reuse a cotton T-shirt instead of buying a new one.

4. **VOCABULARY** Write a paragraph that correctly uses the terms *reduce, reuse,* and *recycle.*

Test Prep

5. **Critical Thinking** Miles reads the newspaper every day. What resource would he conserve if he recycled the newspapers?

Make Connections

 Writing

Expository
Find out about your state and local recycling laws. Write about how those laws protect the environment and how people make sure those laws are followed.

 Music

Recycling Rhyme
Write the words to a song that tells people why recycling is good. Use a tune you already know. Sing your song in class.

Science Spin
From Weekly Reader
TECHNOLOGY

The Great Dam

Thousands of years ago, Chinese workers built the Great Wall of China. Now they are building a great dam called the Three Gorges Dam.

The dam crosses China's mighty river, the Chang Jiang. The dam is 2.4 kilometers (1.5 miles) wide and nearly 182 meters (600 feet) high. When it is finished, it will be the largest dam in the world. Water will pass through the dam and turn 26 giant turbines. A turbine is like a giant engine that produces electricity.

The Three Gorges Dam has changed the face of China. It has created a lake about 643 kilometers (400 miles) long and hundreds of meters deep.

Boat on the Chang Jiang

Officials hope the Three Gorges Dam will help stop flooding by the Chang Jiang. ▼

Why Build a Dam?

To build the dam, workers had to construct a coffer dam. This is a temporary dam that sends the river around the building site.

Chinese officials hope the dam will stop the Chang Jiang from flooding. For centuries, flooding of the river has been a problem. Floods in China have killed more than a million people over the past 100 years.

Not everyone is happy about the dam. Some people say the dam will hurt the environment. They fear that the new giant lake caused by the dam will collect a lot of pollution from towns and factories along the Chang Jiang. Some people think that the dam caused the extinction of the Yangtze River dolphin.

Think and Write

1. How will the Three Gorges Dam change the landscape of China?
2. Why would people want a dam built? Why would people not want a dam built?

Find out more. Log on to www.hspscience.com

Chapter 8 Review and Test Preparation

Vocabulary Review

Use the terms below to complete the sentences. The page numbers tell you where to look in the chapter if you need help.

renewable resource p. 324

reusable resource p. 325

nonrenewable resource p. 326

humus p. 332

clay p. 335

pollution p. 344

conservation p. 352

reduce p. 354

1. Soil with the very tiniest grains of rock is _____.
2. A resource that may be used up one day is a _____.
3. When you use resources wisely, you are practicing _____.
4. When you use less of a resource, you _____ your use of it.
5. A resource that can be replaced in a human lifetime is a _____.
6. A harmful material in the environment is _____.
7. The part of soil that is made up of parts of dead plants and animals is _____.
8. A resource that you can use again and again is a _____.

Check Understanding

Write the letter of the best choice.

9. Which of the following is a reusable resource?
    A. coal
    B. oil
    C. seed
    D. water
10. Which type of resource is the crude oil that gasoline is made from?
    F. nonrenewable resource
    G. refundable resource
    H. renewable resource
    J. reusable resource
11. Which of these is a type of soil?
    A. bedrock
    B. loam
    C. mineral
    D. pebble

12. COMPARE AND CONTRAST Compare the soils in the picture below. Why is Layer A the topsoil?
   F. It has no humus.
   G. It has smaller rocks.
   H. It has humus.
   J. It has large rocks.

Layer A
Layer B
Layer C

13. MAIN IDEA AND DETAILS Which technology can you use to help cut down on air pollution?
   A. a bicycle
   B. a car
   C. a cell phone
   D. a water bottle

14. Which technology is used to clean polluted water?
   F. a car engine
   G. an oil tanker
   H. a tire-recycling plant
   J. a water treatment plant

15. Which of these resources can you conserve by recycling paper?
   A. aluminum   C. gasoline
   B. coal   D. trees

16. Look at the symbol below. What does it represent?
   F. conservation
   G. recycling
   H. renewing
   J. soil formation

Inquiry Skills

17. List the resources you observe on your way to school. List at least one of each kind of resource—renewable, reusable, and nonrenewable.

18. Compare recycling resources with reducing the use of resources.

Critical Thinking

19. Many minerals found in Earth's crust are mined as resources. Are these minerals renewable or nonrenewable resources? Explain.

20. This graph describes a sample of loam. Why is loam the best resource for growing vegetables?

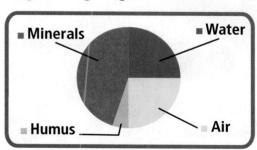

Visual Summary

Tell how each picture shows the **Big Idea** for its chapter.

Big Idea

Rocks are made up of different minerals, and they go through changes during the rock cycle.

Big Idea

Processes on Earth can change Earth's landforms. Some of these changes happen slowly, while others happen quickly.

Big Idea

Living things use Earth's resources to meet their needs. Some of these resources can be recycled or reused.

Weather and Space

UNIT D — EARTH SCIENCE

CHAPTER 9
The Water Cycle 364

CHAPTER 10
Earth's Place in the
Solar System. 402

Unit Inquiry

Space Suits

Scientists at NASA have designed special space suits for astronauts to wear in space. The suits need to be thick to protect the astronauts. How does a thick space suit affect how easily an astronaut can move around in space? For example, can an astronaut make repairs quickly while wearing a space suit? Plan and conduct an experiment to find out.

CHAPTER 9
The Water Cycle

What's the Big Idea?

Water is important to all living things in many different ways.

Essential Questions

Lesson 1
Where Is Water Found on Earth?

Lesson 2
What Is the Water Cycle?

Lesson 3
What Is Weather?

Student eBook
www.hspscience.com

What do you wonder?

Water is just about everywhere on Earth. Many people enjoy surfing waves like this one. What other ways do people use water at a beach? How does this relate to the **Big Idea?**

Cresting wave

LESSON 1

Investigate to find out how much water there is on Earth.

Read and Learn the importance to Earth of both fresh water and salt water.

Essential Question

Where Is Water Found on Earth?

Fast Fact

Water Everywhere! Almost three-fourths of Earth is covered by water! In the Investigate you will observe this by using a model.

Tourists viewing an Arctic glacier

Vocabulary Preview

fresh water [FRESH WAWT•er] Water that has very little salt in it (p. 371)

glacier [GLAY•sher] A huge sheet or block of moving ice (p. 372)

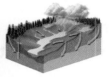

groundwater [GROWND•wawt•er] An underground supply of water (p. 372)

Investigate

Where in the World Is Water?

Guided Inquiry

Start with Questions

Chicago was built next to water, but this isn't an ocean. It's a huge lake!

- Do you see more land or more water in this picture?
- Do you think there is more land or more water on Earth's surface?

Investigate to find out. Then read and learn to find out more.

Prepare to Investigate

Inquiry Skill Tip

You often use a measuring tool, such as a ruler or a thermometer, when you collect data. In this Investigate, you collect data by putting tally marks in the correct spaces of a data table. Making good observations is an important part of collecting data.

Materials

- plastic inflatable globe
- sheet of paper
- pencil

Make a Data Table

Location	Number of Catches	Total
Land		
Water		

Follow This Procedure

1. Work in a group of five. Choose one person to be the recorder.
2. The other four persons toss the globe gently to one another.
3. The catcher catches the globe with open hands. The recorder records the data of whether the catcher's right index finger touches land or touches water.
4. Continue tossing and recording until the globe has been tossed 20 times.

Draw Conclusions

1. Total the catches. How many times did the catcher's right index finger touch water? How many times did it touch land?
2. Where did the catcher's finger land more often? Why do you think this happened?
3. **Inquiry Skill** Scientists use numbers to collect data. Using your data, estimate how much of Earth's surface is covered by water.

Step 2

Step 3

Independent Inquiry

The more data you collect, the more accurate your data will be. How would doing the Investigate 10 more times change your data? Try it! Communicate your results in a bar graph.

Read and Learn

VOCABULARY
fresh water p. 371
glacier p. 372
groundwater p. 372

SCIENCE CONCEPTS
▶ why water is important
▶ where water is found on Earth

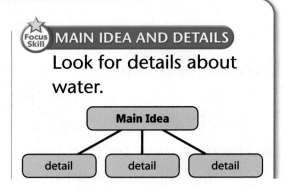

MAIN IDEA AND DETAILS
Look for details about water.

The Importance of Water

You drink water. You use water when you take a bath or a shower. You might even play in water. More than two-thirds of your body is made of water! Animals need water to stay healthy. Plants also need water. Some animals and plants live in water. You and other living things on Earth could not live without water.

Water isn't important only to living things. It's also important to Earth's environment. Water changes Earth's surface. Without water, there would be no rivers, lakes, or oceans. There would be no rain, snow, or clouds.

Deer must drink water to stay alive. ▶

Farmers use a lot of water to grow crops. It takes 11 liters (3 gal) of fresh water to grow a single tomato.

If you look at a picture of Earth taken from space, you will see a lot of blue. Earth looks blue because of the water. Most of the water seen from space is in Earth's seas and oceans. This water is salt water.

Most land plants can't live if salt water is put on them. Many animals, including people, can't drink salt water. Most plants and animals need fresh water to live. **Fresh water** is water that has very little salt in it. Some fresh water has no salt at all. Only a small part of Earth's water is fresh. Most of the fresh water on Earth is frozen.

Math in Science
Interpret Data

Water on Earth

Most of the water on Earth is found in the oceans. For every liter of liquid fresh water, there are 2 liters of frozen fresh water and 97 liters of salty ocean water.

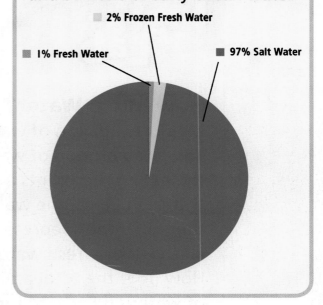

2% Frozen Fresh Water
1% Fresh Water
97% Salt Water

MAIN IDEA AND DETAILS
What are two types of water?

Fresh Water

Almost all the fresh water on Earth is found in one place—Antarctica. Antarctica is the land at the South Pole. Most of the fresh water there is frozen in glaciers. A **glacier** is a huge sheet of moving ice.

Not all fresh water is frozen. Fresh water is also found in the form of rain and melted snow. The water from rain and snow flows into streams, rivers, ponds, and lakes.

Fresh water can also be found under the ground. After a rain, water soaks into the ground. The water moves through the soil until it reaches solid rock. The water collects above the solid rock. This underground supply of water is called **groundwater**. Many people dig wells in the ground and pump up groundwater to supply their homes with water.

MAIN IDEA AND DETAILS

Where is fresh water found on Earth?

Lake Eola is an example of a freshwater lake.

How Much Water?

Measure a gallon of water and a tablespoon of water. If the gallon represents all the water on Earth, the water in the tablespoon represents all the liquid fresh water. How does the total amount of water on Earth compare with the amount of water available for people to drink?

This is a picture of Lake Powell. You can see how high the water has been by looking at the water lines on the rocks.

▲ The state of Florida is almost completely surrounded by salt water.

Salt Water

Almost all water on Earth is salt water. You can find salt water in oceans, seas, and gulfs. Three major saltwater bodies border the United States. These are the Atlantic Ocean, the Pacific Ocean, and the Gulf of Mexico. You can also find salt water in some lakes, such as Great Salt Lake, in Utah.

You can't drink salt water, but salt water is very important. When sunlight hits the ocean, the top part of the water becomes warm. Warm ocean water moves around Earth and helps keep some places warm. People use many things from the ocean, such as fish.

 MAIN IDEA AND DETAILS

Where is salt water found?

Lesson Review

Essential Question

Where is water found on Earth?

In this lesson, you learned that Earth's surface is mostly water. Water is important to living things and the environment.

1. **MAIN IDEA AND DETAILS** Draw and complete a graphic organizer for the main idea: About three-fourths of Earth's surface is covered by water.

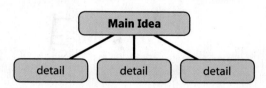

2. **SUMMARIZE** Write a summary of this lesson that tells about fresh water and salt water.

3. **DRAW CONCLUSIONS** Why is it important not to litter in fresh water and salt water?

4. **VOCABULARY** Write a sentence explaining how the terms *fresh water*, *glacier*, and *groundwater* are related.

Test Prep

5. **Critical Thinking** Suppose you are digging a hole in the ground. Right after you start digging, you notice that the soil is very damp. Using your observations, explain what might have happened.

Make Connections

 Writing

Persuasive

Write a **letter** to your city mayor, explaining that there isn't a lot of fresh water on Earth. Give the mayor reasons people should protect the fresh water.

 Art

Animal Mobile

Find out about the kinds of animals that live in water. Make a mobile that shows some of these animals. Hang a label under each animal that says whether the animal lives in salt water or in fresh water.

375

LESSON 2

Essential Question
What Is the Water Cycle?

Investigate to find out how water moves in a terrarium.

Read and Learn about how water moves through the water cycle.

Fast Fact

Sharing Water with Dinosaurs
The water now on Earth is the same water that has been here for millions of years. These raindrops could be water from a pool where a dinosaur once drank. In the Investigate, you will model the water cycle.

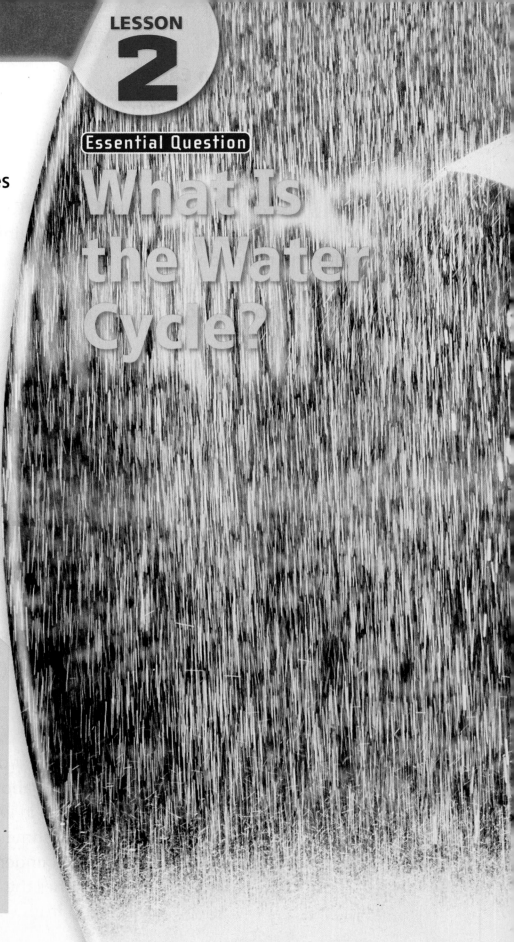

Vocabulary Preview

condensation
[kahn•duhn•SAY•shuhn]
The process by which water vapor changes into liquid water (p. 382)

evaporation
[ee•vap•uh•RAY•shuhn]
The process by which liquid water changes into water vapor (p. 383)

precipitation
[pree•sip•uh•TAY•shuhn]
Rain, snow, sleet, or hail (p. 384)

water cycle [WAWT•er SY•kuhl] The movement of water from Earth's land, through rivers toward the ocean, to the air, and back to the land (p. 384)

Investigate

Condensation in a Terrarium

Guided Inquiry

Start with Questions

What is fog? Fog is a cloud that touches the ground. It is made up of tiny droplets of water and possibly tiny particles of ice.

- How is fog related to weather?
- Where do you think the fog in the picture came from?

Investigate to find out. Then read and learn to find out more.

Prepare to Investigate

Inquiry Skill Tip

A good way to find possible answers in science is to infer. Think about the facts you already know. Then think about your observations. Use what you already know to explain what you have observed.

Materials
- clear plastic salad container
- soil
- packet of seeds
- water
- spray bottle

Make an Observation Chart

Day	Observations

Follow This Procedure

1. To build a terrarium, put about 3 cm of soil in a clear plastic salad container.

2. Plant the seeds. Follow the instructions on the seed package.

3. Using a spray bottle, spray the soil until it is moist. Close the lid of the container, and label the container with your name.

4. Place the terrarium next to a sunny window or under a lamp. **Observe** your terrarium for several days. Write down all the changes you see.

Draw Conclusions

1. What changes happened inside the terrarium?

2. Did anything in the terrarium remind you of weather? If so, what was it?

3. **Inquiry Skill** You watered your terrarium only one time. **Infer** how water may have gotten on the lid of the terrarium.

Independent Inquiry

Do the same Investigate, but don't close the lid. What do you think will happen? **Compare** your observations with the lid closed and not closed.

379

Read and Learn

VOCABULARY
condensation p. 382
evaporation p. 383
precipitation p. 384
water cycle p. 384

SCIENCE CONCEPTS
▶ how water changes forms
▶ how water moves from place to place

SEQUENCE
Look for the sequence in which water changes forms.

Different States of Water

Maybe you have watched water boiling in a pot. If the water boils for a long time, no water will be left in the pot. Even though the water seems to disappear, it really doesn't. The water just changes from a liquid to a gas. The water in the pot was a liquid. The rising vapor that you couldn't see was in the form of a gas.

All water on Earth is in one of three states. It is a liquid, a solid, or a gas. Solid water is ice. Water in gas form is *water vapor.* You can't see water vapor.

The water in this canal is in a solid form during the winter.

Lesson Review

Essential Question

What is the water cycle?

In this lesson, you learned that water can be a solid, a liquid, or a gas. As water evaporates into the air and then condenses as precipitation, it goes through the water cycle.

1. **SEQUENCE** Draw and complete a graphic organizer that shows the sequence of the change of state in water.

2. **SUMMARIZE** Write a paragraph summarizing the states of water.

3. **DRAW CONCLUSIONS** Suppose you leave a glass of water outside early in the morning. The temperature is 29°C (84°F). If there is no rain during the day, what change might you find that night in the glass?

4. **VOCABULARY** Write a paragraph about the water cycle. Include the terms *condensation, evaporation, precipitation,* and *water cycle.*

Test Prep

5. What form does water take at room temperature?
   - **A.** ice
   - **B.** liquid
   - **C.** snow
   - **D.** solid

Make Connections

 Writing

Expository

Water can evaporate slowly or quickly. Research to find out what factors affect how quickly water evaporates. Write a page to tell what you find.

 Math

Construct a Bar Graph

Look at the Interpret Data in Lesson 1. Using the same data, construct a bar graph. Which graph do you think is easier to read? Explain your answer.

385

LESSON 3

Essential Question

What Is Weather?

Investigate to find out how to measure wind.

Read and Learn about what causes weather, how to measure weather, and how to predict it.

Fast Fact

When the Wind Blows

Weather vanes show the direction from which the wind is blowing. Long ago, weather vanes were among the few tools that people used to predict the weather. In the Investigate, you will make a tool to measure the wind.

Weather vane

Vocabulary Preview

atmosphere
[AT•muhs•feer] The air around Earth (p. 390)

oxygen [AHK•sih•juhn] A gas that all living things need, and that plants give off into the air (p. 390)

weather [WETH•er] What is happening in the atmosphere at a certain place and time (p. 392)

temperature [TEM•per•uh•cher] The measure of how hot or cold something is (p. 393)

anemometer [an•uh•MAHM•uht•er] A weather instrument that measures wind speed (p. 393)

Investigate

Measuring Wind

Guided Inquiry

Start with Questions

Wind can be a gentle breeze or a powerful force. These trees are being blown by the wind.

- Is the wind in this picture forceful or gentle?
- How can you measure the strength of the wind?

Investigate to find out. Then read and learn to find out more.

Prepare to Investigate

Inquiry Skill Tip

When you want to compare data, putting it in a bar graph can help. A bar graph shows all the data side by side.

Materials

- 2 cardboard strips
- 3 small white cups
- stapler
- scissors
- wire
- watch
- red cup
- cap of a ballpoint pen

Make a Data Table

Day	Number of Turns
1	
2	
3	
4	
5	
6	
7	

388

Follow This Procedure

1. Make an X with the cardboard strips. Staple them together.
2. **CAUTION:** Carefully use scissors to make a hole in the middle of the X. Push the pen cap into the hole.
3. Cut a slit in both sides of each cup. Attach a cup to the end of each cardboard strip by pushing the cardboard through the slits.
4. Push most of the wire into the ground. Balance the pen cap on the wire.
5. **Observe** the cups for one minute. Count the number of times the red cup spins by. **Record** your data.
6. Repeat your observation at the same time each day for a week.

Draw Conclusions

1. Make a bar graph that shows the number of turns the cup made on each day of the week.
2. **Inquiry Skill** How did making a graph help **compare** your results?

Step 3

Step 5

Independent Inquiry

Predict what would happen if you added 2 more cups. Try it.

Read and Learn

VOCABULARY
atmosphere p. 390
oxygen p. 390
weather p. 392
temperature p. 393
anemometer p. 393

SCIENCE CONCEPTS
▶ what weather is
▶ how weather is measured

 COMPARE AND CONTRAST
Look for ways that clouds are alike and different.

[alike]———[different]

The Air Around You

How do you know there is air around you? You can't see air. You can't hold or taste air. You know that air is all around because you can feel it. Every time the wind blows, you can feel air moving.

The air around Earth is the **atmosphere**. The atmosphere is important because it has oxygen. **Oxygen** is a gas that people need to live. The atmosphere also has gases that plants need to live.

Water vapor is in the atmosphere. Clouds form from water vapor. Rain, snow, and other kinds of precipitation fall from clouds.

 COMPARE AND CONTRAST

How does air compare with a solid object, such as a rock?

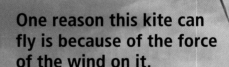

One reason this kite can fly is because of the force of the wind on it.

Types of Clouds

Cirrus Clouds
Cirrus clouds are the highest clouds. They look like wisps of hair. Cirrus clouds do not produce precipitation that reaches Earth.

Cumulus Clouds
Cumulus clouds are lower than cirrus clouds. They look puffy. Cumulus clouds can bring strong storms with rain and hail.

Stratus Clouds
Stratus clouds are the lowest clouds. They look like a sheet or a layer of clouds. A stratus cloud usually brings rain, sleet, or snow.

Weather Patterns

What is today's weather like where you live? Is it sunny? Cloudy? Rainy? Weather both helps and harms people. For example, rain brings water for crops, but too much rain can cause floods that destroy homes.

Weather is what is happening in the atmosphere at a certain place and time. The weather in a place depends on where the place is. In the United States, places in the south are usually warmer than places in the north.

 COMPARE AND CONTRAST

How can weather both help and harm people?

Features on Earth's surface, such as a coastline, can change the weather.

Sea Breeze

▲ On the coast, cooler air over the water moves toward land during the day. This makes a sea breeze.

Land Breeze

▲ During the night, air is cooler over land, so a land breeze moves air toward the water.

Gathering Weather Data

You can learn a lot about the weather by looking out a window. Sometimes, you may want to know facts about the weather that you can't see. Weather instruments can help you gather weather data. For example, you can use a thermometer to find the temperature. **Temperature** is the measure of how hot or cold something is. Another weather instrument is an anemometer. An **anemometer** (an•uh•MAHM•uht•er) measures wind speed.

 COMPARE AND CONTRAST

Compare a thermometer with an anemometer.

These people are using special instruments to collect weather data at weather stations.

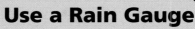

Use a Rain Gauge
Tape a centimeter ruler to the outside of a clear plastic jar. Place your "rain gauge" outside in an open area before a rain. Afterward, use the ruler to measure the amount of rain. How much rain fell?

Severe Weather

Sometimes, weather changes can be dangerous. Dangerous weather changes are sometimes called severe weather. Severe weather can bring heavy rain or snow. Lightning, high winds, and big temperature changes may also happen.

Thunderstorm

Thunderstorms can bring heavy rain and high winds. Hail and lightning can happen in these storms.
Precipitation: rain and hail are possible
Temperature: thunderstorms generally occur during hot weather
Wind: sometimes strong; very strong thunderstorms can produce tornadoes

Tornado

A tornado is a spinning column of air that touches the ground. Tornadoes happen in some strong thunderstorms. They are very dangerous.
Precipitation: rain and hail are possible
Temperature: generally warm
Wind: very strong: can be up to 500 kilometers per hour (310 miles per hour)

Drought

A drought is a time with very dry weather. In a drought, the dryness lasts for a long time. Rivers and lakes can begin to dry up. People may need to be careful about how much water they use.
Precipitation: very little, if any, precipitation
Temperature: may change during drought
Wind: wind speed may change during droughts

Flood

Floods happen when a lot of rain falls in a short amount of time. Rivers and ditches can overflow and cause problems.
Precipitation: periods of heavy rain and some periods with no rain
Temperature: usually warm or mild
Wind: wind speed may change during flooding

The kind of weather depends on the temperature, the speed and direction of wind, and the precipitation. Some weather conditions can last for a long time.

COMPARE AND CONTRAST

How is severe weather different from everyday weather?

Hurricane

Hurricanes are the largest storms on Earth. They form over oceans when the water is warm. Hurricanes can cause tornadoes and thunderstorms to form.
Precipitation: rain—mostly heavy
Temperature: generally warm
Wind: at least 119 kilometers an hour (74 miles an hour)

Blizzard

Blizzards happen in the winter when it is cold and below freezing. They bring heavy snow and strong, cold winds.
Precipitation: snow—mostly heavy
Temperature: below 0°C (32°F)
Wind: at least 56 kilometers per hour (35 miles per hour)

Ice Storm

Ice storms happen in the winter when rain freezes on things. The ice can be heavy enough to break power lines and damage trees. The slippery ice can make walking or driving difficult.
Precipitation: precipitation falls as rain, but freezes when it lands
Temperature: around 0°C (32°F)
Wind: wind speed may change during the storm

Heat Wave

When the temperature is very hot for a long time, a heat wave may be happening. Heat waves last from a few days to several weeks. They can be very dangerous.
Precipitation: rain is possible
Temperature: generally from 35°C (95°F) to 44°C (110°F) and above
Wind: wind speed may vary during a heat wave

Predicting Weather

People use weather instruments to predict weather. One instrument that is used to predict weather is a weather satellite. A weather satellite in space may take pictures of clouds. The satellite sends the pictures to Earth. People use the pictures to see how clouds move. People also use the pictures to make weather maps.

Look at the satellite image below. You can find images like this in newspapers. You can also see weather maps on television news shows. Weather maps show information such as temperature and the chance of rain.

 COMPARE AND CONTRAST

Look at this weather map. How does Orlando's weather differ from Seattle's weather?

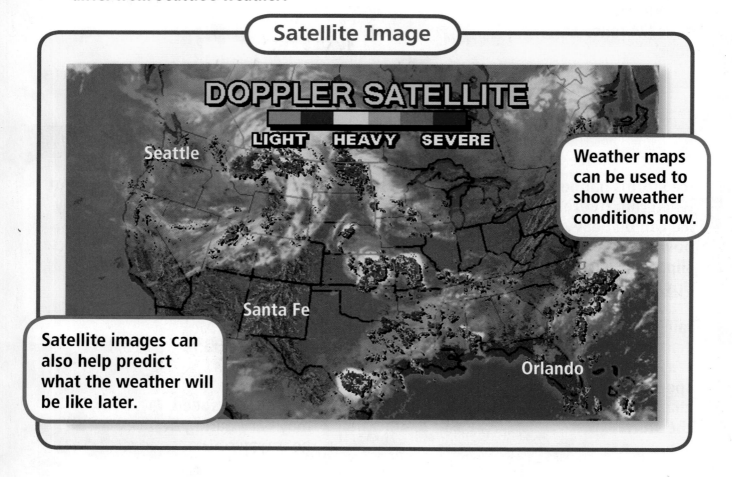

Satellite Image

Weather maps can be used to show weather conditions now.

Satellite images can also help predict what the weather will be like later.

Lesson Review

Essential Question

What is weather?

In this lesson, you learned that weather happens in the atmosphere. The atmosphere contains oxygen, other gases, and water vapor. Changes in temperature, precipitation, and wind speed cause different weather conditions. Instruments can help scientists predict the weather.

1. **COMPARE AND CONTRAST** Draw and complete a graphic organizer that shows how clouds are alike and different.

2. **SUMMARIZE** Write a two- or three-sentence summary for this lesson. Tell about three types of weather.

3. **DRAW CONCLUSIONS** Does the ocean affect weather? Explain.

4. **VOCABULARY** Write a short story about a weather forecaster. Use all the vocabulary terms from this lesson in your story.

Test Prep

5. **Critical Thinking** Examine the weather map on the left-hand page. Describe the weather in Santa Fe, New Mexico.

Make Connections

Writing

Narrative

Write a personal **story** describing some weather you have seen. Describe what the weather was like and tell if it was helpful or harmful to you or your family.

Health

Weather and Health

Weather affects people's health. Choose a type of weather, and draw a poster showing how to stay healthy in that kind of weather. For example, it is important to wear sunscreen on sunny days.

397

People in Science

Adriana Ocampo

You know that Earth's surface is mostly water, but have you ever wanted to know more about Earth's neighbors in space? Adriana Ocampo gets to research planets and their moons. She even helped map Phobos, one of the moons that orbit Mars.

Ocampo works at the Jet Propulsion Laboratory in Pasadena, California. She worked on the Mars Observer Project to learn more about the surface of Mars.

Many scientists want to know if there was ever water on the surface of Mars. They know that living things need water. If there used to be water on Mars, they believe there might have been life on Mars too!

▶ **ADRIANA OCAMPO**
▶ Research Scientist for NASA

Think and Write

1. What might be some signs that water did exist on Mars?
2. If there ever was water on the surface of Mars what might have happened to it?

Phobos

Ocampo compares Earth's surface with those of other planets, such as Mars.

Bin Wang

▶ **BIN WANG**
▶ Meteorologist at the International Pacific Research Center

Bin Wang lives in Hawai`i. Even though the weather there is usually beautiful, Wang studies dangerous storms and changes in Earth's weather.

Bin Wang is a meteorologist, or weather expert. He studies how weather changes over time. He can forecast what the weather will be like in the near future.

Wang has studied Pacific Ocean-area weather for many years. Thanks to him, people can be better prepared for deadly Pacific storms.

Think and Write

1. Why would people want to predict near-future weather?
2. What kinds of storms might Bin Wang study?

Career

Meteorologists don't just predict next week's weather. These scientists use computers and historical records to study weather patterns over many years. Information from weather experts can also help businesses. Farmers, pilots, and ship captains are just some of the people who rely on weather information in their work.

CHAPTER 9 Review and Test Preparation

Vocabulary Review

Use the terms below to complete the sentences. The page numbers tell you where to look in the chapter if you need help.

fresh water   p. 371

groundwater   p. 372

condensation   p. 382

evaporation   p. 383

precipitation   p. 384

water cycle   p. 384

atmosphere   p. 390

weather   p. 392

1. The changing of a liquid into a gas is _____.

2. The movement of water between Earth's air and land is part of the _____.

3. Water that has very little salt in it is _____.

4. The air around Earth makes up the _____.

5. The changing of a gas into a liquid is _____.

6. What is happening in the atmosphere at a certain place and time is _____.

7. Earth's underground supply of water is _____.

8. Rain, snow, sleet, and hail are examples of _____.

Check Understanding

Write the letter of the best choice.

9. Where is most of the water on Earth located?
   A. in freshwater lakes
   B. in glaciers
   C. in oceans
   D. underground

10. **MAIN IDEA AND DETAILS** What are glaciers made of?

    F. gas
    G. ice
    H. groundwater
    J. salt water

11. Which of the following is **not** a type of precipitation?
    A. groundwater
    B. hail
    C. sleet
    D. snow

400

12. What happens to water vapor as it moves upward in the air?
  F. It becomes hot.
  G. It evaporates.
  H. It forms clouds.
  J. It becomes a solid.

13. SEQUENCE Which sequence shows forms of water from the coldest to the hottest temperatures?
  A. ice, liquid water, water vapor
  B. ice, water vapor, liquid water
  C. liquid water, water vapor, ice
  D. water vapor, ice, liquid water

14. If water vapor loses thermal energy, which process will occur?
  F. condensation **H.** freezing
  G. evaporation **J.** melting

15. Which instrument would you use to measure wind speed?
  A. anemometer
  B. rain gauge
  C. thermometer
  D. weather vane

16. Which of the following are the lowest clouds in the atmosphere?
  F. cirrus
  G. cumulus
  H. satellite
  J. stratus

Inquiry Skills

17. LaDonne observes and collects data from her rain gauge each day. Compare her drawings from Tuesday and Wednesday.

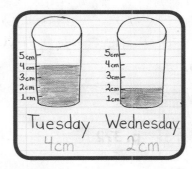

18. Suppose you leave an empty covered jar outside in the evening. When you get up the next morning, there is water in the jar. Infer what happened.

Critical Thinking

19. Scott is a farmer. Why is it important for him to watch the weather report each day?

20. You know that most of the water on Earth is salt water. Explain why it's important to keep water clean, especially fresh water.

CHAPTER 10
Earth's Place in the Solar System

What's the Big Idea? The sun and everything that orbits it make up the solar system. Other objects, such as constellations, are found outside the solar system.

Essential Questions

Lesson 1
What Causes Earth's Seasons?

Lesson 2
How Do Earth and the Moon Interact?

Lesson 3
What Is the Solar System?

Student eBook
www.hspscience.com

What do you wonder?

The moon moves in a path around Earth. It takes about 28 days for the moon to complete this journey. Why does the moon circle Earth instead of Earth circling the moon? How does this relate to the **Big Idea?**

View of Earth from space

Investigate to find out how the sun's rays strike different parts of Earth in different ways.

Read and Learn about how Earth's motion causes the seasons and the cycle of day and night.

Essential Question

What Causes Earth's Seasons?

Fast Fact

Weather Mystery
Earth is farthest from the sun in July and closest to it in January. Yet it is colder in the United States in January than it is in July! In the Investigate, you will learn the reason.

Vocabulary Preview

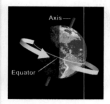

axis [AK•sis] A line—which you cannot see—that runs through the center of Earth from the North Pole to the South Pole (p. 408)

rotation [roh•TAY•shuhn] The spinning of Earth on its axis (p. 408)

revolution [rev•uh•LOO•shuhn] The movement of Earth one time around the sun (p. 409)

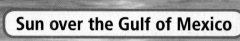

Sun over the Gulf of Mexico

405

Investigate

How Sunlight Strikes Earth

Guided Inquiry

Start with Questions

Have you been up early enough to watch a sunrise? This is what a sunrise looks like from space.

- Why can you see only part of Earth clearly in the picture?
- In the picture, does the sunlight strike Earth evenly?

Investigate to find out. Then read and learn to find out more.

Prepare to Investigate

Inquiry Skill Tip

When you compare things in an investigation, think about the properties you're studying. For this Investigate, you are interested in the direction and strength of the light as it struck the surface.

Materials

- clear tape
- graph paper
- large book
- flashlight
- meterstick
- black marker
- wooden block
- red marker

Make a Data Table

Light Beam	Brightness	Number of Squares
Straight		
Tilted		

Follow This Procedure

1. Tape graph paper to a book. Hold a flashlight about 50 cm above the paper. Shine the light straight down. You will see a circle of light.

2. Have a partner use a black marker to trace the circle of light. **Observe** the light on the paper. **Record** the number of squares inside the black line.

3. Keep the flashlight in the same position. Have a partner put a block under one end of the book. This time, use a red marker to trace the light on the paper. **Observe** the light on the paper, and **record** the number of squares inside the red line.

Draw Conclusions

1. Inside which line was the light brighter? Inside which line are there more squares?

2. **Inquiry Skill Compare** the ways the light rays struck the straight and tilted surfaces. At which time would you infer that a place on Earth has warmer weather—when the sun's rays strike it directly or when they strike it at a slant? Explain.

Independent Inquiry

Form a **hypothesis** about what will happen if the book is tilted more. **Plan and conduct an experiment** to test your idea.

407

Read and Learn

VOCABULARY
axis p. 408
rotation p. 408
revolution p. 409

SCIENCE CONCEPTS
▶ what causes the seasons
▶ what causes day and night

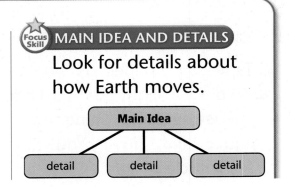
MAIN IDEA AND DETAILS
Look for details about how Earth moves.

How Earth Moves

As you read this book, it may seem as if you're sitting still. In fact, you're moving through space at about 107,000 kilometers (about 66,000 mi) per hour! Although you don't feel the motion, Earth is both traveling and spinning.

Earth spins just the way a top does. Picture a line going through Earth from the North Pole to the South Pole. This imaginary line through Earth is called Earth's **axis**. Look at the picture on this page. As you can see, Earth's axis is not straight up and down. Instead, the axis is tilted a little. The spinning of Earth on its axis is called **rotation**.

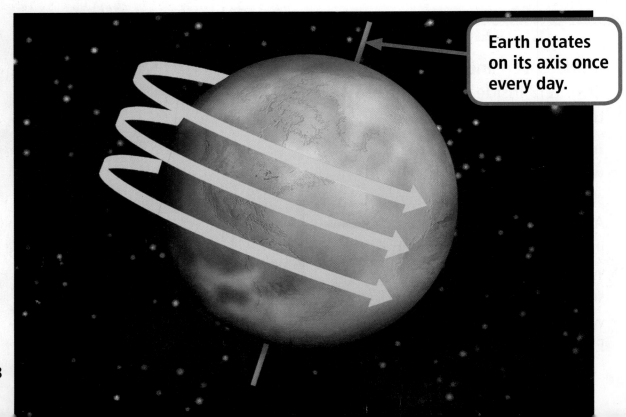
Earth rotates on its axis once every day.

It takes one year for Earth to revolve around the sun and return to this position.

Rotation isn't the only way Earth moves. Earth also revolves around the sun. One trip around the sun is one **revolution**. Each revolution takes about 365 days. People use Earth's movements to measure time. One rotation of Earth takes one day. One revolution of Earth takes one year.

 MAIN IDEA AND DETAILS

What are two ways in which Earth moves?

Insta-Lab

Modeling Motion

Crumple a sheet of yellow paper into a ball to model the sun. Then crumple a sheet of blue paper into a ball to model Earth. Use the models to show the two ways Earth moves. As Earth turns and travels, is the same part always facing the sun? Explain.

409

Seasons

During the year, some parts of Earth have four seasons—winter, spring, summer, and fall—with different weather. One weather difference is temperature. You might wonder what causes temperature differences.

In the Investigate, you learned that when light rays strike directly, the light is brighter than when they strike at a slant. That's because light rays that strike at a slant spread out more.

> When Earth's northern half is tilted away from the sun, light rays strike at their greatest slant and we have winter. At the same time, light rays strike part of Earth's southern half more directly. Summer starts in December there. When light rays strike part of Earth's northern half directly, we have summer. At the same time, light rays strike Earth's southern half at their greatest slant. There, winter starts in June.

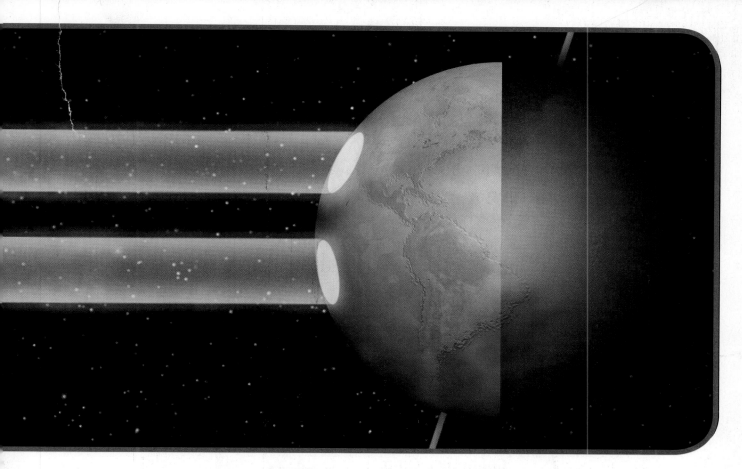

▲ Light rays that strike at a slant have the same amount of energy as light rays that strike directly. Their energy is spread out over a larger area.

As Earth revolves, it is tilted on its axis. This causes the same part of Earth that is tilted toward the sun at one time to be tilted away from it at another time. Places where the sun's rays strike directly are warmer than places where the rays strike at a slant.

Where the sun's rays strike Earth directly, the season is summer. Where they strike Earth at the greatest slant, the season is winter.

 MAIN IDEA AND DETAILS

What causes seasons?

Day and Night

Earth takes one year to revolve around the sun but only one day to rotate. As it rotates, half of Earth faces the sun and has daytime. The other half of Earth faces away from the sun and has nighttime.

Since some places have day while others have night, it can't be the same time everywhere. People have divided the world into time zones so that places near one another have the same time. When it's 12 noon in Baltimore, Maryland, it's 6 o'clock in the evening in Madrid, Spain.

What causes day and night?

When it's daytime in Columbus, Ohio,...

... it's nighttime in Paris, France.

The sun is always shining on one half of Earth while the other half is in darkness.

Lesson Review

Essential Question

What causes Earth's seasons?

In this lesson, you learned that Earth moves in two ways. Earth rotates on its axis once every 24 hours. This rotation causes day and night. Earth revolves around the sun about once every 365 days. This revolution and the tilt of Earth's axis cause the seasons.

1. **MAIN IDEA AND DETAILS** Draw and complete a graphic organizer for this main idea: Earth moves in two different ways.

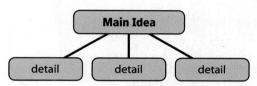

2. **SUMMARIZE** Write a summary of this lesson. Begin with the sentence: Earth moves in two ways.

3. **DRAW CONCLUSIONS** What might happen to seasons if Earth was not tilted on its axis?

4. **VOCABULARY** Draw and label a diagram that shows Earth's revolution, rotation, and axis.

Test Prep

5. **Critical Thinking** Mike lives in Chicago. At 8:00 A.M., Mike calls a friend in Australia. He finds out his friend and the family are sleeping. Why would Mike's friend be asleep?

Make Connections

 Writing

Expository

A day always has 24 hours, but the lengths of daytime and nighttime change during the year. Research why this happens. Then write a **report** about your findings.

 **Math**

Construct a Graph

As the seasons change, so does the weather, including the temperature. Find out the average temperature for each month in your town. Make a bar graph that shows this information.

LESSON 2

Investigate to find out why the moon seems to have different shapes throughout the month.

Read and Learn about the phases of the moon and about eclipses.

Essential Question

How Do Earth and the Moon Interact?

Fast Fact

Earth's Satellite

The moon is about 384,500 kilometers (238,900 mi) from Earth. If you could drive there, the trip would take about 180 days nonstop. In the Investigate, you will find out more about the moon.

Vocabulary Preview

moon phases
[MOON FAYZ•uhz] The different shapes that the moon seems to have in the sky when it is observed from Earth (p. 418)

lunar cycle [LOON•er SY•kuhl] The pattern of phases of the moon (p. 418)

lunar eclipse [LOON•er i•KLIPS] An event in which Earth blocks sunlight from reaching the moon and the Earth's shadow falls on the moon (p. 420)

solar eclipse [SOH•ler ih•KLIPS] An event in which the moon blocks sunlight from reaching Earth and the moon's shadow falls on Earth (p. 422)

A full moon

Investigate

The Moon's Phases

Guided Inquiry

Start with Questions

On some nights, when you look at the moon, it's a full circle. On other nights, it looks like this.

- Do you know what causes a full moon?
- What causes the moon to change from a full moon to a sliver?

Investigate to find out. Then read and learn to find out more.

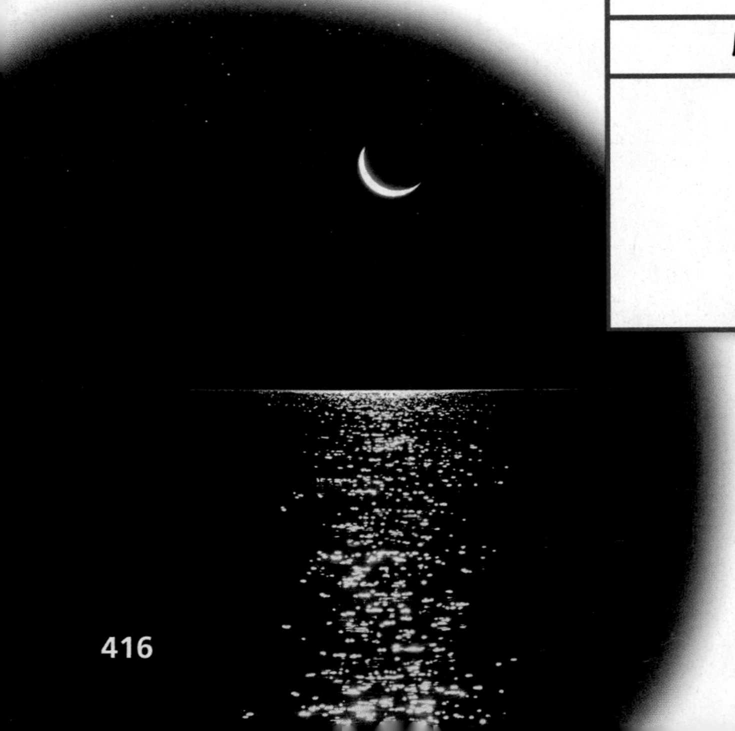

Prepare to Investigate

Inquiry Skill Tip

When you infer, you use your observations and what you already know to figure out something.

Materials

- flashlight
- volleyball

Make an Observation Chart

Position 1	Position 2
Position 3	Position 4

Follow This Procedure

1. Work in groups of three. Use the picture in Step 1 to set up the area. Your teacher will darken the room. One classmate will hold a volleyball at position 1. Another classmate will shine a flashlight on it. The third group member will stand in the middle, observe the ball, and make a drawing of the ball's lighted side.

2. The classmate holding the volleyball moves to positions 2, 3, and 4. The other group members rotate to face the classmate with the volleyball. Observe and record the light at each position.

3. Switch roles so that everyone can observe the pattern.

Draw Conclusions

1. What part of the ball is lighted at each position?

2. What does the ball represent? The flashlight? The person recording?

3. **Inquiry Skill** If the ball represents the moon, what can you infer that the different parts of the lighted ball represent?

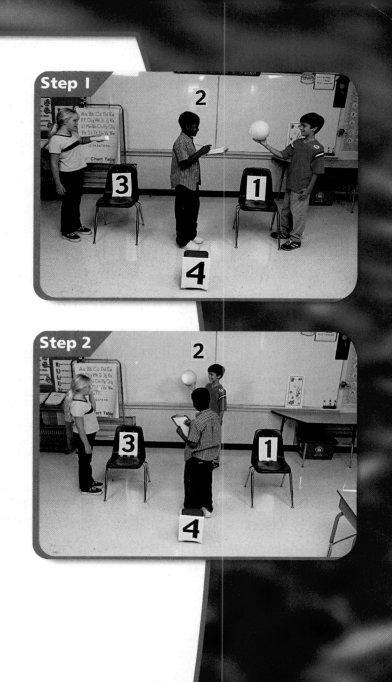

Independent Inquiry

The moon's phases occur in a regular pattern. Predict how long it will take the moon to go through all its phases. Test your prediction.

Read and Learn

VOCABULARY
moon phases p. 418
lunar cycle p. 418
lunar eclipse p. 420
solar eclipse p. 422

SCIENCE CONCEPTS
▶ what the moon's phases are
▶ what causes eclipses

 **SEQUENCE**

Look for the sequences of lunar phases and eclipses.

Phases of the Moon

You may have noticed that over a period of time, the moon doesn't always look the same. It may look like a circle, a half-circle, or just a thin sliver. The different shapes we see are called **moon phases**.

The moon does not really change. As with Earth, half of the moon is always lit by the sun. The moon reflects that light from the sun. Just as Earth revolves around the sun, the moon revolves around Earth. It takes the moon about $29\frac{1}{2}$ days to go through its phases in order. This pattern of phases is called the **lunar cycle**. The phase you see at a certain point in the cycle depends on how much of the sunlit side of the moon is facing Earth at that time.

 **SEQUENCE**

Which phase comes after the new moon phase?

The moon goes through its phases every $29\frac{1}{2}$ days. ▼

418

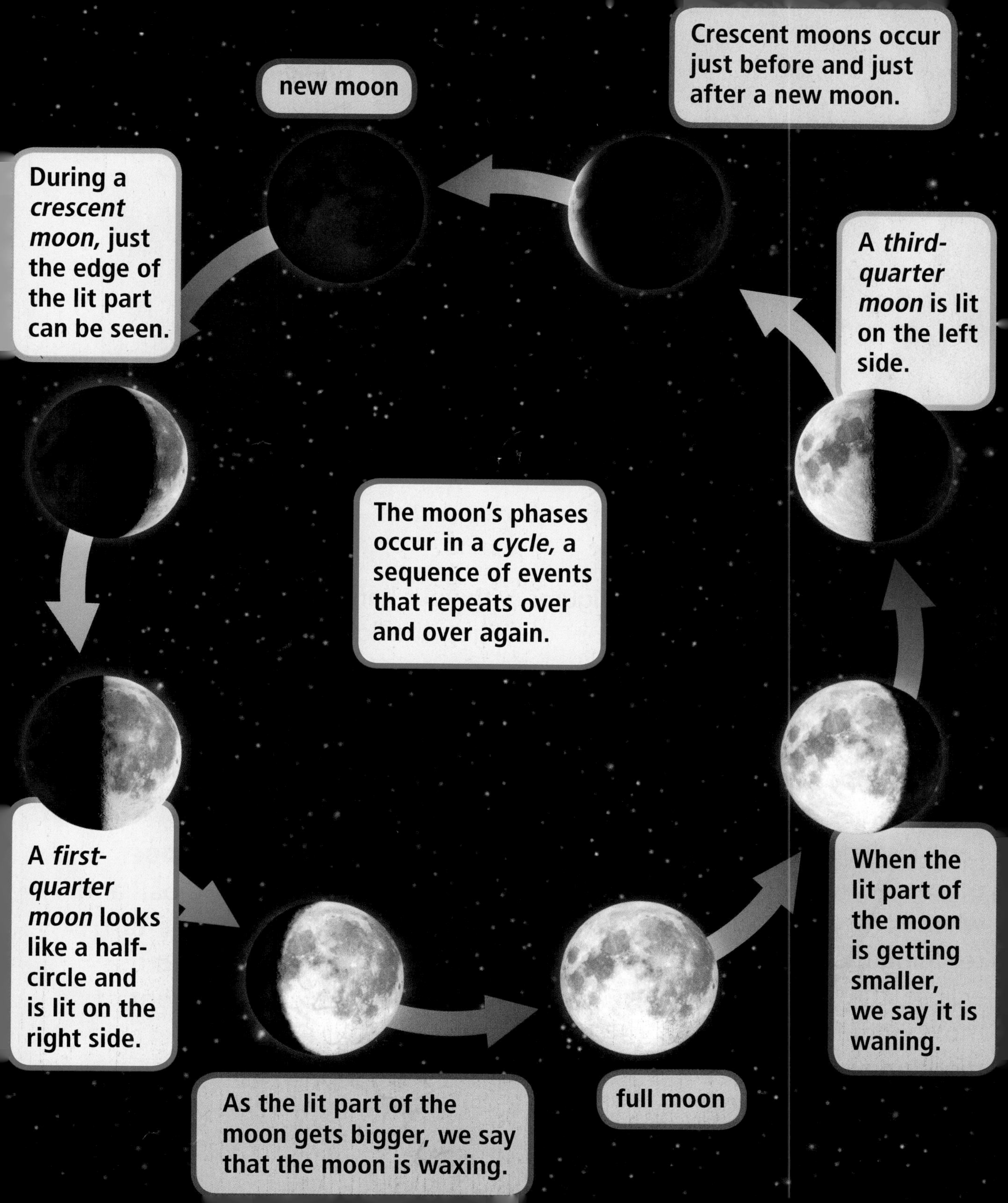

new moon

Crescent moons occur just before and just after a new moon.

During a *crescent moon*, just the edge of the lit part can be seen.

A *third-quarter moon* is lit on the left side.

The moon's phases occur in a *cycle*, a sequence of events that repeats over and over again.

A *first-quarter moon* looks like a half-circle and is lit on the right side.

When the lit part of the moon is getting smaller, we say it is waning.

As the lit part of the moon gets bigger, we say that the moon is waxing.

full moon

During a total lunar eclipse, the moon can appear to be red.

Eclipses of the Moon

Have you ever made shadow figures with your hands? When you hold your hands in front of a light, the light can't pass through them, so they make shadows. In the same way, Earth makes a shadow because the sun's light can't pass through it.

Sometimes the moon moves into Earth's shadow. This causes a lunar eclipse. A **lunar eclipse** happens when Earth blocks sunlight from reaching the moon. In a lunar eclipse, the moon first gets dark as it moves into Earth's shadow. Then the moon moves out of the shadow and becomes bright again.

 SEQUENCE

How does a lunar eclipse happen?

Modeling an Eclipse
Make a medium-size ball and a small ball out of clay. Turn off the lights, and hold the medium-size ball in front of a flashlight. Have a partner pass the small ball behind the medium-size ball. What happens? What does this show?

Science Up Close

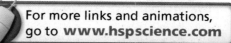

Lunar Eclipses

Lunar eclipses can happen only during the full moon phase. Sometimes the full moon passes through only part of Earth's shadow. This causes a partial lunar eclipse. At other times, the moon passes through all of Earth's shadow. This causes a total lunar eclipse. The stages of a total lunar eclipse are shown below.

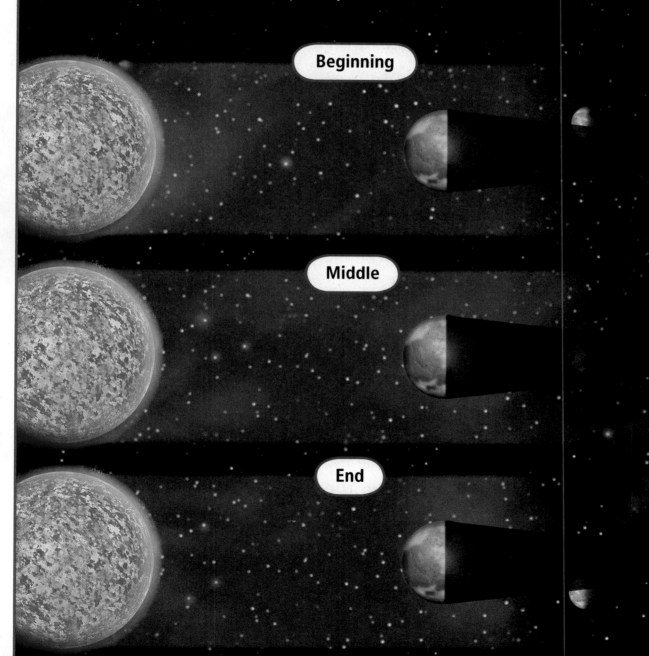

This shows the stages of a solar eclipse.

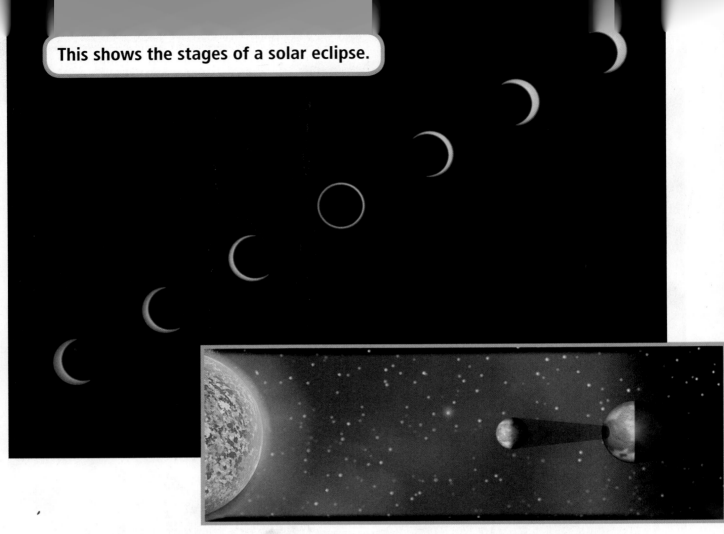

▲ During a solar eclipse, the moon passes between Earth and the sun. Solar eclipses can happen only during the new moon phase.

Eclipses of the Sun

Just as Earth can block sunlight from the moon, the moon can block sunlight from Earth. When the moon's shadow falls on Earth, it is called a **solar eclipse**. This happens when the moon moves between the sun and Earth. During a solar eclipse, the sky gets dark for a few minutes. Then the moon moves on, and daylight returns.

The moon's shadow is not large enough to cover all of Earth. Only a few places on Earth can view a solar eclipse each time. Those places change with every eclipse.

SEQUENCE How does a solar eclipse happen?

Lesson Review

Essential Question

How do Earth and the moon interact?

In this lesson, you learned that the moon goes through phases as it revolves around Earth. A lunar eclipse takes place when Earth comes between the sun and the moon. A solar eclipse takes place when the moon passes between the sun and Earth.

1. **SEQUENCE** Draw and complete a graphic organizer that shows the sequence of the moon's phases.

2. **SUMMARIZE** Write a lesson summary by using the lesson vocabulary terms in a paragraph.

3. **DRAW CONCLUSIONS** Why does a solar eclipse happen only during the new moon phase?

4. **VOCABULARY** Make drawings to illustrate the terms *moon phase, lunar eclipse,* and *solar eclipse.* Be sure to label your drawings.

Test Prep

5. How often does the lunar cycle repeat itself?
   - A. about every day
   - B. about every 29 days
   - C. about every week
   - D. about every year

Make Connections

Writing

Narrative
Astronauts first landed on the moon on July 20, 1969. Do research to learn details about the event. Then write a **story** about the landing from the point of view of one of the astronauts.

Art

Moon Changes
Observe the moon for several months. Draw it as you observe it. How does it change from night to night? Does it look different during the day?

LESSON 3

Investigate to find out the order of the planets in the solar system.

Read and Learn about the kinds of things found in our solar system and the universe.

Essential Question

What Is the Solar System?

Fast Fact

Earth's Neighbors
At the Hayden Planetarium in New York City, you can see models of all the planets. The model of Jupiter, the largest planet, is almost 3 meters (9 ft) across. In the Investigate, you will find out more about the planets.

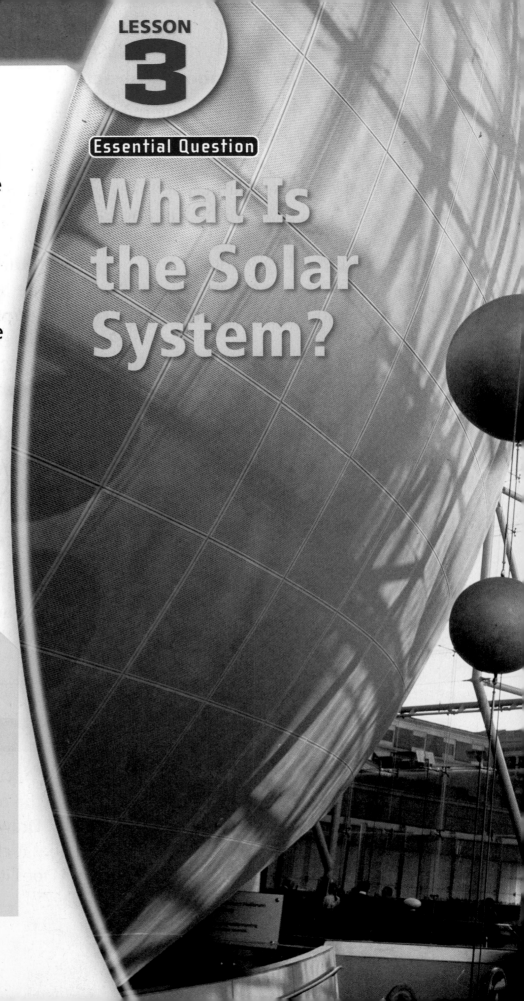

424

Solar system model

Vocabulary Preview

planet [PLAN•it] A large body of rock or gas in space (p. 428)

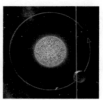

orbit [AWR•bit] The path a planet takes as it revolves around the sun or that a moon takes as it revolves around a planet (p. 428)

solar system [SOH•ler SIS•tuhm] The sun, the planets and their moons, and the small objects that orbit the sun (p. 428)

star [STAR] A hot ball of glowing gases that gives off energy (p. 429)

constellation [kahn•stuh•LAY•shuhn] A group of stars that appears to form the shape of an animal, a person, or an object (p. 434)

425

Investigate

The Planets

Guided Inquiry

Start with Questions

Can you see Venus in the picture? It's the small speck in front of the sun. It's passing between the sun and Earth.

- Is Earth or Venus closer to the sun?
- How can you tell which is closer by looking at the picture?

Investigate to find out. Then read and learn to find out more.

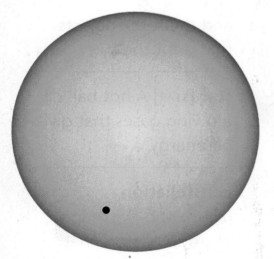

Prepare to Investigate

Inquiry Skill Tip

You often need to compare when you use numbers in an investigation. First notice how many digits the numbers have. A number with fewer digits is lower than a number with more digits. If the number of digits is the same, start with the digits on the left when you compare.

Materials

- pencil
- paper

Make a Data Table

Order of Planets and "Dwarf Planets"	Distance from the Sun (in millions of kilometers)

Follow This Procedure

1. **Use numbers** from the Planet Data table to list the planets in **order** by their distance from the sun, from closest to farthest.

2. Next to the planets' names, **record** their distances from the sun.

Draw Conclusions

1. Which planet is closest to the sun?
2. Which planet is farthest from the sun?
3. How many planets are between Earth and the sun?
4. Which planets are Earth's nearest neighbors?
5. **Inquiry Skill** Scientists sometimes **use numbers** to put things in order. List other ways you could order the planets.

Planet Data	
Planet	Distance From Sun (in millions of kilometers)
Earth	150
Jupiter	778
Mars	228
Mercury	58
Neptune	4,505
Pluto*	5,890
Saturn	1,427
Uranus	2,869
Venus	108

*In 2006 a group of scientists classified Pluto as a "dwarf planet".

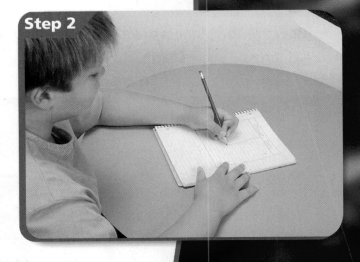

Step 2

Independent Inquiry

How could you **use numbers** to help you plan a **model** of the solar system? **Plan a simple investigation** for the **model**.

427

Read and Learn

VOCABULARY
planet p. 428
orbit p. 428
solar system p. 428
star p. 429
constellation p. 434

SCIENCE CONCEPTS
▶ what the planets are like
▶ what other kinds of bodies in the solar system are like

COMPARE AND CONTRAST
Look for ways to compare details about bodies in the solar system.

alike — different

The Solar System

In Lesson 1, you learned that Earth revolves around the sun. Earth is not the only **planet**, or large body of rock or gases, that does this. All of the planets revolve around the sun. Each planet revolves in a path called an **orbit**. Many small objects revolve around the sun, too. The sun, the planets, and these small objects make up the **solar system**.

COMPARE AND CONTRAST

How are the planets alike? How are they different?

Sun   Mercury   Venus   Earth   Mars   Jupiter

428

The Sun

The sun is the center of the solar system. It is also the biggest object in the solar system. The sun is one of the many, many stars in the universe. **Stars** are hot balls of glowing gases that give off energy.

The sun is the star closest to Earth. That is why it appears so much larger and brighter than the other stars. Light energy from the sun is very important to life on Earth. The sun's energy helps plants grow and keeps Earth warm. Without the sun, Earth would be too cold for anything to live on it.

COMPARE AND CONTRAST

How are stars different from planets?

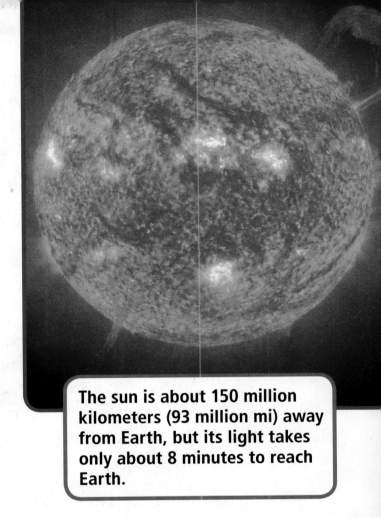

The sun is about 150 million kilometers (93 million mi) away from Earth, but its light takes only about 8 minutes to reach Earth.

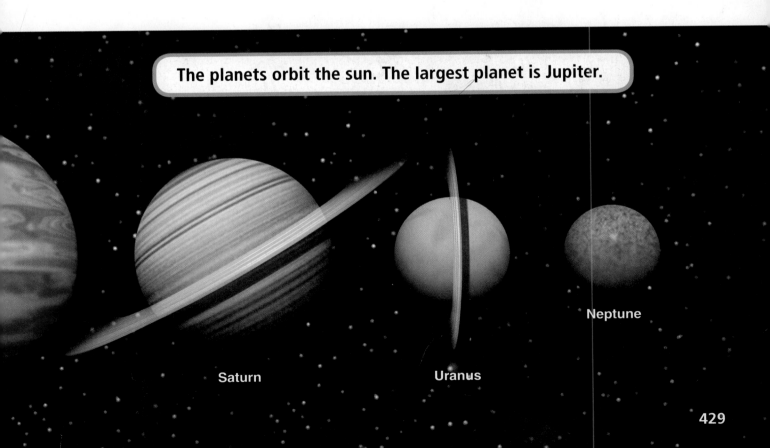

The planets orbit the sun. The largest planet is Jupiter.

Saturn Uranus Neptune

The Inner Planets

The four planets closest to the sun are called the inner planets. The inner planets are Mercury, Venus, Earth, and Mars. You can sometimes see Venus and Mars at night. They look like stars, but their light is steady—it does not twinkle. If you use a telescope, you may be able to see Mercury as well.

Mercury

Fun Fact: In the daytime, much of Mercury's surface is hot enough to melt lead. Even so, there may be ice at its poles.
Length of Day: about 59 Earth days
Length of Year: 88 Earth days
Moons: none
Surface: rocky, with many craters
Distance Across: about 4,900 kilometers (3,100 mi)

Venus

Fun Fact: Venus rotates backward compared with most other planets. Venus may once have had oceans, but they have evaporated.
Length of Day: about 243 Earth days
Length of Year: about 225 Earth days
Moons: none
Surface: rocky, with constant cover of thick clouds
Distance Across: about 12,000 kilometers (7,500 mi)

All of the inner planets have rocky surfaces. They are smaller than the other planets. The inner planets are also warmer than the other planets because they are closer to the sun.

 **COMPARE AND CONTRAST**

How are the inner planets different from the other planets?

Earth

Fun Fact: Earth is the only planet known to have water on its surface. Water is necessary to support life.
Length of Day: 24 hours (1 Earth day)
Length of Year: about 365 days (1 Earth year)
Moons: 1
Surface: $\frac{3}{4}$ water, $\frac{1}{4}$ land
Distance Across: about 12,750 kilometers (7,900 mi)

Mars

Fun Fact: Mars has the largest volcano in the solar system. Mars also has ice under its surface.
Length of Day: about 25 Earth hours
Length of Year: 687 Earth days (almost 2 Earth years)
Moons: 2
Surface: rocky, with red dust and no water
Distance Across: about 6,800 kilometers (4,200 mi)

The Outer Planets

The planets farthest from the sun are called the outer planets. They include Jupiter, Saturn, Uranus, and Neptune. The outer planets are made mostly of frozen gases. Most of them are very large and have many moons.

 **COMPARE AND CONTRAST**

How are most of the outer planets alike?

Jupiter

Fun Fact: Two Earth-sized circles could fit inside Jupiter's Great Red Spot, a huge storm on Jupiter's surface.
Length of Day: about 10 Earth hours
Length of Year: about 12 Earth years
Moons: more than 60
Surface: no solid surface
Distance Across: about 143,000 kilometers (88,900 mi)

Saturn

Fun Fact: Saturn has a large system of rings that reaches about 416,000 kilometers (258,000 mi) from its surface.
Length of Day: about 11 Earth hours
Length of Year: about 29 Earth years
Moons: more than 55
Surface: no solid surface
Distance Across: about 120,000 kilometers (74,600 mi)

Uranus

Fun Fact: Uranus rolls on its side as it orbits the sun.
Length of Day: about 18 Earth hours
Length of Year: about 84 Earth years
Moons: more than 25
Surface: frozen gases
Distance Across: about 51,000 kilometers (31,700 mi)

Neptune

Fun Fact: Neptune is the farthest planet from the sun.
Length of Day: about 19 Earth hours
Length of Year: about 164 Earth years
Moons: more than 10
Surface: frozen gases
Distance Across: about 49,500 kilometers (about 30,758 mi)

Math in Science
Interpret Data

How many Earth years long is one revolution around the sun on the planets Jupiter, Saturn, Uranus, and Neptune?

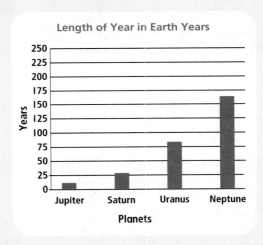

Length of Year in Earth Years

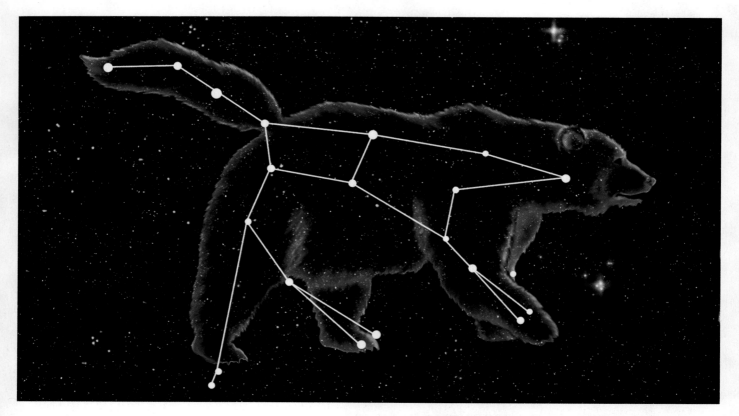

▲ You can see different constellations in different seasons. Some constellations, such as the Big Bear, can be seen all year long.

Patterns of Stars

When you look at stars in the sky, you know that no one could ever count them all. Some are very bright, while others are hard to see. This is because some are farther from Earth than others. How do people tell one star from another? One way is to use constellations. **Constellations** are groups of stars that appear to form the shapes of animals, people, or objects. Looking for constellations helps people find certain stars.

Compare the way people group stars with the way they group the planets.

Finding Constellations
Scatter dried beans on a sheet of paper. Look for shapes made by the beans. Draw them. How might this activity be like what people long ago did when they looked at stars?

Different Seasons, Some Different Stars

During spring and summer, you can see some constellations that you cannot see in fall and winter. You can see other constellations all year long.

In winter, you can see one set of stars. As Earth orbits the sun, the season changes. In summer at the same place, some of the stars you can see will be different from those you saw in the winter. The stars you see in winter are on the other side of the sun in summer.

COMPARE AND CONTRAST How would the night sky be different in summer and winter?

Some stars can be seen only at certain times of the year. Star patterns change with the seasons.

spring

summer

fall

winter

Stars Move at Night

You know that the sun appears to move across the sky from east to west during the day. At night, the stars appear to move, just as the sun does during the day. The rotation of Earth causes stars to seem to move in the nighttime sky.

Like the sun and the moon, most stars seem to rise in the east and set in the west. Stars in the sky near the poles seem to move in a circle.

The shapes of the star patterns we see in the sky do not change. The Big Dipper is a star pattern you might know. It always looks like the Big Dipper, even though it isn't always in the same place in the sky.

 COMPARE AND CONTRAST

How are the moon, the sun, and the stars alike?

This student is observing how the stars appear to move across the sky.

7:00 P.M.   10:00 P.M.

Lesson Review

Essential Question

What is the solar system?

In this lesson, you learned that the sun and everything that revolves around it make up our solar system. The inner planets and outer planets are different. The sun, moon, and stars seem to move because of Earth's movement.

1. **COMPARE AND CONTRAST** Draw and complete a graphic organizer that compares and contrasts the inner and outer planets.

2. **SUMMARIZE** Write a three-sentence summary for this lesson. Tell about the parts of the solar system.

3. **DRAW CONCLUSIONS** How are Earth and Mars alike? How are they different?

4. **VOCABULARY** Write a sentence using the vocabulary terms *planets, orbit,* and *solar system.*

Test Prep

5. **Critical Thinking** Stella likes to look at the stars at night. After several hours, Stella notices that the stars seem to have moved across the night sky. Why do you think the stars seem to move?

Make Connections

 Writing

Descriptive

Observe the stars and moon with a parent or guardian. Write a **description** of what you observe, including any constellations you recognize.

 Health

Lights at Night

Light pollution, or too many lights in the environment, makes it hard to view stars, and can harm human health. Research ways to cut down on light pollution. Make a brochure showing what you learned.

Science Spin — From Weekly Reader
TECHNOLOGY

Ancient Planet Found

Can you guess what has been around for about 13 billion years? It's a planet that scientists recently discovered. If you think that 13 billion years sounds old, you're right. This planet is more than twice as old as Earth. In fact, the new discovery is the oldest planet known to exist. However, don't rush out to try to look at the planet. It is too far away to see with just your eyes. Scientists found the planet only by using the Hubble Space Telescope.

A Faraway Planet

Scientists have found more than 100 planets outside of our solar system. This planet is different because it is so old. It is larger than Jupiter and is most likely made of gases.

The ancient planet is in this group of stars.

The arrow points to the ancient planet.

A Special Telescope

The Hubble Space Telescope is a special kind of telescope. Rather than sitting in a dome on Earth, this telescope floats in space about 600 kilometers (375 mi) above Earth. Since this telescope is in space, it can take clearer pictures of stars and planets than telescopes on Earth can take.

Other Discoveries

Scientists say the Hubble's discovery is important. It means that planets probably began forming much earlier than scientists once thought. This planet wasn't the Hubble's only discovery. Through the years it has taken pictures of many different items from space. It has given scientists closer looks at the stars and even let them see stars they had not seen before. It has also taken pictures of distant galaxies.

Large in Scope

Hubble is a lot bigger than many telescopes. It is 13.1 meters (43.5 ft) long and about 4.27 meters (14 ft) around the outside. This is about the size of a big tractor-trailer truck!

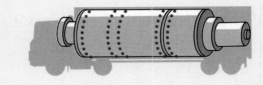

Think and Write

1. What can the Hubble Space Telescope teach scientists about the solar system?
2. Why do you think scientists want to discover new planets?

Find out more. Log on to
www.hspscience.com

CHAPTER 10 Review and Test Preparation

Vocabulary Review

Use the terms below to complete the sentences. The page numbers tell where to look in the chapter if you need help.

axis   p. 408
revolution   p. 409
moon phase   p. 418
lunar cycle   p. 418
solar eclipse   p. 422
planet   p. 428
solar system   p. 428
constellation   p. 434

1. Each stage of the moon's cycle is called a _____.
2. The movement of Earth once around the sun is called a _____.
3. A group of stars that appears to form a picture is a _____.
4. When the moon's shadow falls on Earth, a _____ happens.
5. The imaginary line Earth rotates around is known as its _____.
6. A large body of rock or gases that orbits the sun is a _____.
7. The pattern of lunar phases is called the _____.

8. The sun, the planets, and the other objects that orbit the sun make up the _____.

Check Understanding

Write the letter of the best choice.

9. Which of the following is responsible for Earth's seasons?
   A. Earth's rotation
   B. Earth's tilt
   C. solar eclipses
   D. the lunar cycle

10. Which of the following is responsible for day and night?
    F. Earth's orbit
    G. Earth's revolution
    H. Earth's rotation
    J. Earth's tilt

11. **COMPARE AND CONTRAST** Compare the four areas on the diagram below.

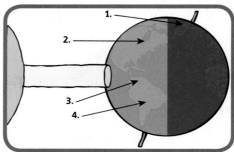

Which area is the warmest?
A. Area 1   C. Area 3
B. Area 2   D. Area 4

440

12. About how long is the lunar cycle?
   F. 29 days
   G. 365 days
   H. 24 hours
   J. 1 week

13. SEQUENCE The steps of a lunar eclipse follow a certain sequence. What happens first?
   A. The moon enters Earth's shadow.
   B. The moon leaves Earth's shadow.
   C. The moon goes dark.
   D. The moon is lighted by the sun again.

14. Which of the following is closest to Earth?
   F. Big Bear
   G. Big Dipper
   H. North Star
   J. the sun

15. Which of the following is true about Mercury?
   A. It has plenty of water.
   B. It is far from the sun.
   C. It is made of frozen gases.
   D. It is very hot.

16. Which of the following is true about Venus?
   F. It has two moons.
   G. It is the closest planet to the sun.
   H. Its rotation is in the opposite direction to the rotations of the other planets.
   J. Its day lasts only 11 hours.

Inquiry Skills

17. Infer why animals and plants are found on Earth but not on other planets of the solar system.

18. Order the inner planets from the one closest to the sun to the one farthest from the sun.

Critical Thinking

19. The sun and moon are part of our solar system. The stars you can see at night are not part of our solar system. Why not?

20. The diagram shows Earth's position for each season. Look at the part of Earth where North America is. Explain why daytime is shorter there in winter than in summer.

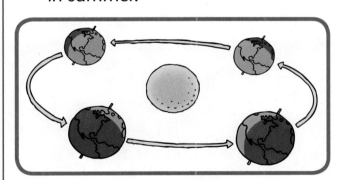

Visual Summary

Tell how each picture shows the **Big Idea** for its chapter.

CHAPTER 9

Big Idea
Water is important to all living things in many different ways.

CHAPTER 10

Big Idea
The sun and everything that orbits it make up the solar system. Other objects, such as constellations, are found outside the solar system.

PHYSICAL SCIENCE

ILLINOIS

Illinois Excursions and Projects
Magic Waters Waterpark 444
Wrigley Field 446
Discovery Center Museum 448
Magnets at a Distance 450

UNIT E
INVESTIGATING MATTER AND ENERGY 451

CHAPTER 11
Properties of Matter 452

CHAPTER 12
Energy 490

CHAPTER 13
Electricity and Magnets 522

CHAPTER 14
Heat, Light, and Sound 552

UNIT F
EXPLORING FORCES AND MOTION . . 593

CHAPTER 15
Forces and Motion 594

CHAPTER 16
Work and Machines 628

Illinois Excursions

Cherry Valley

Magic Waters WATERPARK

Have you ever visited a waterpark? What kind of water slide is your favorite? Do you like the winding tunnel slides best? Or do you like to zip down a steep slide? At the Magic Waters Waterpark in Cherry Valley, Illinois, you can take your pick.

Magic Waters Waterpark from above

Magic Water?

Water is an amazing substance, and science can help you understand it. Remember that water is a type of matter. Matter is made up of tiny particles called molecules. Matter can exist in three states: solid, liquid, and gas.

The water you swim and play in is liquid water. In a liquid, the molecules move around. A liquid flows. It takes the shape of its container. It's also slippery. That's what makes water slides so much fun.

What happens when you put liquid water in a freezer? It gets cold and turns to ice, right? The molecules slow down and hardly move at all. Ice is the solid state of water. A solid has a definite shape and size.

What happens when you heat liquid water in a tea kettle? Adding heat causes the molecules to have more energy. The molecules move faster and spread farther apart. The water boils. Water vapor—a gas—is produced. A gas expands to fill its container.

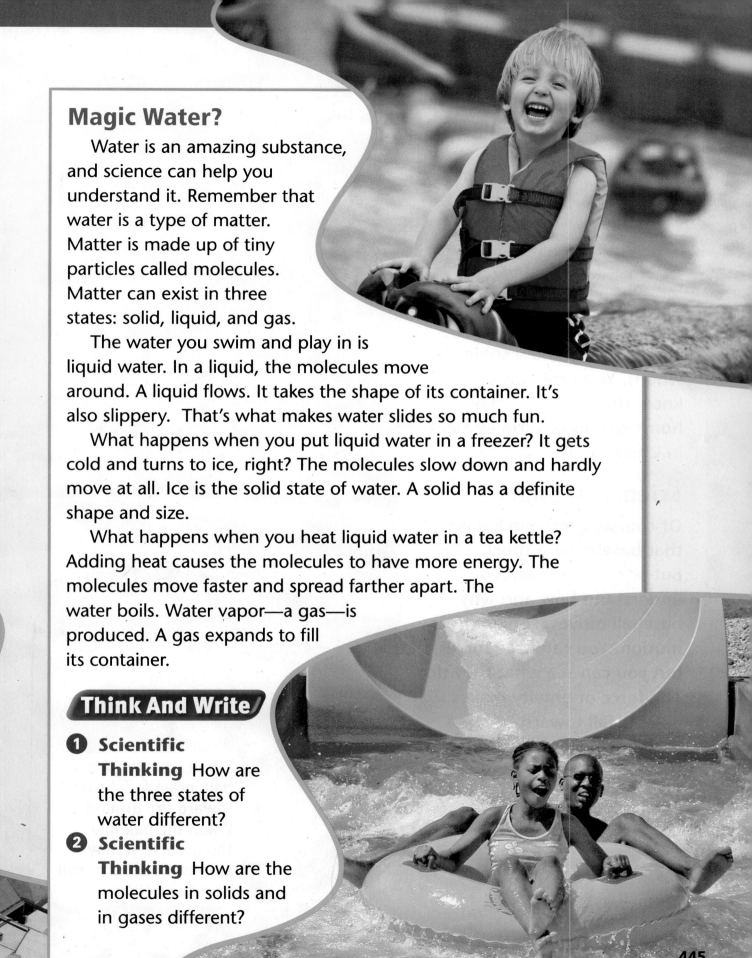

Think And Write

1. **Scientific Thinking** How are the three states of water different?
2. **Scientific Thinking** How are the molecules in solids and in gases different?

Illinois Excursions

Chicago

Wrigley Field

Are you a Chicago Cubs fan? Even if you're not, you probably know someone who is. You probably also know that the Cubs play home games at Wrigley Field, in Chicago.

Science or Sport?

Of course, everyone knows that baseball is a sport. But science can help you understand how and why a baseball moves. Forces affect motion. You can't see forces, but you can see what they do. The force of gravity makes objects fall toward Earth. A baseball player uses a force on the ball to make it move.

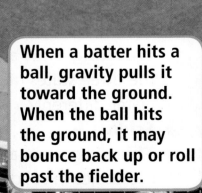

When a batter hits a ball, gravity pulls it toward the ground. When the ball hits the ground, it may bounce back up or roll past the fielder.

Swing, Batter!

What happens when a batter hits a ball? When the bat hits the ball, the bat pushes on the ball. This is called the *action force*. You can see the effect. The ball changes directions.

Forces act in pairs. The action force causes a *reaction force*. The reaction force is the ball's push back on the bat. The effects of this force are harder to see because the bat is moving so fast. But if you were the batter, you would feel the reaction force!

Think And Write

1. **Scientific Thinking** You know that gravity causes objects to fall toward Earth. Describe how gravity affects a baseball during a baseball game.

2. **Scientific Inquiry** Describe how the actions of the pitcher and the batter affect the motion of a baseball.

A pitcher uses force to get the ball from the pitcher's mound to home plate.

Illinois Excursions

Rockford

Discovery Center MUSEUM

Visitors to the Discovery Center Museum in Rockford, Illinois, have a lot of questions. The Discovery Center has a lot of answers!

Something for Everyone

Are you interested in how roller coasters work? Do you wonder where static shocks come from? Are you curious about what colors look like under different types of light? If so, you will enjoy the Amusement Park Science exhibit, the Power House exhibit, and the Color and Light exhibit at the Discovery Center Museum.

What do all these exhibits have in common? They are all about different kinds of energy. Light is a kind of energy. Static electricity is a kind of electrical energy. A speeding roller coaster has kinetic energy, or energy of motion.

Investigating Matter and Energy

CHAPTER 11
Properties of Matter 452

CHAPTER 12
Energy 490

CHAPTER 13
Electricity and Magnets 522

CHAPTER 14
Heat, Light, and Sound. 552

Unit Inquiry

Mold Sand

Playing in the sand at the beach or in a sandbox can be fun. But, turning a pile of sand into a sand sculpture isn't easy. What can you mix with sand to make it easier to mold into different shapes? For example, what is the best amount of water to add to sand to make it keep its shape? Plan and conduct an experiment to find out.

CHAPTER 11 Properties of Matter

What's the Big Idea?

Matter has properties that can be observed, described, and measured.

Essential Questions

Lesson 1
What Is Matter?

Lesson 2
What Are States of Matter?

Lesson 3
How Does Matter Change?

 Student eBook
www.hspscience.com

What do you wonder?

What are some properties of the matter you can see in this picture? How do those properties relate to the **Big Idea?**

LESSON 1

Essential Question
What Is Matter?

Investigate to find out how to measure the volume of a liquid.

Read and Learn about matter and its properties.

Fast Fact

A Lot of Water
An adult human needs about 10 cups of water each day. That sounds like a lot, but it isn't much for a cow. A cow needs at least 400 cups of water each day! In the Investigate, you will measure the volume of water.

Vocabulary Preview

matter [MAT•er] Anything that takes up space (p. 458)

physical property [FIZ•ih•kuhl PRAHP•er•tee] Anything that you can observe about an object by using one or more of your senses (p. 460)

mass [MAS] The amount of matter in an object (p. 462)

volume [VAHL•yoom] The amount of space that matter takes up (p. 463)

density [DEN•suh•tee] The mass of matter compared to its volume (p. 463)

Investigate

Measuring Volume

Guided Inquiry

Start with Questions

These children are sliding down a water slide. Lots of people enjoy water slides on a hot day!

- How much water do you think this water slide uses each day?
- How could you measure the amount of water it uses?

Investigate to find out. Then read and learn to find out more.

Prepare to Investigate

Inquiry Skill Tip

Use data, observations, and facts you have learned to **predict** what you think will happen in the future.

Materials

- metric measuring cup
- water
- 3 clear containers of different shapes
- masking tape

Make an Observation Chart

Container #1	Container #2	Container #3

Follow This Procedure

1. **Measure** 100 mL of water.
2. Pour the water into a clear container.
3. Use a piece of masking tape to mark the level of the water in the container. Put the bottom edge of the tape at the water line.
4. Repeat Steps 1–3 until all three containers have 100 mL of water in them and all three levels are marked.

Draw Conclusions

1. How much water is in each container?
2. Describe the height of the water in each container. Explain why the height of the water looks different in each container.
3. **Inquiry Skill** Scientists use data and observations to **predict** what will happen. What do you **predict** will happen if you pour the water from each container back into the measuring cup?

Independent Inquiry

Fill three containers of different shapes with water. **Predict** how much water you will find in each container. **Measure** the water in each container. How close were your predictions?

Read and Learn

VOCABULARY
matter p. 458
physical property p. 460
mass p. 462
volume p. 463
density p. 463

SCIENCE CONCEPTS
▶ what matter is
▶ how to measure some physical properties of matter

MAIN IDEA AND DETAILS
Find out about the properties of matter.

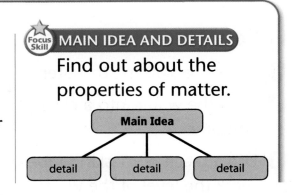

Matter

Ice-skating can be fun. Skaters glide over the ice. They feel the breeze against their faces.

Everything that the skaters see and feel is matter. **Matter** is anything that takes up space. Ice, water, and clouds are matter. The air the skaters breathe is matter. Skaters are matter, too. Since matter takes up space, two objects cannot take up the same space at the same time.

What examples of matter do you see in this picture? ▼

458

▲ No two objects can take up the same space at the same time. This woman's ski takes the place of the water, forcing the water into the air.

Look around you. What matter can you see? Your desk and books and the other objects in the classroom are matter. Your teacher and your classmates are matter. You know they are matter because they take up space.

Now, look outside. Is it raining? Is it sunny? Is it snowing? Rain, snow, and the sun are all types of matter. Is the wind blowing? Air is matter that you can't see. How do you know that air is matter? Air takes up space. You can see air move leaves on trees. You can see that air takes up space when you blow up a balloon.

 MAIN IDEA AND DETAILS

What is matter?

Physical Properties of Matter

Anything you can observe about matter by using one or more of your senses is a **physical property**. Here are some things you observe with your senses.

Sight—Young ducks are small and yellow. You observe their size and color.

Hearing—Bells ring. Rain pings on a metal roof. Wind rustles leaves.

Touch—Ice feels cold and hard. Blankets feel soft. Sandpaper feels rough.

Smell—Baking bread smells delicious. Rotting garbage smells bad.

Taste—One physical property of a food is its flavor. Flavors can be sweet, salty, sour, or bitter.

▲ This pineapple feels rough on the outside. The inside tastes sweet.

Your sense of smell tells you that there is popcorn in the container. ▼

Your sense of touch tells you that the cat is fluffy. ▼

The cymbals are shiny. They also make a different sound than a drum makes. ▼

Color, size, shape, and texture are physical properties matter can have. Some matter, such as rubber, can bounce and stretch. Other matter, such as salt, mixes with water. Some metals bend easily. Others, such as steel, do not. These are just a few examples of the many physical properties of matter. Think of the different physical properties of matter that you can find in your classroom.

 MAIN IDEA AND DETAILS

List two physical properties of a pineapple.

Color is a property you observe with your eyes. What are other physical properties of this glass? ▼

Measuring Matter

Another physical property of matter is mass. **Mass** is the amount of matter in something. You can measure mass by using a balance. Mass is often measured in grams (g) or kilograms (kg). One kilogram equals 1,000 grams.

Suppose you have some apple slices. You know their mass. You also have some orange slices. You know their mass, too. You mix them together to make a fruit salad. How can you find the mass of the salad without using a balance? If you said, "Add the two known masses," you were right! The mass of two or more things together is the sum of their individual masses.

This graduated cylinder and measuring cup measure the volume of liquid in milliliters.

Science Up Close

Measuring Mass

For more links and animations, go to www.hspscience.com

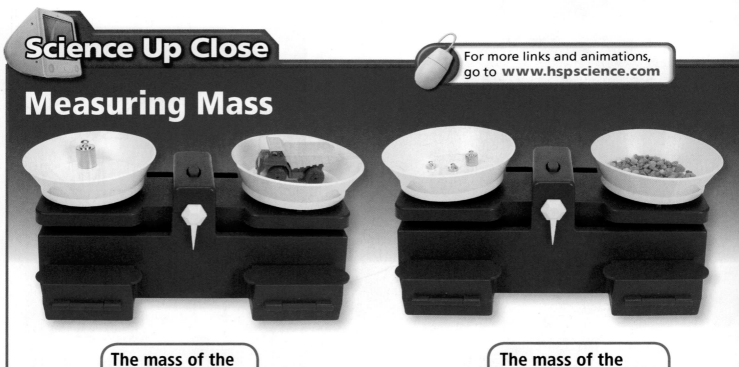

The mass of the truck is 50 grams.

The mass of the pebbles is 14 grams.

Another physical property of matter is volume. **Volume** is the amount of space matter takes up. In the Investigate, you measured 100 mL of water. The volume of the water was 100 mL.

Density is another physical property of matter. **Density** is the mass of matter compared to its volume. Think about two identical boxes. You fill one box with feathers. You fill the other box with rocks. The boxes have the same volume, but the box of rocks has much more mass. Rocks have greater density than feathers.

Insta-Lab

Compare Densities
Fill one sandwich bag with marbles. Fill another with cotton balls. Seal the bags. Measure the mass of each bag. How do their volumes compare? How do their densities compare? Which matter is denser—cotton or marbles?

MAIN IDEA AND DETAILS

What is density?

The mass of the truck and the pebbles is 64 grams. That is the sum of the two masses added together.

463

Sink and Float

How well an object floats is a physical property, too. Steel bars sink. Steel boats float. Why?

Density is the reason. Matter that is less dense than water floats. Matter that is denser than water sinks.

How could you make a steel bar float? You would have to change its shape to increase its volume. Changing a steel bar into a boat shape changes the volume of the steel. The same steel in a boat's shape takes up more space. Yet its mass doesn't change. The same mass that has a greater volume is less dense. If the volume of the steel boat is great enough, the boat floats.

MAIN IDEA AND DETAILS

What is the difference between an object that sinks and one that floats?

Which balls in this picture are less dense than water? Which are denser? How do you know? ▼

Lesson Review

Essential Question

What is matter?

In this lesson, you learned that matter is anything that takes up space. Two objects cannot take up the same space at the same time. Matter has physical properties that can be observed by using the senses. Matter can be measured in terms of volume, mass, and density.

1. **MAIN IDEA AND DETAILS** Draw and complete a graphic organizer for this main idea: Physical properties can be observed and measured.

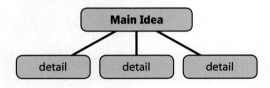

2. **SUMMARIZE** Write a three-sentence summary of this lesson. Tell about the different properties of matter.

3. **DRAW CONCLUSIONS** You have two equal masses of feathers and rocks. Which has the greater volume?

4. **VOCABULARY** How is volume different from mass?

Test Prep

5. A boat loaded with too much cargo sank. Why?
   - A. It became less dense than water.
   - B. Its volume became too great.
   - C. Its density increased.
   - D. All heavy things sink.

Make Connections

 Writing

Expository
Gather some classroom objects made of paper, wood, and metal. Write a **description** of each object's physical properties. Ask a classmate to identify and classify the objects by using only the properties you described.

 Music

Physical Properties Symphony
Use the sounds that different kinds of matter make to perform an original physical properties symphony. Perform your symphony for the class.

LESSON 2

Essential Question

What Are States of Matter?

Investigate to find out how temperature affects matter.

Read and Learn about the states of matter and one way matter changes.

Fast Fact

Water Temperature
The water temperature decreases from the ocean surface to the ocean floor. You will learn about the temperature of ice in the Investigate.

Can you see the three states of matter?

Vocabulary Preview

solid [SAHL•id] A form of matter that has a volume and a shape that both stay the same (p. 471)

liquid [LIK•wid] A form of matter that has a volume that stays the same but a shape that can change (p. 472)

gas [GAS] A form of matter that has no definite shape or volume (p. 473)

evaporation [ee•vap•uh•RAY•shuhn] The process by which liquid water changes into water vapor (p. 474)

condensation [kahn•duhn•SAY•shuhn] The process by which water vapor changes into liquid water (p. 474)

467

Investigate

Temperature and Matter

Guided Inquiry

Start with Questions

These icicles are frozen water. It must be very cold there!

- What causes icicles like these to form?
- What is causing these icicles to melt?

Investigate to find out. Then read and learn to find out more.

Prepare to Investigate

Inquiry Skill Tip

You can use tables, charts, drawings, line graphs, or circle graphs to communicate what happened in an investigation. When deciding which method to use, consider what you need to show. For this investigation, you need to compare temperatures.

Materials

- metric measuring cup
- hot water
- plastic jar or beaker
- thermometer
- 3 ice cubes
- plastic spoon

Make a Data Table

Step	Observations	Temperature (°C)
Hot Water		
After 1 ice cube		
After 2 ice cubes		
After 3 ice cubes		

Follow This Procedure

1. **Measure** 200 mL of hot water from the tap into a measuring cup. Pour the water into a jar or beaker.

2. With a thermometer, **measure** the temperature of the water. **Record** the data.

3. Add an ice cube to the water. Stir with a plastic spoon. **Record** what you **observe**.

4. **Measure** the temperature of the water again. **Record** the data.

5. Repeat Steps 3 and 4 twice.

Draw Conclusions

1. What happened to the ice cubes in the water?

2. What happened to the temperature of the water each time you added an ice cube?

3. **Inquiry Skill** One way scientists can **communicate** data is in a bar graph. Make a bar graph to **communicate** what happened to the temperature of the water in this activity.

Step 2

Step 3

Independent Inquiry

Put 100 mL of water in a freezer. **Measure** its temperature every 10 minutes. **Communicate** the data in a bar graph. **Interpret the data.**

Read and Learn

VOCABULARY
solid p. 471
liquid p. 472
gas p. 473
evaporation p. 474
condensation p. 474

SCIENCE CONCEPTS
▶ what three states of matter are
▶ how temperature affects states of matter

COMPARE AND CONTRAST
Find out how states of matter are alike and different.

alike — different

States of Matter

You have read that matter takes up space. Matter also has different forms called states. Three states of matter are solid, liquid, and gas.

In the Investigate, you watched an ice cube change states. If you had boiled the water, it would also have changed to another state.

The wax of a candle can also change states. To make a candle, wax is melted and poured into a mold. When the wax has cooled and hardened, it has changed states.

COMPARE AND CONTRAST

How are ice and the wax of a candle alike?

What are some solids and liquids in this picture? ▶

470

Name the solids you see in these pictures.

Solids

Think about what an ice cube is like. An ice cube is a solid. A **solid** is matter with a volume and a shape that stay the same.

Solids stay solids unless something, such as heat, changes them. When ice is heated, it melts and becomes a liquid. When you heat matter, the motion of its small particles, or pieces, speeds up.

The opposite happens, too. If you remove enough heat from water, it freezes. When matter cools, the motion of its small particles slows down.

COMPARE AND CONTRAST
How are all solids alike?

Is It Solid?
Take a frozen pat of butter and place it on a dish. Record your observations about the butter. Then place the dish and butter under a lamp. Turn the lamp on and record what happens to the butter every minute. Write a description of how the butter changed.

Liquids

Think about a glass of water. The water in the glass is a liquid. A **liquid** is matter that has a volume that stays the same but a shape that can change.

Like a solid, a liquid has a volume that doesn't change. However, a liquid's shape can change. A liquid takes the shape of whatever container holds it. The volume of water can look large in a tall, slim container. In a short, wide container, it can look small.

You know that water is a liquid. Paint, juice, and shampoo are liquids, too. What are some liquids you see or use every day?

▲ This soap is a liquid.

COMPARE AND CONTRAST

How are liquids and solids different?

Vinegar and oil are liquids that make salads tasty.

Water changes shape as it falls.

472

Fizzy drinks have carbon dioxide gas in them. The bubbles you see are bubbles of gas.

Gases

The helium inside this balloon is a gas. A **gas** has no definite shape or volume. A gas takes up all the space in a container. If you blow up a balloon, you can see that the air spreads out to fill the space inside the balloon.

The air that you breathe is a mixture of gases. Some stoves cook with natural gas. You can't see natural gas, but when it burns, you can see a blue flame.

 **COMPARE AND CONTRAST** How are gases different from solids and liquids?

A gas called helium fills this balloon.

Changes of State

In the Investigate, heat changed solid ice to liquid water. Heat can also change a liquid to a gas. If you leave a cup of water in a warm place, after a day or two, the cup will be empty. The liquid changes into a gas, but it's still water. Water moves into the air in the process of **evaporation**. Boiling also makes water evaporate, but it happens more quickly.

When a gas cools, it changes back to a liquid. The change of a gas to a liquid is **condensation**. You have seen this happen. When the water in air—a gas—changes to a liquid, it rains.

 COMPARE AND CONTRAST How are evaporation and condensation the same? How are they different?

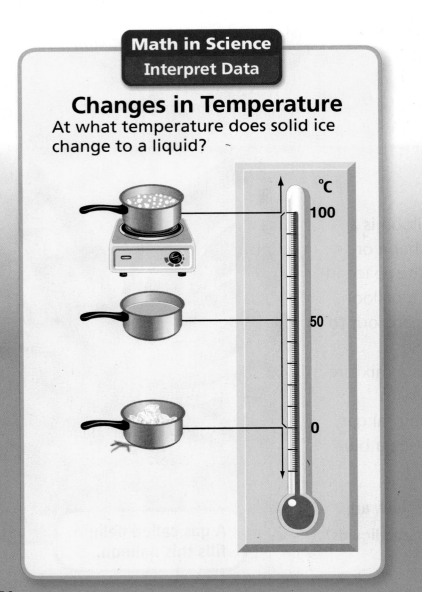

Math in Science
Interpret Data

Changes in Temperature
At what temperature does solid ice change to a liquid?

Lesson Review

Essential Question

What are states of matter?

In this lesson, you learned that matter has three states. Solids have a definite shape and volume. Liquids have a definite volume but take the shape of their containers. Gases have no definite shape or volume and take up all the space in a container. A change in temperature is what causes matter to change its state.

1. **COMPARE AND CONTRAST** Draw and complete a graphic organizer that shows how the different states of matter are alike and different.

2. **SUMMARIZE** Write a summary of this lesson that tells about the different states of matter. Use the terms *solid, liquid,* and *gas.*

3. **DRAW CONCLUSIONS** If matter has a definite volume but no definite shape, what is its state?

4. **VOCABULARY** Write a sentence to explain how evaporation relates to a change of state.

Test Prep

5. **Critical Thinking** A solid object melts to become a liquid. Was heat added or was it removed to cause the change? Explain.

Make Connections

 Writing

Narrative

Write a **story** about an ice cube as temperature changes cause it to change into different states of matter.

 Math

Solve a Problem

Maria took a frozen pop out of the freezer at 3:23 P.M. and placed it in a dish. At 3:35 P.M., the pop was a puddle of liquid. How long did it take the frozen pop to melt?

LESSON 3

Essential Question
How Does Matter Change?

Investigate to find out if certain kinds of matter will mix.

Read and Learn about ways in which matter changes.

Fast Fact

How Big Can It Get?
One of the world's largest sand castles stood more than 20 meters (66 ft) high. Its builders spent three weeks making it, packing the sand into place with water. In the Investigate, you will work with sand and water.

A sand sculpture

Vocabulary Preview

mixture [MIKS•cher] A substance that has two or more different kinds of matter (p. 482)

solution [suh•LOO•shuhn] A mixture in which the different kinds of matter mix evenly (p. 483)

Investigate

Will It Mix?

Guided Inquiry

Start with Questions

These nuts aren't all the same kind. They've been mixed together for people who want to enjoy a snack.

- How many different kinds of nuts do you see in the picture?
- Why are the nuts in the picture a mixture?

Investigate to find out. Then read and learn to find out more.

Prepare to Investigate

Inquiry Skill Tip

If you use drawings to communicate, make the drawings clear and simple. Be sure they show the details you want to communicate. If necessary, use labels to explain the drawings.

Materials

- water
- metric measuring cup
- 2 clear plastic jars
- measuring spoon ($\frac{1}{4}$ teaspoon)
- sand
- hand lens
- plastic spoon
- salt

Make an Observation Chart

Material	Observations
Sand	
Salt	Total number of teaspoonfuls used: _____

Follow This Procedure

1. **Measure** 200 mL of water. Pour the water into a jar.
2. Add $\frac{1}{4}$ teaspoon of sand to the water and stir. Use a hand lens to **observe** the jar's contents. **Record** what you **observe**.
3. Repeat Step 1, using the other jar.
4. Add $\frac{1}{4}$ teaspoon of salt to the water and stir. Use a hand lens to observe the jar's contents. **Record** what you **observe**.
5. Repeat Step 4 until you see salt collect on the bottom of the jar after you stir. **Record** the number of teaspoons of salt you used in all.

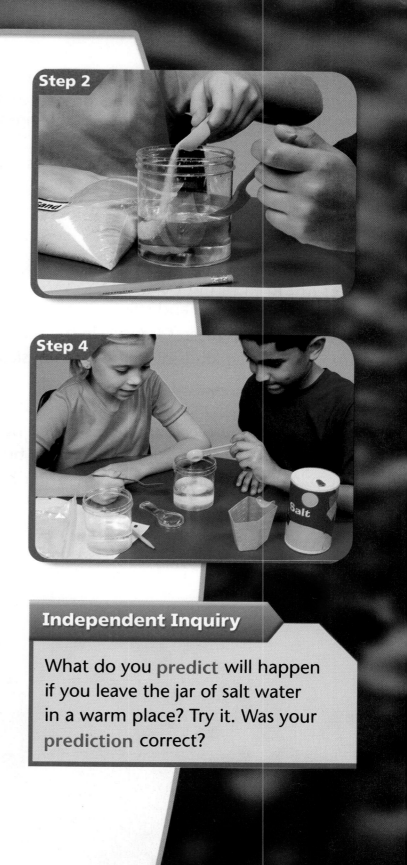

Draw Conclusions

1. What did you **observe** when you stirred in the sand? The salt?
2. **Inquiry Skill** Scientists sometimes use drawings to **communicate**. Make two drawings that **communicate** what happened to the sand and the salt when they were stirred into the water.

Independent Inquiry

What do you **predict** will happen if you leave the jar of salt water in a warm place? Try it. Was your **prediction** correct?

Read and Learn

VOCABULARY
mixture p. 482
solution p. 483

SCIENCE CONCEPTS
▶ what physical and chemical changes are
▶ what mixtures and solutions are

MAIN IDEA AND DETAILS
Find out how physical and chemical changes happen.

Physical Changes

What can change and still be the same? The answer is matter. Changes in matter that don't form new kinds of matter are physical changes.

An example of a physical change is cutting. Cutting makes a piece of paper smaller, but the paper is still paper. Its size changes, but the paper pieces are still the same kind of matter.

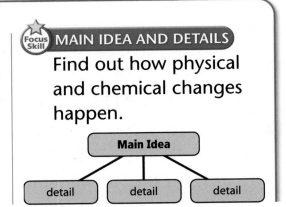

Yarn is packaged and sold.

Sheep grow a thick coat of wool.

▲ A machine spins sheep's wool into yarn.

Knitting a wool cap is another example of a physical change. The thick wool is cut from sheep in spring. This doesn't hurt them, and they grow a new coat before winter. The wool is combed into soft strands, which are pulled into threads and twisted to make yarn. A knitter then knits the wool yarn into a cap. In the cap, the wool looks different from the way it looked on the sheep, but it is still wool. It is the same kind of matter.

MAIN IDEA AND DETAILS
What happens to matter when there is a physical change?

▲ Yarn can be knitted by hand or by machine.

The wool in the cap has been changed physically, but it is still wool.

Mixtures

In the Investigate, you made two mixtures—one of sand and water, and one of salt and water. A **mixture** is a substance that is made up of two or more kinds of matter. Making a mixture is a physical change. You put different types of matter together, but no new types of matter are formed.

Separating the parts of a mixture is a physical change, too. You can separate sand and water by pouring the mixture through filter paper. The water runs through, leaving the sand behind. You can separate salt and water by leaving the mixture in a warm place. The water evaporates, leaving the salt.

MAIN IDEA AND DETAILS

What is a mixture?

▲ **This girl has a mixture of beads.**

This bowl holds a mixture of cereal, strawberries, and bananas.

482

A solution of drink mix and water is a tasty treat.

▲ A solution of detergent and water gets dishes clean.

Solutions

In the Investigate, when you first stirred salt into water, you could not see the salt. The salt dissolved, or mixed with the water. You made a solution. A **solution** is a mixture in which different kinds of matter mix evenly. Your mixture of salt and water was a solution. Since the sand didn't dissolve in the water, that mixture wasn't a solution.

 MAIN IDEA AND DETAILS

Why is a solution a kind of mixture?

483

Chemical Changes

Changes that form different kinds of matter are chemical changes. Cooking causes chemical changes. Suppose you stir flour, sugar, eggs, milk, and butter together to make a cake. After you bake the cake, it has properties that are completely different from the properties of the ingredients.

Burning is also a chemical change. When wood burns, it combines with oxygen in the air. Ashes and smoke form. Those are different kinds of matter than the wood. You can't get the wood back.

MAIN IDEA AND DETAILS
Name at least four examples of chemical changes.

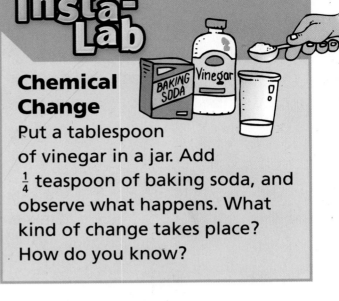

Insta-Lab

Chemical Change
Put a tablespoon of vinegar in a jar. Add $\frac{1}{4}$ teaspoon of baking soda, and observe what happens. What kind of change takes place? How do you know?

Rotting is a chemical change.

Rusting is a chemical change. It happens when oxygen in the air combines with iron in metal.

Lesson Review

Essential Question

How does matter change?

In this lesson, you learned that matter can change. Physical changes don't form a new kind of matter. Chemical changes form new kinds of matter.

1. **MAIN IDEA AND DETAILS**
   Draw and complete a graphic organizer for the main idea: Matter can change in two ways.

   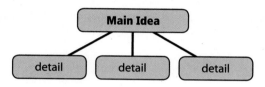

2. **SUMMARIZE** Write a three-sentence summary of this lesson. Tell about two types of physical changes and two types of chemical changes.

3. **DRAW CONCLUSIONS** Mr. Gonzalez put up a wall made of wood and nails. Was the change of the wood physical or chemical? Explain.

4. **VOCABULARY** Explain why all solutions are mixtures but not all mixtures are solutions.

Test Prep

5. Which of these is a chemical change?
   - A. burning gasoline in a car
   - B. putting on fingernail polish
   - C. making a chain from strips of paper
   - D. grinding wheat to make flour

Make Connections

 Writing

Expository
Write two **paragraphs** to compare and contrast changes that happen quickly, such as cutting hair or condensation, and changes that may happen slowly, such as hair growing.

 Math

Solve a Problem
A mixture contains two times as many red beads as white beads. There are six white beads. How many red beads are in the mixture? On paper, show how you solved the problem.

485

People in Science

Steven Chu

▶ **STEVEN CHU**
▶ Physics Professor, Stanford

When Steven Chu was growing up in Garden City, New York, he didn't like school. "Learning seemed like work. I wanted it to be an adventure," he says. So he began asking questions that made learning fun—questions like "Why does that happen?" Chu was on his way to becoming a particle physicist—a scientist who studies atoms and the forces that hold them together or push them apart.

In 1997 Chu received one of the most important awards in science—the Nobel Prize in physics. He won the award for thinking of a way to slow down tiny particles so other scientists can study them more carefully.

Now Chu is a teacher at Stanford University in California. He loves working with students there. "They figure out that textbooks and professors don't know everything, and then they start to think on their own," he says. "Then I begin learning from them."

Think and Write

1. What are you curious about? Pick a question, and think of some experiments that could help you find the answer.

2. Look for a story about science in a newspaper or magazine. What questions were the scientists asking?

Tara McHugh

▶ TARA MCHUGH
▶ Processed Foods Researcher

Imagine you sit down to lunch today and your friend pulls a sandwich out of his backpack. Instead of throwing away the wrap the sandwich came in, he pops it into his mouth and says, "Hmmm! Strawberry."

Thanks to Tara McHugh, edible food wraps are now a reality. McHugh, a scientist who studies food, invented a wrap that you can eat that also helps preserve food. McHugh makes the wraps in her lab.

Oxygen and Food Don't Mix

The wrap helps protect food by keeping oxygen away from it. Oxygen can cause chemical changes to food, making it change color or even rot. McHugh's wraps help prevent this.

Think and Write
1. Why is it important to protect foods?
2. How does keeping oxygen away from food protect it?

Career Materials Scientist

How long do think the rubber soles of your new sneakers will last? Chances are, a materials scientist knows the answer. These scientists study how different materials react to changes. They study how rubber might react to cold or how long it will be before it wears out.

Chapter 11 Review and Test Preparation

Vocabulary Review

Match the terms below to the definitions in items 1–10. The page numbers tell where to look in the chapter if you need help.

matter   p. 458
mass   p. 462
volume   p. 463
density   p. 463
solid   p. 471
liquid   p. 472
gas   p. 473
evaporation   p. 474
mixture   p. 482
solution   p. 483

1. A mixture in which all the parts mix evenly
2. A state of matter with no definite shape or volume
3. The amount of matter in something
4. The process during which water moves into the air
5. The amount of space that matter takes up
6. The state of matter in which volume stays the same but the matter takes the shape of its container
7. The mass of something compared with its volume
8. Anything that takes up space
9. A state of matter with a shape and a volume that don't change
10. A substance with two or more different kinds of matter

Check Understanding

Write the letter of the best choice.

11. **MAIN IDEA AND DETAILS** Which of the following is true of these two jars and their contents?

A. They have the same mass.
B. They have the same volume.
C. They have the same density.
D. They have the same matter.

488

12. COMPARE AND CONTRAST Which states of matter have a volume that doesn't change?
- F. solid and liquid
- G. liquid and gas
- H. solid and gas
- J. solid, liquid, and gas

13. Sarah notices that a metal fence rail feels cold and hard. What is Sarah observing?
- A. chemical changes
- B. densities
- C. physical properties
- D. states of matter

14. Which of these is a solution?
- F. peanut butter and jelly
- G. salt and water
- H. cereal and milk
- J. celery and carrot sticks

15. Which of the following is a chemical change?
- A. dissolving soap in water
- B. filling a balloon with air
- C. grating cheese
- D. burning wood

16. A ball of modeling clay sinks in a pan of water. What change could make the clay float?
- F. Remove some of the water.
- G. Increase the clay's density.
- H. Change its shape.
- J. Add more water to the pan.

Inquiry Skills

17. How could a picture be used to communicate rusting?

18. There are two identical boxes. One box is filled to the top with books. The other box is filled to the top with foam pillows. Which box would you predict has greater mass? Explain your answer.

Critical Thinking

19. If you put sand into a container, it takes the shape of the container. Why is sand still considered a solid?

20. You have a mixture of two kinds of buttons. Some are large. Some are small. How can you separate the mixture without picking out the buttons one at a time?

CHAPTER 12 Energy

What's the Big Idea?

You use many forms of energy every day to grow and live.

Essential Questions

Lesson 1
What Is Energy?

Lesson 2
How Can Energy Be Used?

Lesson 3
Why Is Energy Important?

Go online
Student eBook
www.hspscience.com

What do you wonder?

How many forms of energy can you see in this photo? How do those forms of energy relate to the **Big Idea**?

A field of ripe wheat

LESSON 1

Essential Question

What Is Energy?

Fast Fact

A Real Light Show
People use chemicals to make fireworks. The burning chemicals in the fireworks give off light and sound energy. In the Investigate, you will find out about another kind of energy.

Fireworks over Sydney Harbor in Sydney, Australia

Vocabulary Preview

energy [EN•er•jee] The ability to make something move or change (p. 496)

kinetic energy [kih•NET•ik EN•er•jee] The energy of motion (p. 498)

potential energy [poh•TEN•shuhl EN•er•jee] Energy of position or condition (p. 498)

Investigate

Observing Temperature

Guided Inquiry

Start with Questions

You may have seen panels such as these on the roofs of houses and businesses.

- What are the panels for?
- Why are the panels placed to face the sun?

Investigate to find out. Then read and learn to find out more.

Prepare to Investigate

Inquiry Skill Tip

To infer why something happened, make a list of all the things that changed in your investigation. Decide which of these was a "cause" and which was an "effect."

Materials

- thermometer
- clock

Make a Data Table

Time (hr)	Temperature (°C)
0	
1	
2	
3	

Follow This Procedure

1. With your group, find a place outside that is sunny all day long.
2. In the morning, have a group member place the thermometer on the ground, face up.
3. Wait a few minutes until the temperature reading stops changing.
4. Each member of the group should read and **record** the temperature.
5. **Observe** the thermometer once an hour for several hours. **Communicate** your observations in a line graph that shows time and temperature.

Draw Conclusions

1. What changes did you **observe**? What caused these changes?
2. **Inquiry Skill** Scientists use their observations to **infer** why things happen. The rising temperature reading on the thermometer was caused by energy. Where can you **infer** that the energy came from?

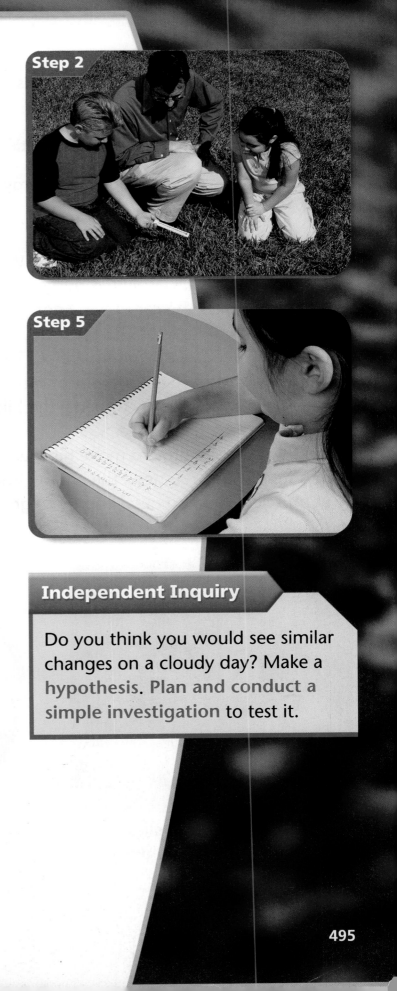

Independent Inquiry

Do you think you would see similar changes on a cloudy day? Make a **hypothesis**. Plan and conduct a simple **investigation** to test it.

Read and Learn

VOCABULARY
energy p. 496
kinetic energy p. 498
potential energy p. 498

SCIENCE CONCEPTS
▶ what energy is
▶ what the relationship between kinetic energy and potential energy is

COMPARE AND CONTRAST
Look for ways in which forms of energy are different.

alike —— different

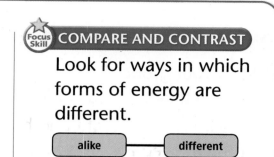

Some Sources of Energy

Think about a moving car. As long as the car is moving, it must have energy. Gasoline is burned to supply the energy that makes the car move. **Energy** is the ability to make something move or change.

Now think about a boy riding a bike. The bike is moving, so the bike must have energy—but from where? The boy's muscles supply the energy to pedal the bike. Moving the pedals makes the bike move.

The boy needs energy to move his muscles. This energy comes from food. You need the energy from food to move your muscles. You also need energy to grow and change.

Moving water supplies the energy to turn the water wheel.

Lightning is a type of electric energy.

496

A seed sprouts and grows into a plant. The energy a plant needs to grow comes from sunlight—plants use it to make food.

Most of the energy we use on Earth comes from the sun. Light and heat from the sun keep Earth warm. The energy in gasoline and other fuels came from the sun, too. The energy was stored by plants and animals that lived many, many years ago.

 COMPARE AND CONTRAST

How is the energy you use to grow different from the energy the plant uses to grow?

You know there is sound energy from this clock, because the clapper is moving.

What types of energy does this picture show?

Electric energy causes this sign to light up.

Forms of Energy

You have just read about different kinds of energy. All kinds of energy can be grouped in two ways. **Kinetic energy** (kih•NET•ik) is the energy of motion. Anything that is moving has kinetic energy. A moving car has kinetic energy. A child moving down a slide has kinetic energy. A leg moving a bike pedal around and around has kinetic energy. If these things didn't have kinetic energy, they wouldn't move.

A child sitting at the top of a slide is not moving but still has energy. The child has potential energy. **Potential energy** (poh•TEN•shuhl) is energy of position or condition. When the child moves, potential energy is changed into kinetic energy.

Turning the crank tightens a spring inside the box.

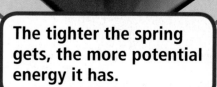

The tighter the spring gets, the more potential energy it has.

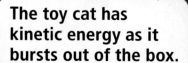

The toy cat has kinetic energy as it bursts out of the box.

A battery has potential energy. This energy can be changed to other forms. When the battery is used to make something move, the potential energy is changed to kinetic energy.

COMPARE AND CONTRAST
What is the difference between kinetic energy and potential energy?

Energy in Motion
Hold a rubber ball 50 cm above the ground. Drop the ball. How high does it bounce? Now drop the ball from 1 m, 1.5 m, and 2 m. Record your observations.

Science Up Close

For more links and animations, go to **www.hspscience.com**

How a Battery Works

Inside the battery are two powders. Each is mixed with a liquid to make a paste. The pastes are kept apart by a tube of fabric.

The two pastes have potential energy. When you put the battery in a circuit, the energy is changed into electric current.

One end of the battery touches a paste. The other end connects to the brass tube.

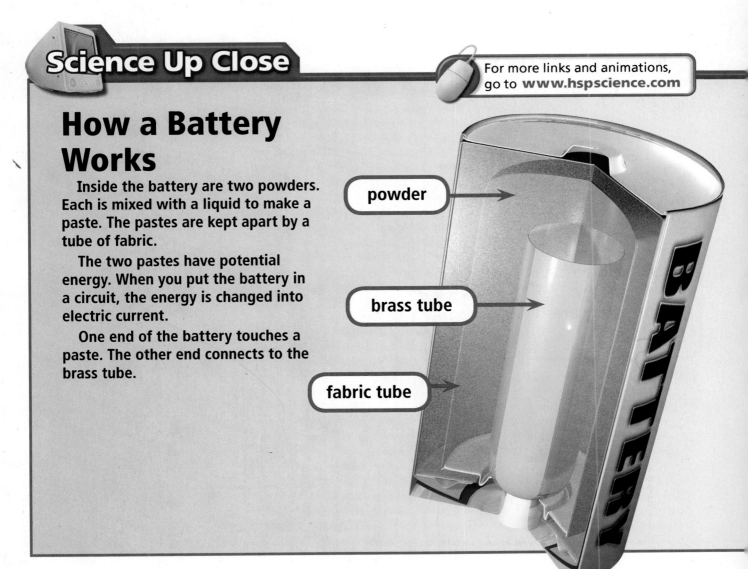

- powder
- brass tube
- fabric tube

Energy Changes

Think about a book on a shelf. It has potential energy. Now suppose the book falls. As it falls, it's moving, so it has kinetic energy. Where did the kinetic energy come from? The potential energy of the book's position was changed to kinetic energy.

Now you put the book back on the shelf. As you lift it, it's moving. It has kinetic energy. When it's sitting on the shelf again, it has potential energy. The kinetic energy was changed to potential energy.

Focus Skill: COMPARE AND CONTRAST

How is the energy of a book sitting on a shelf different from the energy of the book that is falling?

These children have kinetic energy, because they are jumping.

Lesson Review

Essential Question

What is energy?

In this lesson, you learned that energy is the ability to make something move or change. Most of the energy on Earth comes from the sun in the form of heat and light. Kinetic energy and potential energy are also forms of energy.

1. **COMPARE AND CONTRAST** Draw and complete a graphic organizer that shows how potential and kinetic energy are alike and different.

2. **SUMMARIZE** Write a summary of this lesson that tells the difference between kinetic and potential energy.

3. **DRAW CONCLUSIONS** A bike resting on its kickstand has potential energy. How could you change the energy to kinetic energy?

4. **VOCABULARY** Use each of the lesson vocabulary terms in a sentence that correctly shows the term's meaning.

Test Prep

5. Which type of energy do we get from the sun?
   - **A.** potential
   - **B.** electrical
   - **C.** light
   - **D.** kinetic

Make Connections

 Writing

Expository
Light is a type of energy. Write a **paragraph** in which you compare and contrast different sources of light energy.

 Physical Education

Using Energy
Make up a 30-second physical fitness routine that uses kinetic and potential energy. Do your routine for the class, and identify the form of energy you have during each part of the routine.

LESSON 2

Essential Question

How Can Energy Be Used?

Investigate to find out how sunlight affects ice.

Read and Learn about how we use and measure energy.

Fast Fact

Something's Cooking

People have been using thermal energy to cook food for tens of thousands of years. In the Investigate, you will observe thermal energy at work.

Vocabulary Preview

combustion
[kuhm•BUS•chuhn]
Another word for *burning*
(p. 507)

temperature
[TEM•per•uh•cher] The measure of how hot or cold something is (p. 508)

Investigate

The Heat Is On

Guided Inquiry

Start with Questions

Lots of people like to make shapes from snow. This snow isn't going to last much longer!

- What does it have to be like outside to make a snow figure?
- Where is the energy that is melting these figures coming from?

Investigate to find out. Then read and learn to find out more.

Prepare to Investigate

Inquiry Skill Tip

You may observe many changes during an investigation. To infer why things happen, look for a pattern of changes. A change in temperature means that energy has moved from one place to another. Think about what would cause this movement of energy.

Materials
- clear plastic cup
- ice cubes
- thermometer

Make a Data Table

Time (hr)	Temperature (°C)
0	
½	
1	
1½	
2	
2½	
3	

Follow This Procedure

1. Fill a clear plastic cup with ice cubes. Place a thermometer in the cup. Place the cup in sunlight.
2. After several minutes, **record** the temperature inside the cup.
3. Continue to **record** the temperature inside the cup every half hour for three hours.
4. **Communicate** the data from your table by making a bar graph.

Draw Conclusions

1. What did you **observe** about the temperature and the ice? What caused these changes?
2. **Inquiry Skill** Scientists use observations and data to **infer** why things happen. What can you **infer** happens to an object placed in sunlight?

Step 1

Step 4

Independent Inquiry

What do you think happens to the temperature of water placed in a freezer? Make a **hypothesis**. Then test it.

Read and Learn

VOCABULARY
combustion p. 507
temperature p. 508

SCIENCE CONCEPTS
▶ how people use energy

MAIN IDEA AND DETAILS
Look for details about ways people use energy.

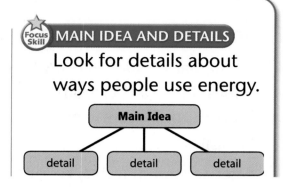

Using Energy

Energy makes things move or change. Without it, we could not live.

The sun provides energy to plants to grow. You then get energy from eating the plants for food. After you eat, energy from the food moves through your body. The energy keeps you healthy and helps you grow.

Energy in your muscles lets you move. When you run or jump, energy moves from your leg muscles to make your body move.

It takes a lot of energy to make this big, heavy train move.

Machines get energy in many ways. One way to get energy is from combustion. **Combustion** (kuhm•BUS•chuhn) is another word for "burning." When fuels burn, they give off heat.

A car engine burns gasoline. The gasoline gives off energy that makes the car move. Some stoves burn gas. The gas gives off energy that cooks your food. Wood burning in a fireplace also gives off energy.

What happens during combustion?

Where does the energy that moves the train come from?

Insta-Lab

Energy from Food
Think about all the activities a student like you could have done this morning, such as brushing teeth or packing a backpack. Food gives a person the energy to do these activities. List at least 10 activities someone might do each morning. Explain how someone might feel if he or she didn't have the energy from the foods.

507

Measuring Energy

You have read that there are different types and uses of energy. There are different ways to measure energy, too. The way energy is measured depends on the type of energy.

You have probably seen a thermometer like the one shown on this page. Thermometers are used to measure temperature. **Temperature** is the measure of how hot or cold something is. To find thermal energy, you measure both temperature and mass. To find the energy of a moving object, you measure both the speed and mass of the object.

MAIN IDEA AND DETAILS
What are some kinds of energy that can be measured?

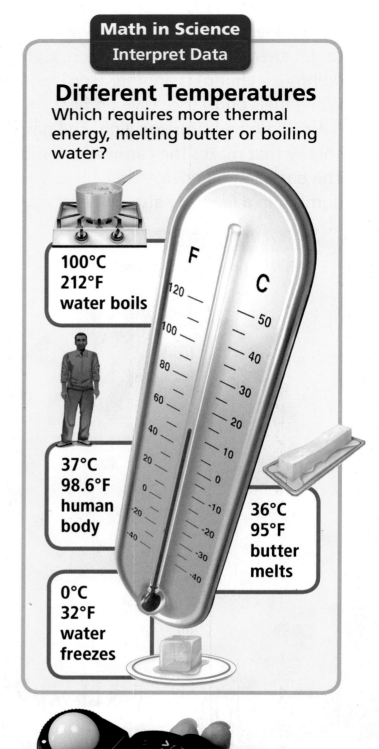

Math in Science
Interpret Data

Different Temperatures
Which requires more thermal energy, melting butter or boiling water?

100°C / 212°F water boils

37°C / 98.6°F human body

36°C / 95°F butter melts

0°C / 32°F water freezes

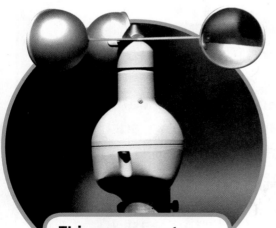

This anemometer measures wind speed. What else do you need to know to find the energy of the wind?

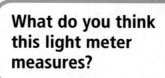

What do you think this light meter measures?

508

Lesson Review

Essential Question

How can energy be used?

In this lesson, you learned that without energy, we could not live. We need energy to move and grow. Machines use energy to move and do work. How energy is measured depends on the type of energy.

1. **MAIN IDEA AND DETAILS** Draw and complete a graphic organizer for this main idea: Without energy people could not live.

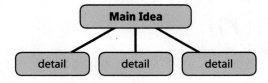

2. **SUMMARIZE** Write a summary of this lesson by writing the most important idea from each page.

3. **DRAW CONCLUSIONS** Matt tells his friend that he uses the sun's energy, even at night. How can this be?

4. **VOCABULARY** What word is a synonym of *combustion*?

Test Prep

5. Which type of energy does a thermometer measure?
    A. electric energy
    B. thermal energy
    C. light energy
    D. sound energy

Make Connections

 Writing

Expository
Find out how a light meter measures energy. Then write a **description** of what you learn. Share the description with a family member.

 Math

Make a Graph
The average temperature of the human body is 37°C. Find the average body temperatures of three other animals of your choice. Make a bar graph to show your findings.

509

LESSON 3

Essential Question
Why Is Energy Important?

Investigate to find out how windmills work.

Read and Learn about why energy is important, how it can be saved, and what types of energy resources exist.

Fast Fact

Lighting Up the Night

Places using the most electricity are the brightest in this satellite picture. Most electricity is produced by burning oil, coal, and natural gas. In the Investigate, you will explore another source of energy.

Follow This Procedure

1. Draw a 12-cm square with dotted lines and dots.
2. Use scissors to cut out the square.
3. Cut along each of the dotted lines to within 1 cm of the center.
4. Take each corner that has a dot, and fold it toward the center of the square to make a vane.
5. **CAUTION: Pushpins are sharp!** Put the pushpin through the center of all the folded corners and into the eraser of the pencil. Be sure the vanes turn freely.
6. You have just made a model of a windmill. First, blow gently on the windmill. Then blow more forcefully. Record what you observed each time.

Draw Conclusions

1. What did you observe when you blew on the windmill each time?
2. **Inquiry Skill** Scientists use models to help them understand processes. How does a windmill work? What kind of energy turns a windmill?

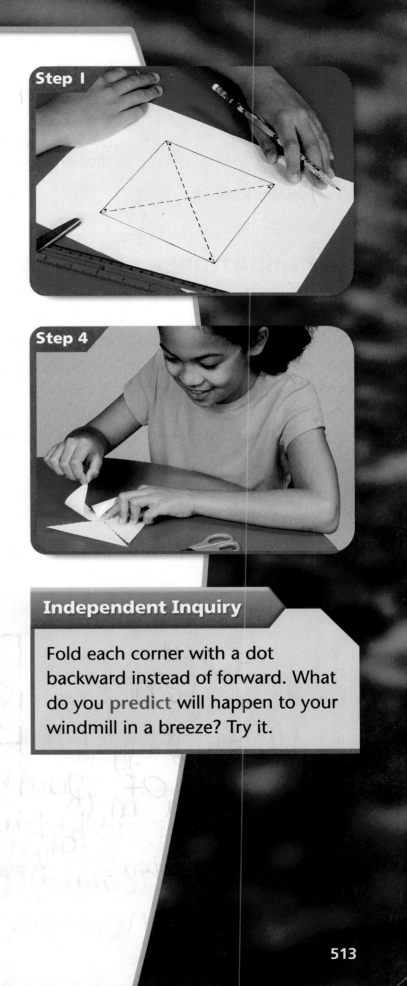

Independent Inquiry

Fold each corner with a dot backward instead of forward. What do you predict will happen to your windmill in a breeze? Try it.

Read and Learn

VOCABULARY
resource p. 515
fossil fuel p. 515
nonrenewable resource p. 516
renewable resource p. 516

SCIENCE CONCEPTS
- how important energy is
- why it's important to save energy

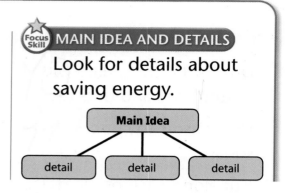

MAIN IDEA AND DETAILS
Look for details about saving energy.

The Importance of Energy

Every living thing needs energy. You are still growing. You're becoming bigger and taller, which takes energy. Walking, running, and playing use energy, too. Even when you are asleep, you're still breathing and your heart is still beating. That takes energy, too.

Communities also need energy. We use energy to cook our food. We use it to light our homes. We use it to run our cars.

 MAIN IDEA AND DETAILS
What are three ways communities use energy?

These people are eating foods that will give them energy.

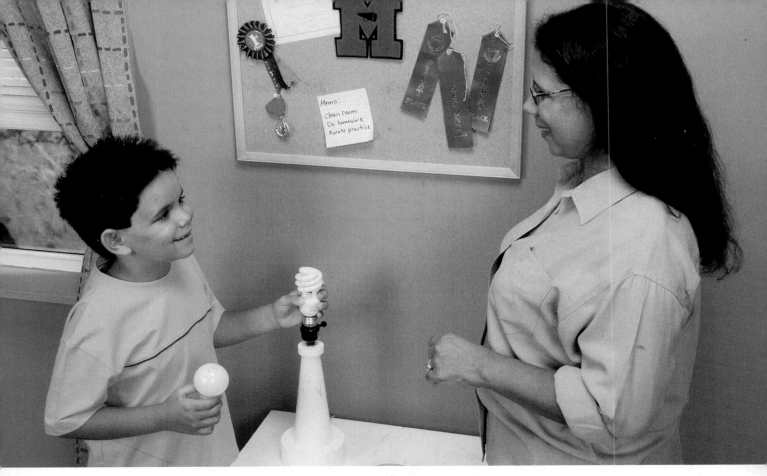

Ways to Save Energy

Coal, oil, and natural gas are all energy resources. A **resource** is something in nature that people can use. Energy resources are used to produce energy.

Coal, oil, and natural gas are called **fossil fuels**. They come from the remains of long-ago plants and animals. Fossil fuels can't be replaced. When they are used up, there will be no more. So it's important to save them and make them last longer. You save resources when you reduce the amount of energy you use.

**MAIN IDEA AND DETAILS**

Why is it important to save energy?

▲ This boy is replacing an old light bulb with a new one. The new kind of bulb gives off the same amount of light as the old bulb but uses less energy.

Save Fossil Fuels

With a partner, design and make a table about fossil fuels. List as many fossil fuels as you can. Then list all the ways you can think of for people to use less of each fossil fuel. Share your table with your classmates.

Other Energy Resources

It took millions of years for fossil fuels to form. They are said to be **nonrenewable resources**, because no more fossil fuels can be made in a human lifetime. *Nonrenewable* means "unable to be replaced."

Some resources are **renewable resources**, or resources that can be replaced. In the Investigate, you saw that wind can move things. Wind can turn windmills to produce electricity. Solar energy, or the sun's energy, can also be used to produce electricity.

 MAIN IDEA AND DETAILS

What are three renewable resources?

This water is warmed by geothermal (jee•oh•THER•muhl) energy, or heat from deep within Earth. That same heat provides power to a nearby city.

Lesson Review

Essential Question

Why is energy important?

In this lesson, you learned that people need energy for many things, including living and growing, cooking food, lighting homes, and transportation. You can save resources by reducing the amount of energy you use.

1. **MAIN IDEA AND DETAILS** Draw and complete a graphic organizer for this main idea: There are different types of energy resources.

   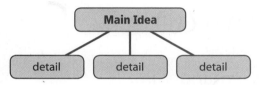

2. **SUMMARIZE** Write a three-sentence summary for this lesson. Tell why energy is important and how we can save resources.

3. **DRAW CONCLUSIONS** As water plunges over Niagara Falls, it turns machines that produce electricity. Is moving water a renewable energy resource or a nonrenewable energy resource? Explain.

4. **VOCABULARY** Use each lesson vocabulary word in a sentence that correctly shows its meaning.

Test Prep

5. Which of the following is a nonrenewable energy resource?
   A. electricity C. water
   B. coal D. wind

Make Connections

 Writing

Persuasive
Some people think we should rely more on renewable energy resources. Write a **letter** to your local newspaper. Tell your opinion.

 Literature

Saving Resources
Read a book about ways to save resources. Write a summary of the book to share with a first grader.

People in Science

Enrico Fermi

On December 2, 1942, Enrico Fermi conducted a successful experiment that made it possible to develop the atomic bomb. That bomb was a result of the work of many scientists over several years. Fermi had become interested in atoms after reading about the research done by other scientists.

▶ ENRICO FERMI
▶ Physicist

In 1938, Fermi had won the Nobel Prize in physics. His wife and their two children went with him to Sweden to accept the prize. Fermi's wife, Laura, was Jewish. She was in danger in Italy because of the hatred against Jews then. Instead of returning to Italy, the Fermi family came to the United States.

Fermi taught at Columbia University. A few years later, he went to the University of Chicago, where many of his experiments were conducted. In 1943, Fermi went to Los Alamos, New Mexico, to help develop the first atomic bomb. All of the major countries involved in World War II were racing to make this bomb. After the war ended, Fermi returned to the University of Chicago.

Think and Write

1. Why did all the major countries in the war want to make an atomic bomb?
2. Why is it important for scientists to write about their discoveries?

Evangelista Torricelli

▶ **EVANGELISTA TORRICELLI**
▶ Inventor of the barometer

Evangelista Torricelli is best known for an invention that helps predict the weather. That invention is the barometer.

A barometer measures the weight of the air and tells when its weight has changed. This weight pushing on an area is known as *air pressure.*

When energy from the sun heats air, that air has less weight than the cooler air. The warmer air rises as the cooler air falls. This change in air pressure can cause the weather to change, too. Barometers tell people when the air pressure is changing.

▲ Modern Barometer

Think and Write

1. How do barometers help predict the weather?
2. What causes the air to weigh less?

Career Energy Manager

Energy managers help people decide on the best way to use energy in homes, offices, and factories. They make suggestions about what kinds of lights to use, when to turn air conditioning or heating on and off, and what kinds of machines are the most efficient.

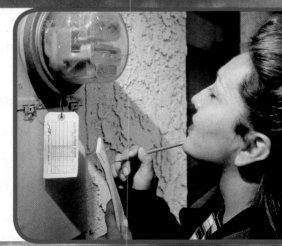

Chapter 12 Review and Test Preparation

Vocabulary Review

Use the terms below to complete the sentences. The page numbers tell you where to look in the chapter if you need help.

energy   p. 496
kinetic energy   p. 498
potential energy   p. 498
combustion   p. 507
resource   p. 515
fossil fuels   p. 515
nonrenewable resource   p. 516
renewable resource   p. 516

1. Oil, coal, and natural gas are resources called _____.
2. A resource that can be replaced is said to be a _____.
3. Energy of position is _____.
4. Gasoline gives off energy during _____.
5. Something in nature that people can use is a _____.
6. The ability to make something move or change is _____.
7. A resource that can't be replaced is a _____.
8. Energy of motion is _____.

Check Understanding

Write the letter of the best choice.

9. Where does the energy you need to grow come from?
   A. from potential energy
   B. from the food you eat
   C. from fossil fuels
   D. from wind energy

10. Which kind of energy does a book sitting on a shelf have?
    F. potential energy
    G. light energy
    H. speed energy
    J. kinetic energy

11. Which CD has kinetic energy?
    A. CD 1   C. CD 3
    B. CD 2   D. CD 4

520

12. **MAIN IDEA AND DETAILS** Which tool measures how hot or cold something is?
    F. an anemometer
    G. a light meter
    H. a sound meter
    J. a thermometer

13. Which kind of energy do people need to see things?
    A. sound energy
    B. light energy
    C. kinetic energy
    D. electrical energy

14. **COMPARE AND CONTRAST** Which resource can't be replaced?
    F. air
    G. oil
    H. water
    J. wind

15. Which resource is constantly being renewed?
    A. coal
    B. fossil fuels
    C. natural gas
    D. water

16. Which of the following takes millions of years to form?
    F. a plant
    G. a fossil fuel
    H. an energy need
    J. a hot spring

Inquiry Skills

17. Suppose you want to know how a water wheel works. How could you use a model to learn about a water wheel?

18. Suppose you have two ice cubes. You place one of the ice cubes outside in the direct sunlight and the other ice cube outside in the shade. Infer what will happen to each ice cube.

Critical Thinking

19. What changes can solar energy cause to a person lying in the sun?

20. List five ways you use energy in your home.

The Big Idea

CHAPTER 13 Electricity and Magnets

What's the Big Idea?

Electricity and magnetism are related and are part of things you use every day.

Essential Questions

Lesson 1
What Is Electricity?

Lesson 2
What Are Magnets?

Lesson 3
How Are Electricity and Magnets Related?

Go online
Student eBook
www.hspscience.com

What do you wonder?

How are electricity and magnets used at an amusement park? How does this relate to the **Big Idea?**

LESSON 1

Essential Question
What Is Electricity?

Investigate to find out how static electricity affects objects.

Read and Learn about types of electricity and how electricity moves.

Fast Fact

A Hair-Raising Experience

This Van de Graaff generator builds up a charge of static electricity. While the girl is touching the metal ball, she gets a charge, too. In the Investigate, you will see static electricity at work.

Vocabulary Preview

static electricity [STAT·ik ee·lek·TRIS·uh·tee] An electric charge that builds up on an object (p. 528)

current electricity [KER·uhnt ee·lek·TRIS·i·tee] Electricity that moves through a wire (p. 529)

circuit [SER·kuht] A path that electricity follows (p. 529)

525

Investigate

Looking for Static Electricity

Guided Inquiry

Start with Questions

This balloon is staying on the wall with no strings. It's not magic; it's static electricity!

- Have you ever used static electricity to stick a balloon to the wall?
- Where does the static electricity that holds the balloon to the wall come from?

Investigate to find out. Then read and learn to find out more.

Prepare to Investigate

Inquiry Skill Tip

A **hypothesis** is sometimes called an "educated guess." It is based on your ideas, on what you observe, and on things you already know. By combining these, you can decide what is probably the reason something happens. The data you collect may not support your hypothesis. Instead, the data may lead you to a new hypothesis.

Materials

- tissue paper
- comb
- piece of wool (sweater or blanket)

Make an Observation Chart

Action	Observation
Comb not passed through hair or rubbed with wool	
Comb passed through hair	
Comb rubbed with wool	

Follow This Procedure

1. Tear a small piece of tissue paper into tiny bits. Make the pieces smaller than your fingertips. Put the pieces in a pile.

2. Hold the comb just above the pile. What happens? **Record** what you **observe**.

3. Pass the comb through your hair several times. Repeat Step 2.

4. Rub the comb with your hand. Repeat Step 2.

5. Rub the comb with a piece of wool. Repeat Step 2.

Draw Conclusions

1. How did the comb change after you passed it through your hair? What happened when you rubbed the comb with wool?

2. **Inquiry Skill** When you **hypothesize**, you use observations or data to give a reason something happens. State a hypothesis to explain what happened in this investigation.

Independent Inquiry

Do you get the same result if you rub the comb with other things, such as silk or plastic wrap? **Plan and conduct an experiment** to find out.

527

Read and Learn

VOCABULARY
static electricity p. 528
current electricity p. 529
circuit p. 529

SCIENCE CONCEPTS
▶ what electricity is
▶ how electricity moves

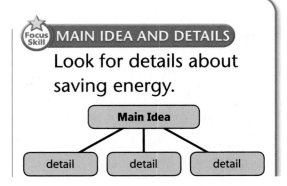

MAIN IDEA AND DETAILS
Look for details about saving energy.

Static and Current Electricity

On a cold day, you may see a person's hair stand up straight. The reason the hair stands up is static electricity. **Static electricity** is an electric charge that builds up on an object.

Lightning is one example of static electricity. A static charge builds up in a cloud, and then current electricity moves to the ground in the form of a lightning flash. You may even hear a crackling noise.

Math in Science
Interpret Data

Which of these states has the most deaths caused by lightning? Which have the fewest?

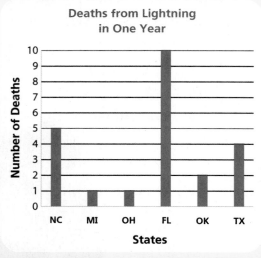

Deaths from Lightning in One Year

A lightning flash has enough electricity in it to light a 100-watt bulb for a little more than a month.

528

In the mixer, current electricity runs a motor, which makes the beaters turn.

What happens when you plug a lamp into a wall outlet? Electricity flows from wires in the wall through the plug and into the lamp's wires. Electricity that moves through a wire is **current electricity**.

A lamp's bulb glows only when there is a closed circuit. A **circuit** is a path that electricity follows. Electricity moves from the wall outlet, through the lamp, and back to the wall.

People use current electricity to light, heat, and cool homes, to run motors, and to cook food. What other things do you know of that use electricity?

MAIN IDEA AND DETAILS
What is current electricity?

Make It Light
Put a night-light bulb in a bulb holder. Attach two wires to the bulb holder. Touch one wire to one knob of the battery. Close the circuit by touching the other wire to the other knob on the battery. What happens?

529

How Electricity Moves

Electricity moves through some things but not through others. Something that electricity moves freely through is called a conductor. Copper wire is a good conductor.

Something that electricity will not move through easily is called an insulator. Wood and plastic are examples of insulators. Electricity can be dangerous, so people use insulators to stay safe.

 MAIN IDEA AND DETAILS

What is a conductor? What is an insulator?

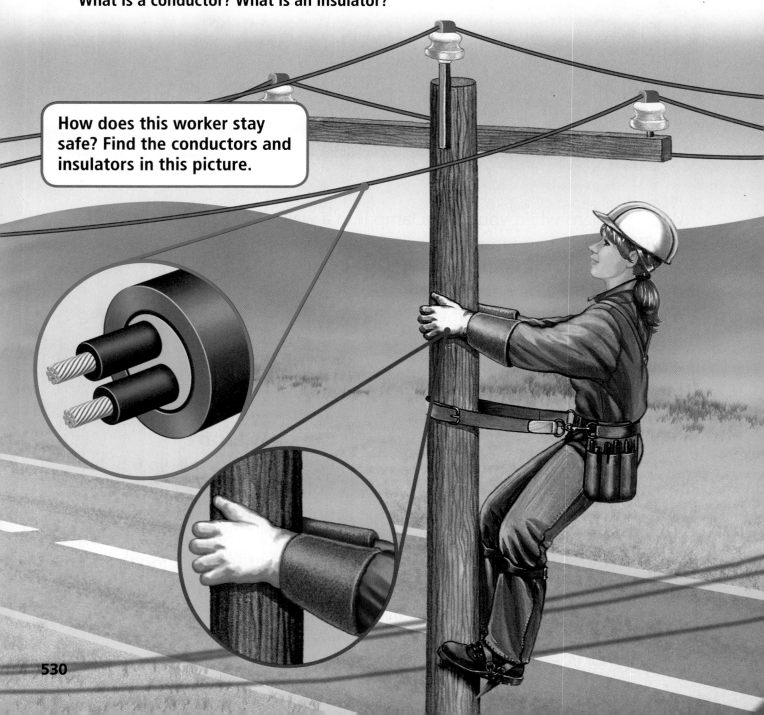

How does this worker stay safe? Find the conductors and insulators in this picture.

Lesson Review

Essential Question
What is electricity?

In this lesson, you learned that static electricity is an electric charge that builds up on an object. Current electricity is electricity that moves through a wire. Conductors are materials that electricity can move freely through. Materials that electricity cannot move easily through are insulators.

1. **MAIN IDEA AND DETAILS** Draw and complete a graphic organizer for this main idea: There are two kinds of electricity.

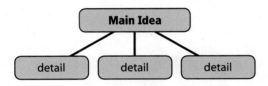

2. **SUMMARIZE** Write three sentences to summarize this lesson. Tell how static electricity and current electricity are different.

3. **DRAW CONCLUSIONS** Why do wires often have plastic wrapped around them?

4. **VOCABULARY** Write a sentence to tell what *current electricity* is.

Test Prep

5. Which of the following makes an oven work?
   A. a magnet
   B. current electricity
   C. an open circuit
   D. static electricity

Make Connections

 Writing

Expository
Electricity that is used the wrong way can be dangerous. Research ways to be safe around electricity. Use your findings to write a **how-to** booklet.

 Math

Solve Problems
Suppose Sam's computer is on a desk that's centered exactly between two wall outlets. The wall outlets are 3 meters apart. What is the least amount of cord Sam can use to plug his computer into one of the outlets?

531

LESSON 2

Essential Question
What Are Magnets?

Investigate to find out which are stronger, bar magnets or horseshoe magnets.

Read and Learn about magnets and their uses.

Fast Fact

How Strong Is It?
Some magnets can pick up only small objects, such as these fish. Others are strong enough to lift steel beams for buildings. In the Investigate, you will observe and compare the strengths of two magnets with different shapes.

Vocabulary Preview

magnetic [mag•NET•ik]
Able to attract objects that have iron in them
(p. 536)

Investigate

Which Magnet Is Stronger?

Guided Inquiry

Start with Questions

Many people keep magnetic toys on their desks at work. You may have seen one similar to the one below. The magnet is in the base, below the metal pieces.

- What shape is the base?
- Does the shape of a magnet affect how strong it is?

Investigate to find out. Then read and learn to find out more.

Prepare to Investigate

Inquiry Skill Tip

Graphs are a good way to communicate the data you collect during an investigation. Choose the graph style that best shows the type of data you collect. A bar graph is useful if you want to compare data. A circle graph can be used to show parts of a whole. Line graphs are a good way to communicate changes over time.

Materials

- bar magnet
- 10 to 20 steel paper clips
- horseshoe magnet

Make a Data Table

Kind of Magnet	Observations	Number of Paper Clips
Bar		
Horseshoe		

Follow This Procedure

1. Hold the bar magnet near a paper clip. **Record** what happens.
2. Hold the horseshoe magnet near a paper clip. **Record** what happens.
3. Pick up as many paper clips as one end of the bar magnet will hold. Count them. **Record the data.**
4. Pick up as many paper clips as one end of the horseshoe magnet will hold. Count them. **Record the data.**

Draw Conclusions

1. Which magnet is stronger? How can you tell?
2. **Inquiry Skill** Scientists can **communicate** data in graphs. Make a bar graph to show how many paper clips each magnet held.

Independent Inquiry

If you hold two magnets together, can they lift as many paper clips as each one can separately? **Plan an investigation** to find out.

Read and Learn

VOCABULARY
magnetic p. 536

SCIENCE CONCEPTS
▶ what magnets do

COMPARE AND CONTRAST
Find out how magnets of different shapes work.

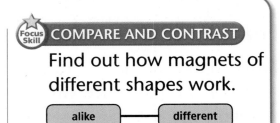

Magnets

In the Investigate, you used magnets to pick up paper clips. Magnets are made from metal that is magnetic. **Magnetic** things attract objects that have iron in them. Paper clips are made of steel, but steel has iron in it. That's why the magnets lift and hold steel paper clips. A magnet won't pick up plastic paper clips, because they don't contain iron.

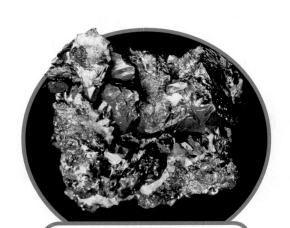

This rock is magnetite. It's naturally magnetic.

How are magnets being used here?

Science Up Close

For more links and animations, go to www.hspscience.com

Magnets will attract only when the north-seeking and south-seeking poles match up. Which magnets are attracting? Which are repelling?

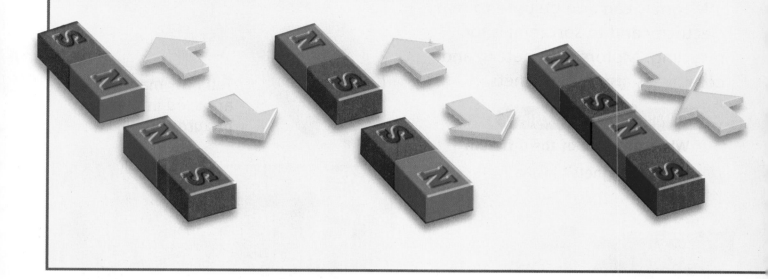

All magnets have two ends, called poles—a north-seeking (N) pole and a south-seeking (S) pole. To see how a magnet works, use two bar magnets. Hold the two N poles together. Can you push them together? You probably can't. Poles that are alike repel each other. *Repel* means "push away." What do you think will happen if you put the two S poles together?

Now hold opposite poles—an N pole and an S pole—together. They pull toward each other, or *attract.* The opposite poles of a magnet attract.

How are all magnets alike?

Are Horseshoe Magnets Like Bar Magnets?
Try to push two horseshoe magnets together. What happens? Turn one over and try again. What happens? Explain what you observe.

537

Some Uses of Magnets

Magnets have many uses. They can keep cabinets closed or hold papers on a refrigerator door. Some of your favorite games might use magnets.

Magnets can also be used to make electricity and to sort metals for recycling. Motors, computers, and compasses also use magnets.

COMPARE AND CONTRAST
What are some of the different uses of magnets?

Tell how magnets are used in these pictures.

Lesson Review

Essential Question

What are magnets?

In this lesson, you learned that magnets are made from magnetic metal. They attract objects that have iron in them. Magnets have two poles. Poles that are alike repel each other. Opposite poles attract each other. Magnets can be used for producing electricity.

1. **COMPARE AND CONTRAST** Draw and complete a graphic organizer to compare and contrast magnets.

   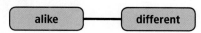

2. **SUMMARIZE** Write a summary of this lesson. Tell how magnet poles react to each other.

3. **DRAW CONCLUSIONS** Juan is using a bar magnet to pick up items in his room. Explain why the magnet he is using won't pick up a rubber ball.

4. **VOCABULARY** Write a sentence to explain what *magnetic* means.

Test Prep

5. What must an object be made of to be attracted to a magnet?
   A. iron
   B. plastic
   C. rubber
   D. wood

Make Connections

 Writing

Expository
Some objects are naturally magnetic. Research lodestone. Write a two-paragraph **description** of your findings.

 Art

Uses for Magnets
List at least five ways that magnets are used in your home or classroom. Draw an illustration for each use.

LESSON 3

Investigate to find out how magnets can be helpful.

Read and Learn about how electromagnets and generators work.

Essential Question

How Are Electricity and Magnets Related?

Fast Fact

Recycling
Americans recycle more than 70 million tons of scrap metal each year. Electromagnets are used to sort scrap metal. In the Investigate, you will find out why an electromagnet is useful.

Vocabulary Preview

generator [JEN•er•ayt•er]
A device that uses a magnet to produce a current of electricity (p. 546)

Investigate

Simple Sorting

Guided Inquiry

Start with Questions

The globe in this picture looks as if it's floating! Magnets are keeping it up.

- How do you think the magnets are able to keep the globe in the air?
- What are some other ways magnets are useful?

Investigate to find out. Then read and learn to find out more.

Prepare to Investigate

Inquiry Skill Tip

Good observations are important when you infer information. Some of the most important things you learn from an investigation are not specifically found in the results. You have to think about what the results mean.

Materials

- steel paper clips
- bowl or paper plate
- plastic beads
- magnet
- stopwatch

Make a Data Table

Method	Time (s)
Without a magnet	
With a magnet	

Follow This Procedure

1. Put a handful of paper clips in the bowl. Add a handful of beads. Mix them up.

2. Remove the paper clips from the bowl by hand, making sure not to pick up any of the beads. Use a stopwatch to measure and record how long this takes.

3. Return the paper clips to the bowl of beads. Mix them up again.

4. Now use a magnet to remove the paper clips from the bowl. Use a stopwatch to measure and record how long this takes.

Draw Conclusions

1. How do the two times compare? Which is the quicker way to separate the steel paper clips from the plastic beads?

2. **Inquiry Skill** Scientists use what they know to infer, or conclude. Look at the picture on pages 540 and 541. What are some things that you can infer about the use of an electromagnet?

Independent Inquiry

Mix various small metal objects together. Predict which objects you can separate by using a magnet. Test your prediction.

Read and Learn

VOCABULARY
generator p. 546

SCIENCE CONCEPTS
▶ how an electromagnet works

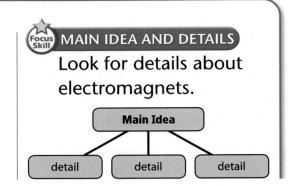

MAIN IDEA AND DETAILS
Look for details about electromagnets.

Electromagnets

At a salvage (SAL•vij) yard, there are many recycled cars. The cars don't run anymore, so a person can't drive or move them. However, a very strong magnet attached to a crane can move a car. Together, the crane and magnet can pick up a car and put it somewhere else. There is one problem. After the car is on the magnet, how is it removed? A person cannot pull the car from the magnet. The magnet's pull is too strong.

How does the boy know that the screw isn't magnetic?

Wouldn't it be handy if there were a magnet whose magnetism could be turned on and off? Well, there is—an electromagnet (ee•lek•troh•MAG•nit). This metal object is made of iron or steel, and you can turn its magnetism on or off.

It's easy to make an electromagnet. A plain screw isn't a magnet. However, the screw can be made into an electromagnet. First, wrap wire around the screw. Then, run an electric current through the wire. The screw becomes a magnet that you can turn on and off. When the electric circuit is open, the screw is no longer a magnet.

MAIN IDEA AND DETAILS

How does an electromagnet work?

What is the source of the electric current for this electromagnet?

Generators

An electromagnet uses an electric current to make a magnet. A generator (JEN•er•ayt•er) does the opposite. A **generator** uses a magnet to make electricity.

Do you remember the wire that was important in an electromagnet? It's important in a generator, too. If you move a coil of wire near a magnet, current electricity flows in the wire. That's one way electricity is made.

Get Ready for an Emergency
Find out the plan that your school would use if the electricity stopped working. Write about it. Does your school have a generator?

MAIN IDEA AND DETAILS
How does a magnet make an electric current?

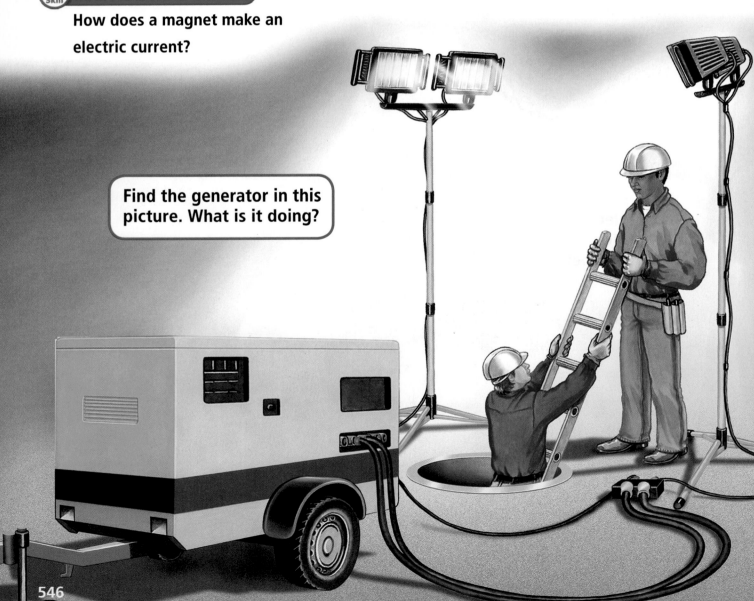

Find the generator in this picture. What is it doing?

Lesson Review

Essential Question

How are electricity and magnets related?

In this lesson, you learned that you can make an electromagnet with iron and an electric current. Electricity can also be generated with a magnet and coiled wire.

1. MAIN IDEA AND DETAILS Draw and complete a graphic organizer for this main idea: Electromagnets

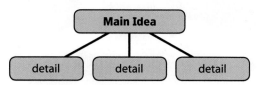

2. SUMMARIZE Write a three-sentence summary for this lesson. Tell how electromagnets and generators are similar.

3. DRAW CONCLUSIONS Why is it important for hospitals to have emergency generators?

4. VOCABULARY Write a sentence to explain how a generator works.

Test Prep

5. Critical Thinking What does a magnet have to do to produce an electric current?

Make Connections

 Writing

Persuasive
Write an **e-mail** to a business owner. Explain the reasons that the business should have a generator.

 Social Studies

Inventing the Generator
Research Nikola Tesla's contributions to the invention of the generator. Share your findings in a short report. Be sure to add drawings to your report.

Science Spin
From Weekly Reader
TECHNOLOGY

Batteries Included

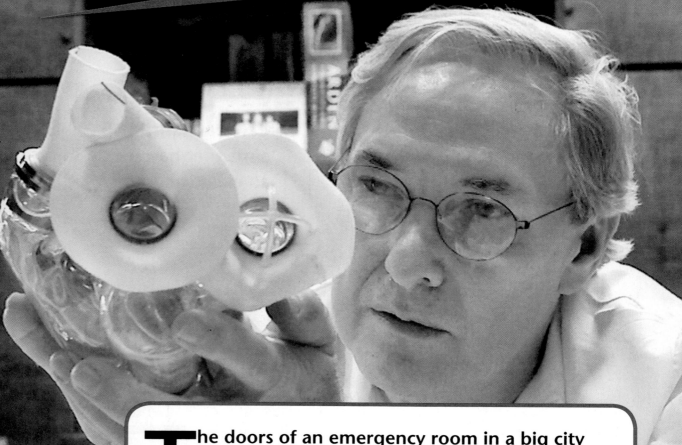

The doors of an emergency room in a big city hospital burst open. "Heart attack!" shouts the medic. Soon the patient is in the operating room. But doctors don't want to wait for a heart transplant from another patient. Instead, doctors insert an artificial heart to save the patient's life.

Although it hasn't happened yet, this type of scene could happen if the makers of the *AbioCor* are successful. The *AbioCor* is a battery-operated artifical heart that weighs about two pounds.

A Life-Saver

Scientists spent about 30 years designing the *AbioCor*. The replacement heart is different from earlier types of devices. This device fits completely inside the body of a patient. There is no need for wires to connect through the skin to an outside power source. This way a patient can be free to walk around without being hooked up to any machines.

The heart received the government's appoval in September of 2006. It is being tested in some patients. The first patient received an *AbioCor* about six years ago.

Think and Write

1. What other lifesaving devices might batteries be used in?
2. Why would patients want a replacement heart that runs on batteries instead of one that is plugged into a power source?

Find out more. Log on to www.hspscience.com

How It Works

The *AbioCor* gets its power from two lithium batteries. One battery is inside the patient and the other battery is outside the patient. Each battery has a set of coils. The outside battery pack can constantly recharge the inside battery. The outside battery does this by sending electricity from its coil, through the skin, to the coil of the inside battery.

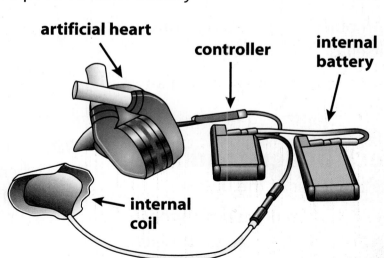

Chapter 13 Review and Test Preparation

Vocabulary Review

Use the terms below to complete the sentences. The page numbers tell where to look in the chapter if you need help.

static electricity p. 528
current electricity p. 529
circuit p. 529
magnetic p. 536
generator p. 546

1. To run a machine, electricity must move along a closed _____.
2. A machine that uses a magnet to make a current of electricity is called a _____.
3. Current that moves in a wire is called _____.
4. An object that pulls iron toward itself is _____.
5. A charge that builds up in an object is _____.

Check Understanding

Write the letter of the best choice.

6. Which term describes copper?
   A. circuit
   B. conductor
   C. insulator
   D. magnet

7. **MAIN IDEA AND DETAILS** What is lightning an example of?
   F. a magnet
   G. an electromagnet
   H. static electricity
   J. a circuit

8. **COMPARE AND CONTRAST** What can an electromagnet do that other magnets can't do?
   A. be turned on and off
   B. coil around a wire
   C. lift metal objects
   D. generate electricity

Use the picture for questions 9–10.

9. What are the ends of the magnets called?
   F. circuits
   G. conductors
   H. insulators
   J. poles

10. What happens to the magnets when they are held like this?
    A. They attract each other.
    B. They cause static electricity.
    C. They make an electromagnet.
    D. They repel each other.

11. Magnet A picks up six paper clips. Magnet B picks up eight paper clips. What conclusion can you draw?
    F. The circuit in Magnet A has been broken.
    G. Magnet B is an electromagnet.
    H. Magnet A is weaker than Magnet B.
    J. Both magnets are equally weak.

12. A coil of wire moves near a magnet. What is produced?
    A. an outlet
    B. electricity
    C. an electromagnet
    D. static electricity

13. Miguel plugs his desk lamp into an outlet. What type of electricity is the lamp using?
    F. current   H. magnetic
    G. lightning   J. static

14. What does a generator make?
    A. a magnet
    B. an electromagnet
    C. a circuit
    D. an electric current

15. A magnet attracts an object. What must the object contain?
    F. iron   H. rubber
    G. plastic   J. wood

16. Plastic is an example of which of the following?
    A. conductor
    B. generator
    C. insulator
    D. magnet

Inquiry Skills

17. Dana wants to separate steel paper clips from steel safety pins. She uses a magnet. Predict what will happen, and tell why.

18. As hard as Brooke tries, she can't push two horseshoe magnets together. Infer what she could do to get them to attract.

Critical Thinking

19. Carson built an electromagnet. He used a battery, a nail, and some wire. His electromagnet doesn't work. Give one reason it might not work.

20. Miguel held a magnet 2 cm above some steel safety pins. His magnet picked up eight pins. What will happen if Miguel holds the magnet 4 cm above the pins? Why?

CHAPTER 14
Heat, Light, and Sound

What's the Big Idea? Heat, light, and sound are different forms of energy that can move from place to place.

Essential Questions

Lesson 1
What Is Heat?

Lesson 2
What Is Light?

Lesson 3
How Are Light and Color Related?

Lesson 4
What Is Sound?

GO online Student eBook
www.hspscience.com

Marching band

What do you wonder?

What forms of energy are present in the photo of the marching band? How do you know? How do these forms of energy relate to the **Big Idea?**

LESSON 1

Essential Question
What Is Heat?

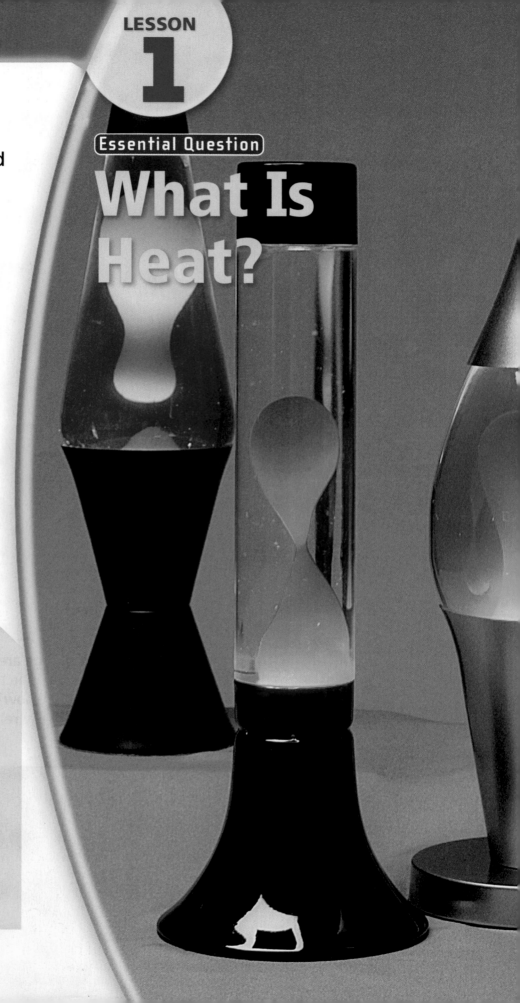

Investigate to find out how different materials conduct heat.

Read and Learn about how heat is produced and how conductors and insulators are used.

Fast Fact

Up and Down
Each lamp has a light bulb in its base. Energy from the bulb warms the goo, which rises. When the goo cools at the top, it falls. That's one way thermal energy moves from one place to another. In the Investigate, you will observe another way.

Lava lamps

Vocabulary Preview

temperature [TEM•per•uh•cher] The measure of how hot or cold something is (p. 558)

thermal energy [THER•muhl EN•er•jee] The form of energy that moves particles of matter (p. 558)

heat [HEET] The movement of thermal energy from hotter to cooler objects (p. 558)

conduction [kuhn•DUK•shuhn] The movement of heat between objects that are touching each other (p. 559)

conductor [kuhn•DUK•ter] An object that heat can move through easily (p. 559)

insulator [IN•suh•layt•er] An object that doesn't conduct heat well (p. 560)

555

Investigate

Getting Warmer?

Guided Inquiry

Start with Questions

You may have used an oven mitt to remove something hot from an oven. In the picture below, a person is being careful not to get burned.

- Why would someone need to use an oven mitt to remove something from an oven?
- How does an oven mitt help protect a person from getting burned?

Investigate to find out. Then read and learn to find out more.

Prepare to Investigate

Inquiry Skill Tip

If you have trouble drawing a conclusion, try drawing a picture! You can use a drawing or a diagram to summarize the results of an investigation and better understand what happened.

Materials

- safety goggles
- wooden spoon
- plastic spoon
- metal spoon
- 3 plastic foam cups
- hot water
- ceramic mug with handle
- plastic mug with handle
- metal mug with handle

Make an Observation Chart

Temperatures of Objects			
	Wooden Spoon	Plastic Spoon	Metal Spoon
Dry			
After 1 min			

	Ceramic Mug	Plastic Mug	Metal Mug
Dry			
After 30 sec			
After 60 sec			
After 90 sec			
After 120 sec			

Follow This Procedure

1. **CAUTION: Put on safety goggles.**
2. Touch the three spoons. **Record** your **observations**.
3. **CAUTION: Be careful with hot water.** Fill three plastic foam cups with hot water. Place one spoon in each cup. Wait 1 minute.
4. Gently touch each spoon. **Record** your **observations**.
5. Touch the three mugs. **Record** your **observations**.
6. Fill each mug with hot water. Carefully touch each handle every 30 seconds for 2 minutes. **Record** what you **observe**.

Draw Conclusions

1. **Compare** your **observations** of the spoons and the mugs before and after the water was used.
2. **Inquiry Skill** Draw a **conclusion** about the way thermal energy travels through different substances. Write down your conclusion and compare it with a classmate's.

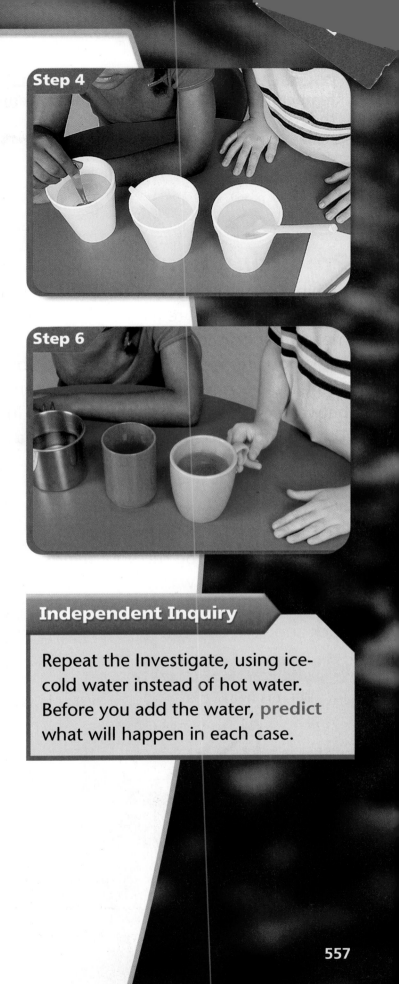

Independent Inquiry

Repeat the Investigate, using ice-cold water instead of hot water. Before you add the water, **predict** what will happen in each case.

Read and Learn

VOCABULARY
temperature p. 558
thermal energy p. 558
heat p. 558
conduction p. 559
conductor p. 559
insulator p. 560

SCIENCE CONCEPTS
▶ how thermal energy moves

MAIN IDEA AND DETAILS
Look for details about the movement of heat.

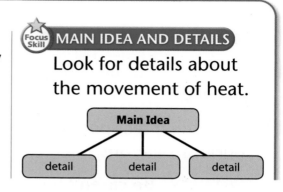

Producing Heat

Think about sitting by a campfire. If you sit too close, your skin temperature might go up. **Temperature** is the measure of how hot or cold something is. You feel hot when you are close to the fire because burning wood gives off heat. **Thermal energy** is a form of energy that moves between objects because of differences in temperature. This movement of thermal energy is **heat**. If you get too hot from the fire's heat, you can move away from it or pour water on the fire to stop the heat.

Heat can be produced in other ways, too. When you rub your hands together, they get warm. If you use a heat pack, the chemicals mix in the pack and give off heat.

MAIN IDEA AND DETAILS
What is heat?

Math in Science
Interpret Data

When you're outside, you can measure the temperature without using a thermometer. Count the number of cricket chirps you hear in one minute, and use this table. What is the temperature when you count 140 chirps?

Number of Chirps per Minute	Temperature in Celsius
10	4°
40	8°
80	14°
120	19°
160	25°
200	31°

558

Conductors

To cook an egg, you put the egg in a pan and put it on the stove. The heat from the burner makes the pan hot. Soon the heat moves through the pan to the egg. This movement of thermal energy between objects touching each other is **conduction**.

Cooking pans are made of metal, such as iron or aluminum. Heat moves easily through most metals. An object that heat can move through easily is called a **conductor**.

What are two conductors?

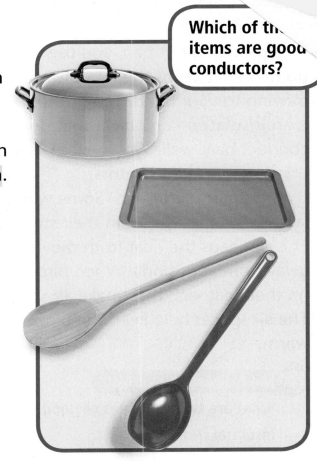
Which of the items are good conductors?

Heat moves from warm objects—such as the burner—to cooler objects—such as the eggs.

Insulators

To pick up a hot pan, you use a potholder. It keeps the pan's heat from moving to your hand. The potholder is an **insulator**—an object that doesn't conduct heat well. Wood, cloth, and plastic are good insulators.

Air is an insulator, too. Some winter jackets have air spaces in their stuffing. The air keeps the heat from moving away from your body. When birds fluff up their feathers, they make air spaces. The air spaces help keep the birds warm.

MAIN IDEA AND DETAILS

What are two examples of good insulators?

Feeling the Heat

Place small pieces of butter on top of a sponge, a piece of wood, and a metal jar lid. Float each object in a shallow pan of hot water. What happens to the butter? Why?

Water seeps in between the wet suit and the surfer's skin. This water warms up and acts as an insulator along with the wet suit. ▼

The cover wrapped around this pot is an insulator.

Lesson Review

Essential Question
What is heat?

In this lesson, you learned that heat is the movement of thermal energy between two objects. Heat can be produced in several ways, including burning wood, rubbing your hands together, and chemical reactions. Conductors are items that heat moves through easily. Insulators do not allow heat to pass through them easily.

1. **MAIN IDEA AND DETAILS** Draw and complete a graphic organizer for this main idea: Some things allow heat to pass through them better than others.

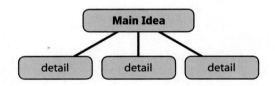

2. **SUMMARIZE** Write two sentences that tell the most important ideas about heat.

3. **DRAW CONCLUSIONS** Tasha had a foam cup of hot cocoa. The cocoa stayed hot, and the cup didn't get too hot to hold. Explain why.

4. **VOCABULARY** Use the vocabulary words in this lesson to make a crossword puzzle.

Test Prep

5. Which of these is a good conductor?
   - **A.** air
   - **B.** cloth
   - **C.** copper
   - **D.** wood

Make Connections

 Writing

Descriptive
Write a **description** of what it feels like to sit by a campfire. Include adjectives to make your description exciting.

 Health

Putting Out a Fire
Research some ways to put out a fire. Make a booklet to show your findings. Include pictures.

LESSON 2

Essential Question

What Is Light?

Investigate to find out how light travels.

Read and Learn about how light moves and how that affects objects.

Fast Fact

Finding the Correct Lighthouse

Sailors sometimes use the light from a lighthouse to know where they are when they're close to shore. In the Investigate, you will find out how light travels.

Follow This Procedure

1. Place a small object in the middle of your desk.

2. Place a flashlight on its side on your desk. Point the flashlight in the direction of the object.

3. Have your partner stand a sheet of poster board on the desk. The poster board should be between the flashlight and the object.

4. Turn on the flashlight. **Observe** the object. Does the light shine on it?

5. Have your partner slide the poster board across your desk just until the light shines on the object. **Observe** and **record** the positions of the object, flashlight, and poster board.

Draw Conclusions

1. What kind of line would you draw to connect the object and the light?

2. **Inquiry Skill** Scientists **infer**, based on their observations. **Infer** what the path of light is like from the flashlight to the object.

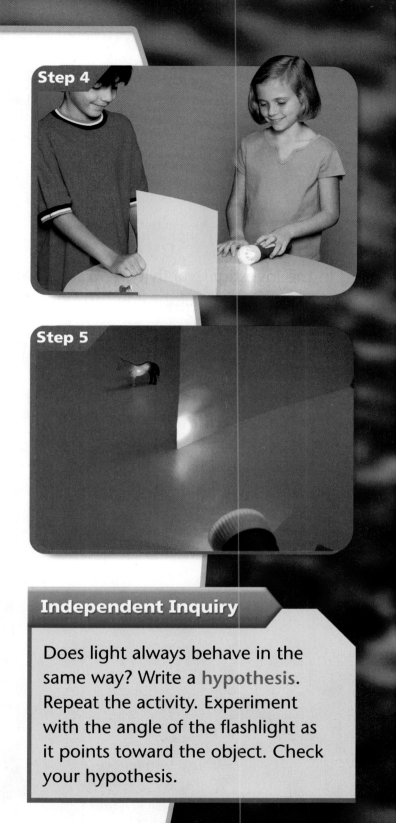

Independent Inquiry

Does light always behave in the same way? Write a **hypothesis**. Repeat the activity. Experiment with the angle of the flashlight as it points toward the object. Check your hypothesis.

Read and Learn

VOCABULARY
reflection p. 566
refraction p. 567
shadow p. 568

SCIENCE CONCEPTS
▶ what the path of light is like and how it can be changed

 **SEQUENCE**
See what happens next when light travels.

Ways Light Moves

In the Investigate activity, you saw that light moves in a straight line through space. When light strikes an object, some of the light bounces off the object. The bouncing of light off an object is called **reflection**. If the object is smooth and shiny, the light reflects in a pattern that you can see.

The reflection in a mirror is reversed from left to right. Everything looks "backward."

You can see trees and mountains reflected in the lake.

Refraction makes this piggy bank look broken.

The bank in the picture appears to be broken, but it isn't. Light bends when it moves from the water to the air. The bending of light as it moves from one material to another is **refraction**. Light from the top of the bank goes straight to your eyes. Light from the bottom of the bank goes through water first. The light bends, or refracts, as it leaves the water. You see the bank in two parts.

 SEQUENCE

What happens to light after it strikes a smooth object?

What Do You See?

Write the capital letters ABC on a card. Hold the card in front of a mirror, and draw what you see in the mirror. Repeat, using BIRD and MOTH. How are these words different?

Shadows

Think back to the Investigate. When the poster board was in front of the light, part of your desk was dark. The poster board blocked the light and made a shadow. A **shadow** is a dark area that forms when an object blocks the path of light.

When an object moves, its shadow moves, too. After you moved the poster board, the shadow moved away from part of your desk.

An object blocks the path of light. What happens next?

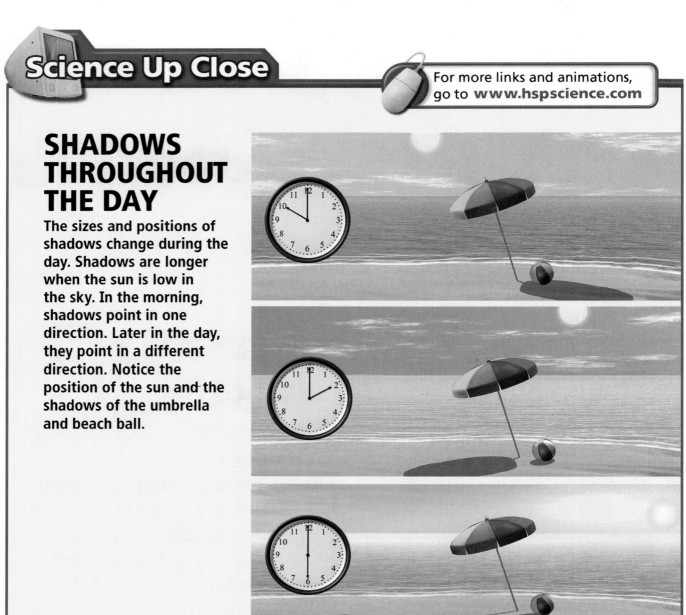

SHADOWS THROUGHOUT THE DAY

The sizes and positions of shadows change during the day. Shadows are longer when the sun is low in the sky. In the morning, shadows point in one direction. Later in the day, they point in a different direction. Notice the position of the sun and the shadows of the umbrella and beach ball.

For more links and animations, go to www.hspscience.com

568

Lesson Review

Essential Question

What is light?

In this lesson, you learned that light travels in a straight line. Light can be reflected and bent. The bending of light is called refraction. A shadow forms when an object blocks light.

1. **Focus Skill SEQUENCE** Draw and complete a graphic organizer that sequences light striking an object and light traveling through water.

2. **SUMMARIZE** Write a summary of this lesson. Start with this sentence: Light travels in a straight line.

3. **DRAW CONCLUSIONS** Why can you see your reflection in a mirror but not on a brick wall?

4. **VOCABULARY** Draw a picture to illustrate each vocabulary word in the lesson. Exchange pictures with a classmate. Write captions for your classmate's pictures.

Test Prep

5. **Critical Thinking** Would the word WOW look the same in a mirror? Explain.

Make Connections

 Writing

Narrative

Write a humorous **story** about a four-year-old who has just discovered his or her shadow. Include facts you have learned about shadows in your story.

 Social Studies

Using Shadows to Tell Time

Research how people have used shadows and sundials to tell time. Write a report to share your research with your classmates.

LESSON 3

Essential Question

How Are Light and Color Related?

Investigate to find out how rainbows form.

Read and Learn about color and about what happens when light hits different objects.

Fast Fact

All the Colors
The colors of the rainbow are always in the same order. Red is on top and violet is on bottom. In the Investigate, you will use a prism to make a spectrum of colors.

Vocabulary Preview

absorbed [ab•SAWRBD] Taken in by an object (p. 574)

opaque [oh•PAYK] Relating to objects that don't let light pass through (p. 575)

transparent [trans•PER•uhnt] Relating to an object that lets most light pass through (p. 575)

translucent [trans•LOO•suhnt] Relating to an object that lets some light pass through (p. 575)

Investigate

Making Rainbows

Guided Inquiry

Start with Questions

The photo below may look like a rainbow, but it's not a true rainbow. Some people call this a fire rainbow.

- How is a fire rainbow different from a true rainbow?
- What causes a fire rainbow to form?

Investigate to find out. Then read and learn to find out more.

Prepare to Investigate

Inquiry Skill Tip

When you predict what will happen in an investigation, think about the pattern of things that have happened before. Then decide what would happen if that pattern were to continue with just one variable changed.

Materials

- prism
- clear tape
- sheet of white paper
- sheet of black paper
- sheet of red paper

Make an Observation Chart

Position of the Prism	Observation
Over white paper	
Over red paper	

Follow This Procedure

1. Cut a narrow slit in a sheet of black paper. Tape the paper to the bottom of a window. Pull down the blinds to make a narrow beam of sunlight.

2. Hold a prism in the beam of light over a sheet of white paper. Slowly turn the prism until it makes a rainbow on the paper.

3. Look closely at the paper. What do you see? **Record** your **observations**.

4. Repeat Steps 2 and 3, using a sheet of red paper.

Draw Conclusions

1. How does a prism change sunlight?

2. How was the light you saw on the white paper different from the light you saw on the red paper?

3. **Inquiry Skill** Scientists **predict** what might happen, based on patterns or experiences. What do you **predict** you would see if you used the prism to shine sunlight on a piece of blue paper?

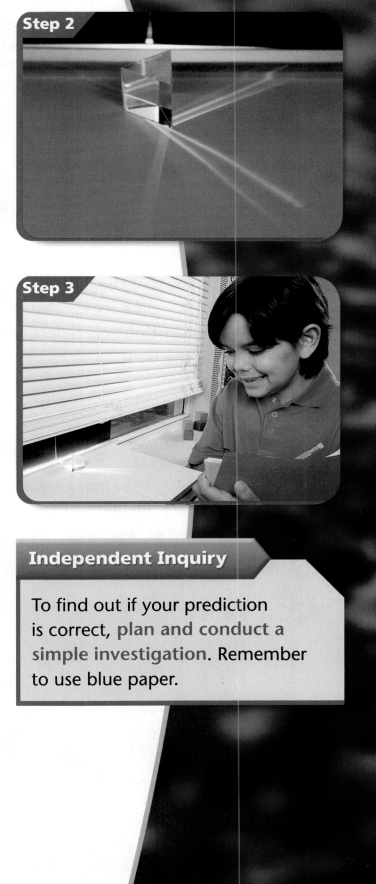

Independent Inquiry

To find out if your prediction is correct, **plan and conduct a simple investigation**. Remember to use blue paper.

573

Read and Learn

VOCABULARY
absorbed p. 574
opaque p. 575
transparent p. 575
translucent p. 575

SCIENCE CONCEPTS
▶ how light affects the way things look

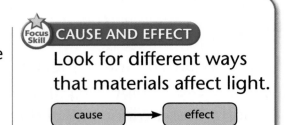

CAUSE AND EFFECT
Look for different ways that materials affect light.

cause → effect

How to Stop Light

When light strikes an object, some of the light is reflected. What happens to the rest of the light? It is **absorbed**, or taken in by the object. Shiny objects, such as mirrors, reflect most of the light that strikes them. Dull, dark objects, such as some rocks, absorb most of the light that strikes them. Most objects reflect some light. They absorb the rest of the light.

The stones are opaque.

The frosted marbles are translucent.

574

Objects that don't let light pass through them are **opaque** (oh•PAYK). Mirrors, rocks, books, wooden desks, and people are opaque.

Some objects let most of the light that strikes them pass through. Objects that let most light pass through them are **transparent**. The clear glass in a window is transparent.

Objects that let some light pass through them, such as frosted light bulbs, are **translucent**. Light passes through the glass of the bulb. However, you can't see what is inside the bulb.

CAUSE AND EFFECT

What effect does an opaque object have on light?

The clear marbles are transparent.

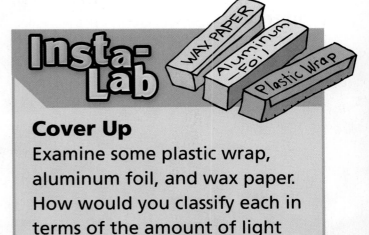

Cover Up
Examine some plastic wrap, aluminum foil, and wax paper. How would you classify each in terms of the amount of light that passes through?

Light and Color

What do you see when you hold white paper in white light, such as sunlight? Since the light that hits the paper is reflected, the paper looks white.

In the Investigate, when sunlight went into the prism, colored light came out. Where did the colors come from? The white light of sunlight is a mixture of colors. In the activity, did you notice that light doesn't go straight through the prism? The light bends as it passes through the prism. This bending separates light into its different colors.

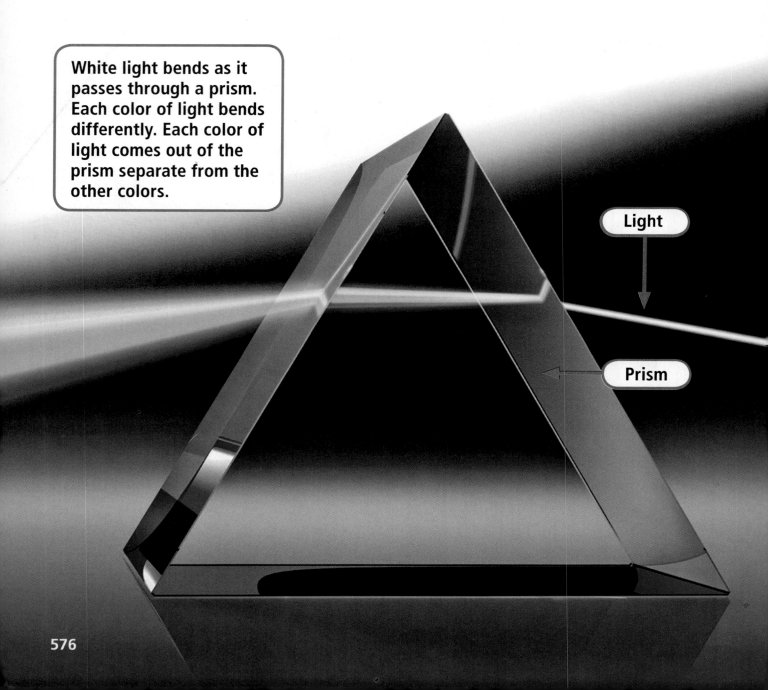

White light bends as it passes through a prism. Each color of light bends differently. Each color of light comes out of the prism separate from the other colors.

You need sunlight and rain to see a rainbow outside. These two things don't often happen at the same time. That's why you don't often see rainbows. When it rains and then the sun comes out, drops of water in the air act like tiny prisms. They break up white light into its colors.

 **CAUSE AND EFFECT**
What effects does a prism have on white light?

How Rainbows Form

A drop of water is like a tiny prism.

White light from the sun strikes drops of water in the air. Light enters each water drop, refracts, and then it's reflected inside the drop. When the light leaves the drop, it separates into many colors.

◀ The colors that most people see in a rainbow are red, orange, yellow, green, blue, and violet.

Making White Light

Do you know that you can make white light? You can do this by shining lights together onto one surface. The picture shows three lights shining on a white surface. Each flashlight has a different-colored filter that the light shines through. Find the place where all three colors shine on the same spot. Red, green, and blue light together make white light.

CAUSE AND EFFECT What effects do red light, blue light, and green light have when you shine them together onto a white surface?

The filters on these flashlights each let only one color pass through. These three colors can combine to make white light. ▼

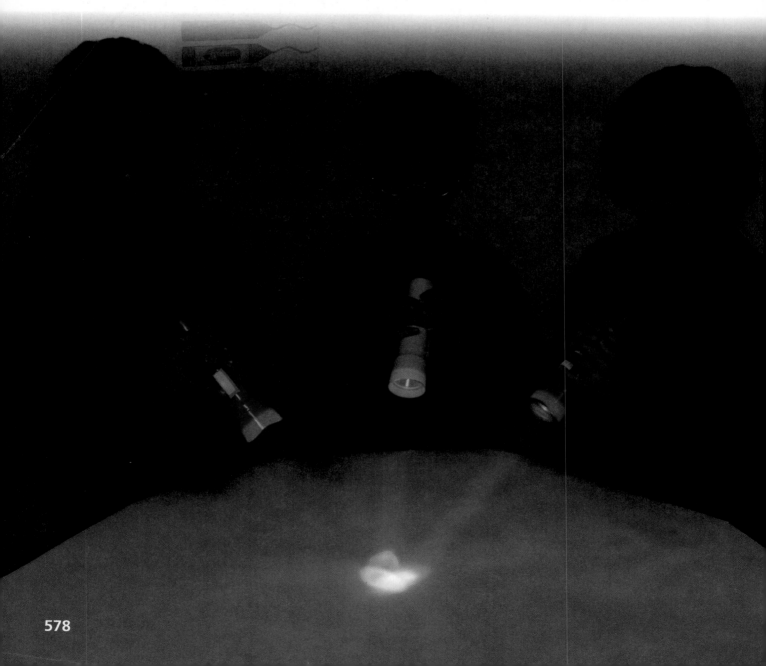

Lesson Review

Essential Question
How are light and color related?

In this lesson, you learned that some objects reflect light and other objects absorb it. White light is a mixture of colors. If you combine red, green, and blue light, you can make white light.

1. **CAUSE AND EFFECT** Draw and complete a graphic organizer to show the effect of light striking an object.

2. **SUMMARIZE** Write a three-sentence summary of this lesson. Use the terms *opaque, transparent,* and *translucent* in your summary.

3. **DRAW CONCLUSIONS** Katie is watering her garden. She is using a hose with a fine spray. She sees a rainbow over her garden. Explain why.

4. **VOCABULARY** Write two sentences that use the vocabulary terms from this lesson.

Test Prep

5. What color will you see if you shine red, green, and blue lights on the same spot on a sheet of white paper?
   A. black
   B. purple
   C. white
   D. yellow

Make Connections

 Writing

Expository
Write an **explanation** of how a rainbow forms.

 Math

Graph Favorite Colors
Using the six main colors of the rainbow, ask 10 classmates which color is their favorite. Place your findings in a pie chart. Color the sections to match the colors chosen.

LESSON 4

Essential Question
What Is Sound?

Investigate to find out how maracas work.

Read and Learn about how sound is made, how it travels, and how you hear it.

Fast Fact

Percussion Section
Percussion instruments include maracas, drums, and tambourines. They keep the beat of the music. In the Investigate, you will make a model of a maraca.

percussion instruments

Vocabulary Preview

vibration [vy•BRAY•shuhn] A series of back-and-forth movements (p. 584)

loudness [LOUD•nuhs] The amount of energy a sound has (p. 584)

pitch [PICH] The highness or lowness of a sound (p. 585)

Investigate

Make a Maraca

Guided Inquiry

Start with Questions

You may have seen a speaker like this one at a party or concert. It can make the music really loud! Loud music can damage your ears.

- Have you ever seen parts of a speaker vibrate when it plays music?
- Why do you think speakers like the one in the picture vibrate when they play music?

Investigate to find out. Then read and learn to find out more.

Prepare to Investigate

Inquiry Skill Tip

When you compare sounds, it may help to hear the sounds several times each, one after the other. Think about whether the sounds are soft or loud. Is each sound high, like the chirping of a bird? Or is it low, like a foghorn?

Materials

- empty paper-towel roll
- dried rice and beans
- stapler
- gravel
- masking tape

Make an Observation Chart

How Shaken	Sound Made by the Maraca
Shaken Gently	
Shaken Harder	

582

Follow This Procedure

1. Flatten one end of a paper-towel roll. Fold it over and staple it closed. Put tape over the staples to protect your hands.

2. Put a handful of dried rice and dried beans into the roll.

3. Flatten the open end of the roll. Don't let the rice and beans leak out. Fold over the end of the roll and staple it closed. Tape over the staples. You have made a maraca!

4. Shake the maraca gently a few times. If it leaks beans or rice, check the ends and tape them again. **Observe** the sound it makes.

5. Shake the maraca with more force. **Observe** the sound it makes now.

Draw Conclusions

1. How does a maraca make sound?

2. **Inquiry Skill** Scientists **compare** to learn how things are alike or different. How did the sounds **compare** in Step 4 and Step 5?

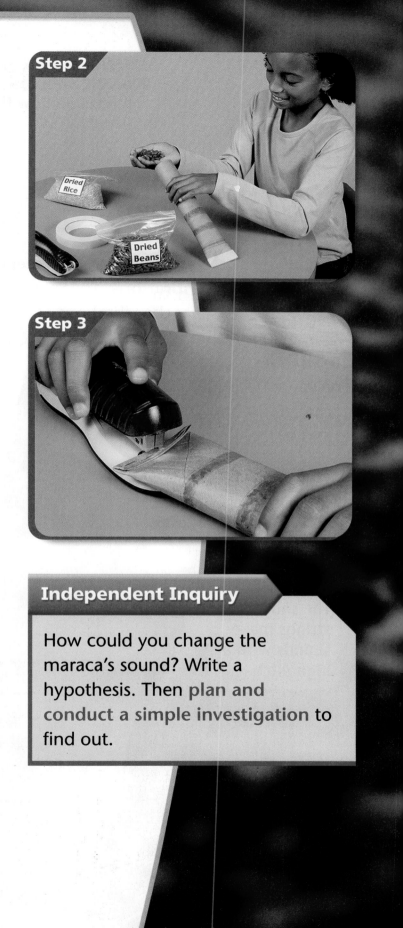

Independent Inquiry

How could you change the maraca's sound? Write a hypothesis. Then **plan and conduct a simple investigation** to find out.

Read and Learn

VOCABULARY
vibrations p. 584
loudness p. 584
pitch p. 585

SCIENCE CONCEPTS
▶ how sounds are made and how they travel

 **CAUSE AND EFFECT**
Look for different ways to cause sound.

Sound

When you shake a maraca, you can feel back-and-forth movements called **vibrations**. The maraca makes the air vibrate, too. You hear the vibrations as sound.

When you shake the maraca harder, the vibrations have more energy. The sound is louder. **Loudness** is a measure of how much energy sound has.

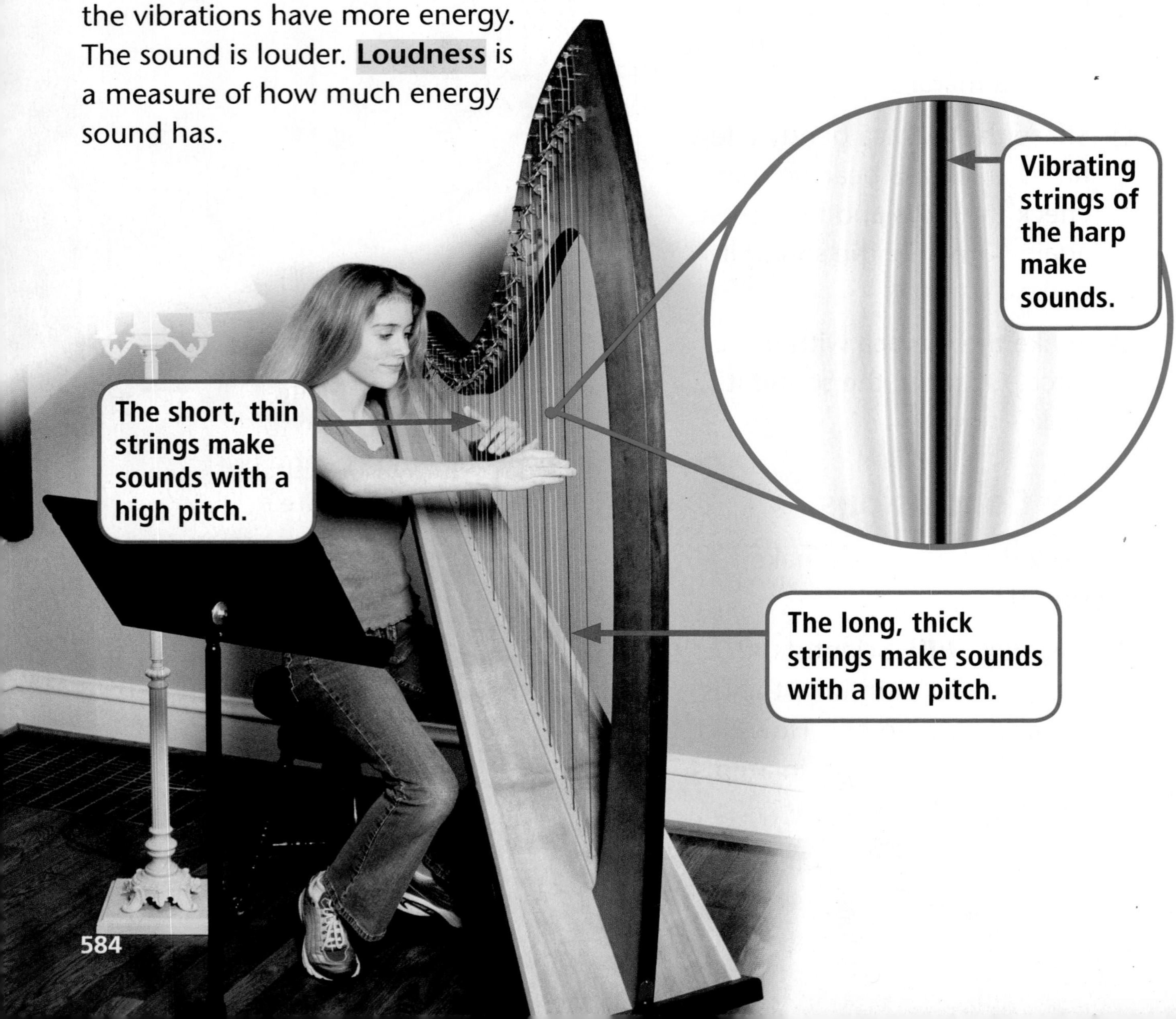

The short, thin strings make sounds with a high pitch.

Vibrating strings of the harp make sounds.

The long, thick strings make sounds with a low pitch.

584

Vibrations can also be made when a musical instrument is plucked, blown, or tapped. When harp strings are plucked, they can make high sounds or low sounds. **Pitch** is how high or low a sound is.

Air isn't the only kind of matter that sound can move through. Other kinds of matter carry sound better than air does. Sound moves farther and faster through liquids and solids than it does through air.

 **CAUSE AND EFFECT**

What causes sound?

When you're swimming underwater, the sound of the boat is much louder.

Hearing Sounds

You have read how sounds come from objects that vibrate. Sounds can also make objects vibrate. You may have heard windows rattle from a loud boom of thunder. The sound made the windows vibrate. It might have made you vibrate, too.

You hear sounds when vibrations move through the air to your ears. The vibrating air makes your eardrums vibrate, and you hear sound.

CAUSE AND EFFECT

What effect does vibrating air have when it reaches your ears?

Big Ears
Stand about 3 meters (10 ft) from a partner. Whisper to each other. Can you hear the soft sound? Now roll up a piece of paper to make a cone. Hold the cone close to but not directly in your ear. Listen again while your partner whispers. Can you hear better now? Explain any difference that you hear.

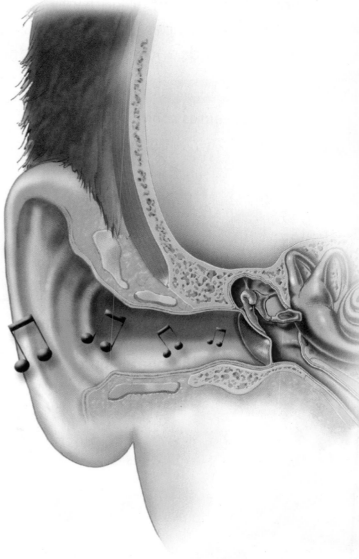

Lesson Review

Essential Question
What is sound?

In this lesson, you learned that sound waves are made by vibrations. Sound waves can travel through gases, solids, and liquids. When sound waves reach your ear, your eardrum vibrates, enabling you to hear the sound.

1. 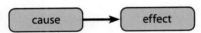 **CAUSE AND EFFECT** Draw and complete a graphic organizer that shows cause and effect between actions and the sound they make.

   cause → effect

2. **SUMMARIZE** Write a two-sentence summary of this lesson that tells what makes sound and how sound travels.

3. **DRAW CONCLUSIONS** If a 20-cm tightly held piece of string is plucked, would it have a higher or lower pitch than a 50-cm tightly held piece of string?

4. **VOCABULARY** For each vocabulary term, list three examples that help explain what the term means.

Test Prep

5. Which of these would **not** produce vibrations?
   - **A.** blowing into a trumpet
   - **B.** listening to an orchestra
   - **C.** plucking a guitar
   - **D.** tapping a piano key

Make Connections

 Writing

Expository
Write a **letter** to a friend, telling about a piece of music that you like. Use some of the vocabulary words you learned in this lesson.

 Music

Make Your Own Musical Instrument
Using an empty tissue box and rubber bands of different thicknesses, make a musical instrument. Pluck the "strings" of your instrument, and record your observations.

Science Spin
From Weekly Reader
TECHNOLOGY

A New Source of Energy?

DANGER
Generator X
Most generators work by burning fuels such as oil or coal. Burning those fuels gives off gases that pollute the air.

Switchgrass

Energy companies make electricity at places called power plants. The plants have huge machines called generators. These machines use thermal energy to generate electricity and send it through wires to homes.

A company in Georgia is testing a fuel that is a mixture of grass and coal. The grass, called switchgrass, is a kind of prairie grass that grows in the South.

To make the fuel, the company mixes switchgrass and coal. The mixture is then formed into cubes. As the cubes burn, they produce almost as much energy as the same amount of coal would. However, the cubes give off less pollution than other fuels, which also helps conserve resources.

Heating Homes

Some people use switchgrass instead of wood to keep their homes warm. A small amount of switchgrass can produce a lot of heat. One acre of this grass can be enough to heat an average home in Canada for a year. Switchgrass must be burned in special heaters, but many people are considering using it to heat their homes.

Think and Write

1. Would you use switchgrass to heat your home?
2. How does using switchgrass help the environment?

Find out more. Log on to
www.hspscience.com

Chapter 14 Review and Test Preparation

Vocabulary Review

Use the terms below to complete the sentences. The page numbers tell where to look in the chapter if you need help.

conductor p. 559
insulator p. 560
reflection p. 566
shadow p. 568
absorbed p. 574
opaque p. 575
translucent p. 575
vibrations p. 584
loudness p. 584
pitch p. 585

1. The bouncing of light off an object is _____.
2. Light is taken in, or _____, by some objects.
3. An object that doesn't conduct heat well is an _____.
4. An object that lets some light pass through it is _____.
5. Back-and-forth movements of matter are _____.
6. A dark area that forms when an object blocks light is a _____.
7. A sound can have a high or a low _____.
8. The amount of energy that a sound has is its _____.
9. An object that doesn't let light pass through it is _____.
10. A material that lets heat move through it easily is a _____.

Check Understanding

Write the letter of the best choice.

11. What does this picture show?
    A. conduction
    B. reflection
    C. refraction
    D. vibration

12. Tamika has two pictures of a tree, taken on different days. By comparing the pictures, what can she look at to tell if they were taken at the same time of day?
    F. the brightness of the reflections
    G. the size of the reflections
    H. the direction of the shadows
    J. the darkness of the shadows

590

13. In which of these materials does sound travel the slowest?
   A. air   C. steel
   B. glass   D. water

14. Which of these is transparent?

F.

G.

H.

J.

15. MAIN IDEA AND DETAILS Which of these is a good insulator?
   A. aluminum   C. copper
   B. cloth   D. steel

16. CAUSE AND EFFECT When light passes through drops of water in the air, what appears next in the sky?
   F. lightning
   G. a shadow
   H. a rainstorm
   J. a rainbow

Inquiry Skills

17. Jason was watching a play. He saw an actor's white shirt change color from white to red to green. What can you infer about the stage lights?

18. You stand in front of a mirror and wave your right hand. Predict what you will see in the mirror.

Critical Thinking

19. Maria is holding an ice cube. Is heat moving from her hand to the ice cube or from the ice cube to her hand? Explain.

The Big Idea

20. Most pots and pans are made of some type of metal. Explain why they are made of metal and not wood or plastic.

Visual Summary

Tell how each picture shows the **Big Idea** for its chapter.

Chapter 11 Big Idea
Matter has properties that can be observed, described, and measured.

Chapter 12 Big Idea
You use many forms of energy every day to grow and live.

Chapter 13 Big Idea
Electricity and magnetism are related and are part of things you use every day.

Chapter 14 Big Idea
Heat, light, and sound are different forms of energy that can move from place to place.

Exploring Forces and Motion

UNIT F
PHYSICAL SCIENCE

CHAPTER 15
Forces and Motion 594

CHAPTER 16
Work and Machines 628

Unit Inquiry

Make an Obstacle Course

An obstacle course is full of forces and motion. To get through, you have to move up and down, back and forth, and left and right. Try building an obstacle course for a ball. Can you control how quickly a ball moves through your course? How does the ball move in different ways? Plan and conduct an experiment to find out.

CHAPTER 15
Forces and Motion

What's the Big Idea?

Movement is caused by forces acting on an object.

Essential Questions

Lesson 1
What Is Motion?

Lesson 2
What Are Forces?

Lesson 3
How Do Waves Move?

Student eBook
www.hspscience.com

What do you wonder?

What different forces work together to make a ride like this one work? How do those forces relate to the **Big Idea?**

State fair swing ride

LESSON 1

Essential Question
What Is Motion?

Investigate to find out in which directions an object can move.

Read and Learn about motion, distance, and speed.

Fast Fact

Speedy Pinball
In the game of pinball, a ball can reach a speed of 145 kilometers (90 mi) per hour! The ball moves in many directions. In the Investigate, you will experiment with several kinds of motion.

Pinball machine

Vocabulary Preview

motion [MOH•shuhn] A change of position (p. 601)

distance [DIS•tuhns] How far one location is from another (p. 602)

speed [SPEED] The distance that an object moves in a certain period of time (p. 604)

Investigate

Make It Move

Guided Inquiry

Start with Questions

Have you ever seen hang gliders in the sky? These gliders ride on air currents.

- Do you think the person in the picture below is moving quickly or slowly?
- In which directions can a hang glider or another object travel?

Investigate to find out. Then read and learn to find out more.

Prepare to Investigate

Inquiry Skill Tip

When you interpret data, look for patterns. Are there any number patterns? Do any patterns repeat? Can you sort the data into similar groups?

Materials
- clay
- string (about 25 cm long)

Make an Observation Chart

Path of Motion	How the Object Was Pushed or Pulled
straight and fast	
straight and slow	
zigzag	
back and forth	
round and round	

Follow This Procedure

1. Mold a piece of clay into a ball. Mold another piece into a ring.
2. Make the ball move in a straight line. Make it move at different speeds. **Record** your observations.
3. Make your ball zigzag. **Record** your observations.
4. Thread the string through the hole in the ring. Tie the string to the ring. Hold the string by the end. Make the ring swing back and forth and then in a circle. **Record** your observations.
5. **Communicate** your observations by making drawings of each movement.

Draw Conclusions

1. Make a table like the one shown. **Record** how you made the objects move.
2. **Inquiry Skill** When you interpret data, you explain what the data means. What did you do to make the objects move in different directions?

Independent Inquiry

Plan an experiment with a different object, by itself and on the string. Try to move it each way you moved the ball and the ring.

599

VOCABULARY
motion p. 601
distance p. 602
speed p. 604

SCIENCE CONCEPTS
▶ what motion is
▶ how distance and time affect speed

COMPARE AND CONTRAST
Look for different ways to describe how objects move.

alike — different

Types of Motion

Suppose you are on the playground. What kinds of motion do you see? A girl throws a ball forward. It curves up and then falls down. A boy bounces a basketball. It goes up and down. Some children run fast and zigzag as they play tag. Others go back and forth on swings or up and down on seesaws. Still others go round and round on a merry-go-round.

Straight-line motion

You can observe many kinds of motion on a playground. The next time you are on a playground, think about all the ways you move.

Back-and-forth motion

600

Zigzag motion

Round-and-round motion

Fast motion and slow motion

An object can keep traveling in one direction, or it can change direction. Objects can move fast or slowly.

Every object has a position, or location. A school bus stops at the same position every day, so you know where to wait for it. You know the position of the cafeteria in your school. The cafeteria's position does not change. What is your position right now?

Motion is a change of position. To get to the cafeteria, you would need to change your position. You would need to *move*. When you ride a bike down the street, the bike changes position. It is in motion. If you park the bike in a bike rack, the bike is no longer changing position. It is no longer in motion.

Focus Skill COMPARE AND CONTRAST

What is one way in which each kind of motion is different from the others? How are all kinds of motion the same?

Distance, Direction, and Time

Suppose you are on a trip with your family. You ask, "How far do we have to drive to get to the next city?" What you want to know is the distance. **Distance** is how far it is from one location to another.

Distance is often measured in inches, yards, or miles. Scientists use units of centimeters, meters, and kilometers. You can use a ruler to measure distances.

An important thing to know about motion is its direction. Your family's car is going *east* from Chicago to New York. Your friend throws a ball *up* into the air.

Math in Science
Interpret Data

The graph shows the average speeds of different types of dogs. How much faster is the cocker spaniel than the basset hound?

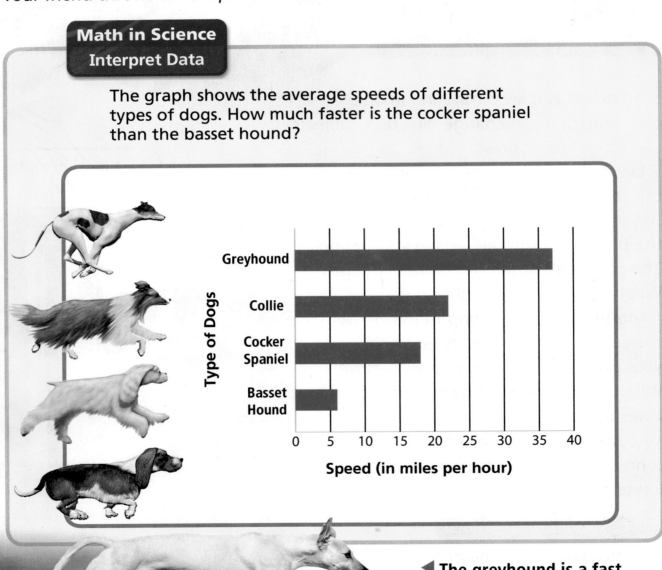

◀ The greyhound is a fast runner. It can cover a long distance in a short time.

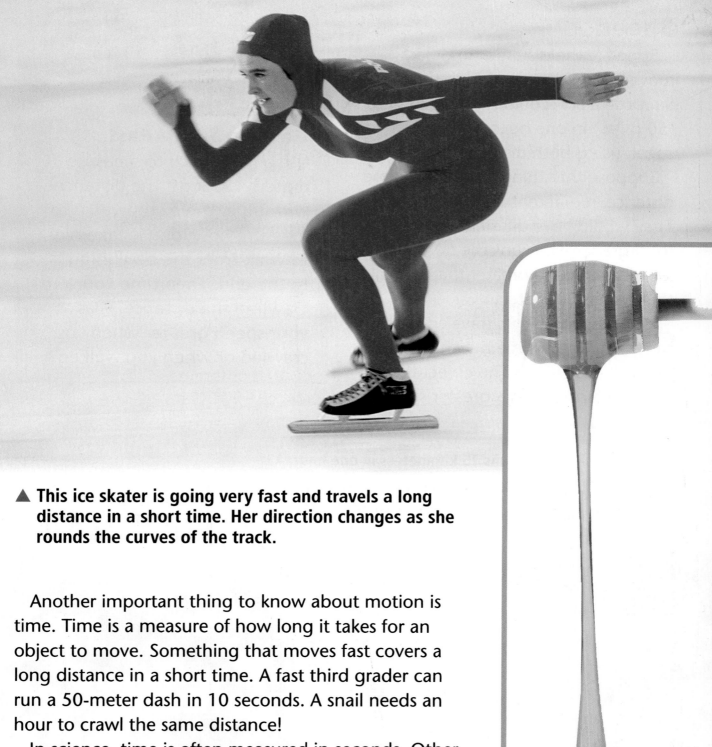

▲ This ice skater is going very fast and travels a long distance in a short time. Her direction changes as she rounds the curves of the track.

Another important thing to know about motion is time. Time is a measure of how long it takes for an object to move. Something that moves fast covers a long distance in a short time. A fast third grader can run a 50-meter dash in 10 seconds. A snail needs an hour to crawl the same distance!

In science, time is often measured in seconds. Other units of time include minutes, hours, days, and years.

COMPARE AND CONTRAST

You are walking home from school. Your parent is driving from your home to school. How is your motion different from your parent's?

▲ The motion of this honey is slow. Its direction is straight down.

603

Speed

Speed is the distance an object moves in a certain period of time. Suppose a lion could run 80 kilometers (50 miles) in one hour. You find its speed using both distance and time.

Suppose that things travel the same distance in different amounts of time. They would have different speeds. A third grader who runs the 50-meter dash has a speed greater than a snail that travels 50 meters.

What if two objects traveled for the same amount of time but they went different distances? The object that traveled farther had a greater speed.

Getting There Fast

Use a meterstick to measure 10 meters. Mark off the distance with tape. Use a stopwatch to measure how long it takes you to walk from the beginning to the end. Then time yourself crawling the distance. Was your speed greater when you crawled or when you walked?

COMPARE AND CONTRAST

Suppose a zebra runs 15 kilometers in one hour. A grizzly bear runs 10 kilometers in one hour. Which animal has the greater speed?

A single beat of a hummingbird's wing takes less than a second.

Plants turn toward the sun at such a low speed that you cannot see them move. This motion is measured in days.

Lesson Review

Essential Question
What is motion?

In this lesson, you learned that motion is a change in position. Motion can be described by direction, including straight, back-and-forth, and zigzag.

1.  **COMPARE AND CONTRAST** Draw and complete a graphic organizer that compares the different types of motion.

   alike — different

2. **SUMMARIZE** Write a four-sentence summary of the lesson. Tell how *motion, distance,* and *speed* are related.

3. **DRAW CONCLUSIONS** Would it be faster to zigzag or to walk in a straight line from one location to another? Why?

4. **VOCABULARY** Write a quiz question that uses all the vocabulary terms.

Test Prep

5. What changes when an object moves?
   A. its direction
   B. its height
   C. its position
   D. its shape

Make Connections

 Writing

Expository
Write a paragraph that **describes** the kinds of motion you see when you go outside at recess. Describe the speeds and distances of the people and things you observe.

 Math

Solve Problems
Use a meterstick to measure the distance from one side of your classroom to the other. If you cross the classroom four times, how far will you walk?

LESSON 2

Essential Question
What Are Forces?

Investigate to find out which items travel fastest down a ramp.

Read and Learn about forces and how they affect motion.

Fast Fact

Super Sledding
Olympic bobsledders zoom down the track at speeds of up to 145 kilometers (90 mi) per hour! These children aren't going quite as fast, but they're having lots of fun! In the Investigate, you will see how the height of a starting point affects speed.

606

Vocabulary Preview

force [FAWRS] A push or a pull (p. 610)

gravity [GRAV•ih•tee] A force that pulls two objects toward each other (p. 614)

weight [WAYT] The measure of the force of gravity on an object (p. 614)

Investigate

Speed Ramp

Guided Inquiry

Start with Questions

Riding a bike downhill is easier than riding one uphill. Going downhill is much faster.

- Is it more fun to ride a bike downhill or uphill?
- Why is going downhill easier?

Investigate to find out. Then read and learn to find out more.

Prepare to Investigate

Inquiry Skill Tip

Check that an inference fits the data and makes sense. *Ask yourself:* Does my inference agree with the measurements or data? Does it agree with what I know about how the world works?

Materials

- books
- cookie sheet
- block
- metric ruler
- penny
- rubber eraser

Make a Data Table

Object	Height of Books (cm)	Speed
Penny		
Block		
Eraser		

Follow This Procedure

1. Work with a partner. Make a table like the one shown.

2. Stack books about 5 cm high. **Record** how high the pile is.

3. Lay one end of the cookie sheet on the books to make a ramp.

4. Place a penny, a block, and an eraser at the top of the cookie sheet. Let the objects go at the same time.

5. **Record** how fast each item traveled. Use words such as *fastest, slowest,* and *did not move.*

6. Add books to make the stack about 10 cm high. Repeat Steps 2–5.

7. Add books to make the stack about 15 cm high. Repeat Steps 2–5.

Draw Conclusions

1. **Compare** the speeds of the sliding objects. **Record** your **observations**.

2. **Inquiry Skill** When you **infer**, you make a guess based on what you observe. **Infer** why the speeds of the objects changed.

Independent Inquiry

Predict how your results would change if you coated the cookie sheet with oil. Try it. **Record** your **observations**.

Read and Learn

VOCABULARY
force p. 610
gravity p. 614
weight p. 614

SCIENCE CONCEPTS
▶ what the kinds of forces are
▶ what forces do

MAIN IDEA AND DETAILS
Look for details that describe forces.

```
      Main Idea
    /     |     \
detail  detail  detail
```

Types of Forces

To move a ball, you can throw it, kick it, or hit it with a bat. Any kind of push or pull is a **force**. You must apply a force to make an object move. An object will keep moving until another force stops it. When you catch a moving ball, the force from your hand stops the ball.

Friction is one force that stops things or slows them down. When two objects rub together, there is friction between them. Most rough surfaces make more friction than smooth surfaces. You can slide farther on ice than you can on dirt or grass.

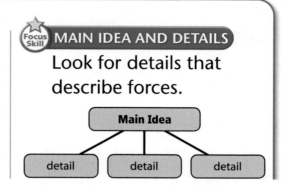

The rocket is pushed upward by the force of its engines.

▲ When you open a drawer, you are using a pulling force.

Science Up Close

For more links and animations, go to www.hspscience.com

Forces at Work

When the basilisk lizard runs fast across the water, it looks as if it is running above the water. If the forward force were stopped, the lizard would start to sink into the water.

Forces also make objects change directions. You apply a force to the handlebar to turn your bike. When you push one end of a seesaw down, the other end comes up.

Some forces do not even have to touch the object. A magnet can be used to push or pull objects made of iron or steel. This is called magnetic force. A magnet can pull a nail.

MAIN IDEA AND DETAILS

A magnet picks up paper clips. What force is at work?

Insta-Lab

Amazing Paper Clips

Put some water in a cup. Carefully lay a paper clip on the water's surface. Observe. Research surface tension. Then explain why the paper clip doesn't sink.

Ways That Forces Change Motion

Three things affect the motion of an object.

- the strength of the force
- the direction of the force
- the mass of the object

The stronger the force is, the greater the change in motion. If you toss a ball gently, it doesn't move fast or far. If you throw a ball hard, the ball moves faster and farther. You use more force on the ball when you throw it as hard as you can.

If an object is not moving and a force pushes or pulls it, the object will move in the direction of the push or pull. When you pull up on an object, it comes up. When you push an object to the right, it moves to the right.

The smaller the mass of an object is, the easier it is to move. It takes less force to pick up a pencil than to pick up a book, since a pencil has less mass.

The jogger pushes the stroller in the direction in which he wants it to go.

net force

▲ Each of these people is using force to push the boat. When forces act in the same direction, they add up to make a greater force. The sum of all the forces is called the net force.

You can add all the forces that push or pull on an object. The sum is called the net force. Suppose you and a friend push the same way on a door. The net force on the door equals your push plus your friend's push.

MAIN IDEA AND DETAILS

Name three things that affect the motion of an object, and tell what each one does.

When equal forces act in opposite directions, they cancel each other out. The net force is zero.

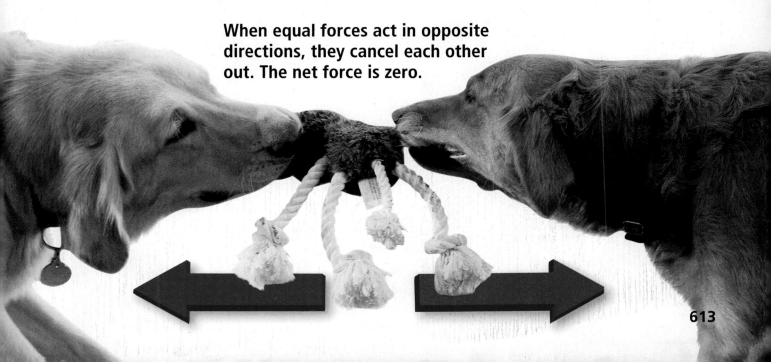

Gravity

When you throw a ball into the air, you know that it will come back down. **Gravity** is a force that pulls two objects toward each other. The ball comes back to Earth because Earth's gravity pulls on it.

Gravity is a very weak force between most objects. You don't feel the force of gravity between you and your desk. But Earth's gravity pulls very strongly. This is because Earth is so large.

The force of gravity depends on how much mass an object has. **Weight** is a measure of the force of gravity on an object. Objects with a large mass weigh more.

MAIN IDEA AND DETAILS

Does gravity pull a horse or a puppy harder? How do you know?

Why does the water fall back down to Earth after shooting up from the fountain?

The force of gravity pulls the roller coaster back to Earth for a thrilling ride!

Lesson Review

Essential Question

What are forces?

In this lesson, you learned that forces are any kind of push or pull. Force is required to make an object move.

1. **MAIN IDEA AND DETAILS** Draw and complete a graphic organizer for the main idea: Things that affect the motion of an object.

   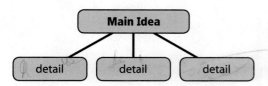

2. **SUMMARIZE** Write a summary of this lesson. Begin with the sentence *An object's motion changes because a force pushes or pulls it.*

3. **DRAW CONCLUSIONS** Why is riding a bike downhill easier than riding it uphill?

4. **VOCABULARY** Use the terms *force, gravity,* and *weight* in sentences.

Test Prep

5. **Critical Thinking** Suppose you kick a ball sideways. In which direction will it move? Explain.

Make Connections

 Writing

Narrative

The moon is smaller than Earth, so the moon's gravity is weaker than Earth's. Write a **short story** about a day on the moon. How would the weaker gravity affect your motion?

 Physical Education

Experiment with Forces

Gather several different sports balls. Throw them with different amounts of force. Push them in different directions. Then write a paragraph that describes your observations.

615

LESSON 3

Investigate to find out what kinds of waves there are.

Read and Learn about the different types of waves and how to measure them.

Essential Question

How Do Waves Move?

Fast Fact

Waves!
Water waves can be tiny, like these ripples in a pond. Waves in the ocean are much larger. The largest wave measured was 520 meters (1,700 ft) high. In the Investigate, you will see how two different kinds of waves move.

Ripples in water

Vocabulary Preview

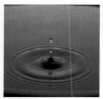

wave [WAYV] A disturbance that travels through matter or space (p. 620)

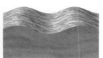

crest [KREST] The highest point of a wave (p. 622)

trough [TRAWF] The lowest point of a wave (p. 622)

wavelength [WAYV•length] The distance from one point of one wave to the same point on the next wave (p. 622)

617

Investigate

Two Kinds of Waves

Guided Inquiry

Start with Questions

At the beach, you're likely to see waves like this one. Sometimes the waves are larger and sometimes they're not very big at all.

- Which do you think have more energy, larger or smaller waves?
- Do you think there are other types of waves than those in this picture?

Investigate to find out. Then read and learn to find out more.

Prepare to Investigate

Inquiry Skill Tip

When you compare two objects, such as diagrams, pictures, or other representations, lay them side-by-side. When you compare two events, look at tables, charts, or other summaries that show the data or observations of the events.

Materials
- rope about 2 meters long
- spring toy

Make an Observation Chart

Diagram of Rope Movement	Diagram of Spring Movement

Follow This Procedure

1. Hold one end of the rope. Your partner will hold the other end. Let the rope hang loosely between you.

2. Move one end of the rope gently up and down as your partner holds the other end still. Then move the rope faster. **Observe** what happens.

3. Put the coiled spring toy on a table or on the floor. Have your partner hold one end still. Push the other end of the toy about 10 cm toward your partner. Then pull and push that end backward and forward. **Observe** what happens.

4. **Record** your **observations** by making diagrams for Steps 2 and 3.

Draw Conclusions

1. How did the force you used on the rope affect it? What happened when you moved it faster?

2. **Inquiry Skill** When you **compare** things, you look at how they are alike. **Compare** the movements of the waves in the rope with the waves in the toy.

Independent Inquiry

What do you **predict** will happen when you move the coiled spring toy in the same way you moved the rope? Try it and see.

619

Read and Learn

VOCABULARY
wave p. 620
crest p. 622
trough p. 622
wavelength p. 622

SCIENCE CONCEPTS
▶ what the types of waves are
▶ how to measure waves

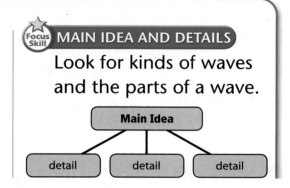

MAIN IDEA AND DETAILS
Look for kinds of waves and the parts of a wave.

Types of Waves

Even if you live far from the ocean, waves are all around you. There are many kinds of waves. Light travels in waves, and so does sound. Microwaves cook your food. A doctor uses X-ray waves to take pictures of the inside of your body.

A **wave** is a disturbance that travels through matter or space. Waves disturb matter by causing it to move. This is because waves carry energy. Waves can travel through solids, liquids, and gases. Some waves can travel through empty space.

Waves disturb the water particles. ▼

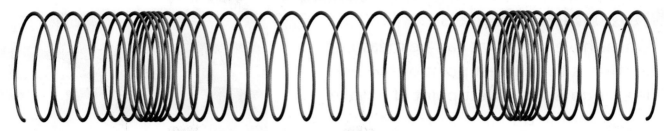

▲ A spring toy that is moved in and out makes a back-and-forth wave.

Sound waves from this xylophone (ZY•luh•fohn) bring sound energy to your ears. Sound waves are back-and-forth waves. ▶

Waves are made when something vibrates. Things that vibrate move back and forth. When you speak, your vocal cords vibrate quickly to make sound waves.

There are two types of waves. Some waves move up and down. Examples are radio waves and light waves. Some waves, like sound waves, move back and forth.

MAIN IDEA AND DETAILS

What is a wave?

Move It!
Sprinkle some tiny bread crumbs into a foil pie plate. Hold a cookie sheet next to the pie plate, and bang a large wooden spoon against the cookie sheet. Observe the crumbs. What makes them move?

Measuring Waves

Waves have parts that can be measured. The **crest** is the highest point of a wave. The **trough** is the lowest point of a wave. The greater the distance between these two points, the larger the wave is and the more energy it carries.

Forces affect the amount of energy that waves carry. A rock thrown into a pond will make bigger waves than a pebble will. Suppose you beat hard on a drum. The sound is louder than if you beat softly.

Scientists also measure another characteristic of waves. **Wavelength** is the distance from one point of one wave to the same point of the next wave.

MAIN IDEA AND DETAILS

What measurement tells how much energy a wave carries?

You can see crests on the waves in this pool.

Wavelength can be measured as the distance from one crest to the next crest or from one trough to the next trough. ▼

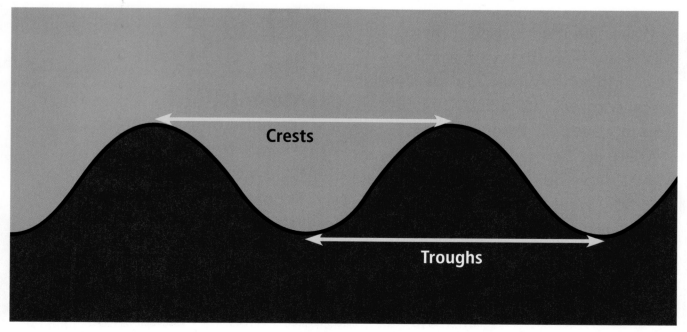

Lesson Review

Essential Question

How do waves move?

In this lesson, you learned that a wave is a disturbance that travels through matter or space. Light, sound, and water waves are all types of waves.

1. **MAIN IDEA AND DETAILS** Draw and complete a graphic organizer for this main idea: Types of waves

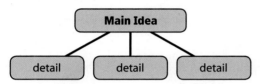

2. **SUMMARIZE** Write three sentences to tell how waves are made and how they are measured.

3. **DRAW CONCLUSIONS** Why can ocean waves wash away rocks on a beach?

4. **VOCABULARY** Make a quiz question for each vocabulary term in this lesson. Provide the answers.

Test Prep

5. How can the length of a wave be measured?
   - A. from trough to crest
   - B. from crest to wavelength
   - C. from crest to trough
   - D. from crest to crest

Make Connections

 Writing

Narrative

Suppose you are a wave. Write a **description** of yourself. Tell what type of wave you are, how you move, and how your energy is measured. Describe how the amount of energy you carry can change.

 Math

Measure Wavelength

Look at the two pictures of waves. Measure their wavelengths. How much longer is Wave A than Wave B?

623

People in Science

Percy Spencer

Have you ever made popcorn in a microwave oven? That's how microwave ovens were invented. In 1945, Percy Spencer was touring a laboratory at Raytheon Company, where he worked. He was standing close to a device that powered a radar set. Suddenly, he realized that the chocolate bar in his pocket was melting. He used some unpopped popcorn to test his theory that the machine could cook food.

▶ **PERCY SPENCER**
▶ Inventor of the microwave

Percy Spencer was awarded 120 patents during his lifetime. His patent for the microwave oven was given in 1952. Today, microwaves are used to do many things besides heat food. Some researchers are working to use microwaves to sterilize food. They are also used to bake tiles for space shuttles. One of the most exciting possibilities is that microwaves might be used to treat some kinds of cancer. Cancer tissue begins to die at 109°F. Researchers hope they can use microwaves to kill the cancer and not harm the healthy tissue.

 Think and Write

1. How do you know that Percy Spencer was creative?
2. Have you ever made an unexpected discovery? What was it?

Christine Darden

▶ CHRISTINE DARDEN
▶ Engineer

One of Christine Darden's childhood loves was fixing things. If her bicycle broke, she tried to fix it herself. Another of her loves was math. She wanted to make her living with math, so she became a math teacher.

After teaching for several years, Darden went back to school. She became a mathematician at the National Aeronautics and Space Administration (NASA).

Darden then decided to become an engineer, so she returned to school and earned a degree in engineering.

In 1992, Darden won the Women in Science and Engineering (WISE) Award for her achievements in science.

 Think and Write

1. Which is more important—finding a job easily or doing work you love to do?
2. How did Christine Darden show her determination?

Career Civil Engineer

Civil engineers design buildings, bridges, roads, airports, and tunnels and make sure these things are safe to use.

Civil engineers have at least a bachelor's degree in engineering. They study physics, chemistry, and math so that they can test and design structures.

Chapter 15 Review and Test Preparation

Vocabulary Review

Use the terms below to complete the sentences. The page numbers tell you where to look in the chapter if you need help.

motion p. 601
distance p. 602
speed p. 604
force p. 610
gravity p. 614
weight p. 614
wave p. 620
crest p. 622
trough p. 622
wavelength p. 622

1. The highest point of a wave is the _____.
2. A change in position is _____.
3. The measure of the force of gravity on an object is the object's _____.
4. A push or a pull is a _____.
5. The distance between one crest and the next crest is the _____.
6. The distance an object moves in a certain period of time is its _____.
7. A disturbance that travels through matter or space is a _____.
8. The lowest point of a wave is the _____.
9. The force that pulls two objects toward each other is _____.
10. How far an object moves is _____.

Check Understanding

Write the letter of the best choice.

11. A boy pushes a box across the floor. The box moves to the right. In which direction is the boy probably pushing?
    A. toward the left
    B. toward the right
    C. downward
    D. upward

12. What do waves carry with them from place to place?
    F. energy   H. speed
    G. motion   J. wavelength

626

13. COMPARE AND CONTRAST Two horses pull a wagon in the same direction with the same force. How does the net force compare with the force of each horse?

   A. The net force is twice the force of each horse alone.
   B. The net force is half the force of each horse alone.
   C. The net force is equal to the force of each horse alone.
   D. The net force is zero.

14. MAIN IDEA AND DETAILS To figure speed, what do you need to know besides time?

   F. distance   H. motion
   G. force   J. wavelength

15. In the picture, the two girls start pushing the trunk with equal force. How does it move?

   A. It moves toward the right.
   B. It moves toward the left.
   C. It doesn't move.
   D. It moves slowly.

16. Which force holds your book on your desk?

   F. electricity
   G. gravity
   H. magnetism
   J. surface tension

Inquiry Skills

17. Interpret the data shown in the pictures. Which wave has the shortest wavelength? How do you know?

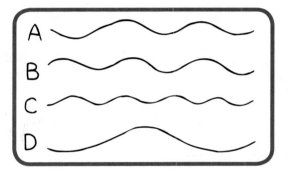

18. Compare two types of waves.

Critical Thinking

19. Look around the room. Find three objects that can move, and describe their motions. Tell what forces make these objects move.

The Big Idea

20. Weigh some objects in your classroom. Record the weights. Which object does gravity pull on the hardest? How do you know?

CHAPTER 16
Work and Machines

What's the Big Idea? Simple machines allow people to change the direction or size of a force.

Essential Questions

Lesson 1
What Is Work?

Lesson 2
What Are Some Simple Machines?

Lesson 3
What Are Some Other Simple Machines?

GO online
Student eBook
www.hspscience.com

Skiing in Alaska

What do you wonder?

How do the simple machines used in skiing help the skiers change their direction? How do these simple machines relate to the **Big Idea?**

LESSON 1

Essential Question
What Is Work?

Investigate to find out what work is.

Read and Learn about the different types of work and how to measure it.

Fast Fact

Hard Work!
This girl weighs less than the dog. However, she does work and uses simple machines to pull him through the park. In the Investigate, you will also use work to move an object.

Vocabulary Preview

work [WERK] The use of a force to move an object
(p. 636)

Investigate

Work with Me

Guided Inquiry

Start with Questions

In places where there is a lot of snow, people have to clear paths to walk. Some people use shovels to move the snow. Others use snowblowers like the one in the picture.

- Would you rather use a shovel or a snowblower to move snow?
- How could you measure how far the snow moved?

Investigate to find out. Then read and learn to find out more.

Prepare to Investigate

Inquiry Skill Tip

You can use many tools to measure the size of objects. Any tool that you use should have units that are equal in size.

Materials

- safety goggles
- graph paper
- checker
- drinking straw

Make an Observation Chart

Action	Observation
First time blowing through straw	
Second time blowing through straw	

Follow This Procedure

1. **CAUTION: Put on the safety goggles.** Work with a partner. On a sheet of graph paper, make a start line. Place the checker on the graph paper behind the line.

2. Put one end of the straw in your mouth, and touch the other end to one edge of the checker. Blow hard through the straw.

3. Place the checker back at the same point on the paper. Have your partner press down on the checker while you repeat Step 2. **Record** your observations.

Draw Conclusions

1. Was the force of blowing on the checker the same or different each time? Explain.

2. Was the result the same or different each time? Explain.

3. **Inquiry Skill** Scientists often **measure** things during an experiment. How could you use the graph paper to measure how far the checker moved?

Independent Inquiry

Predict how using a stack of two checkers might affect your results each time. Try it!

633

Read and Learn

VOCABULARY
work p. 636

SCIENCE CONCEPTS
▶ what scientists mean by work

MAIN IDEA AND DETAILS
Look for details about work.

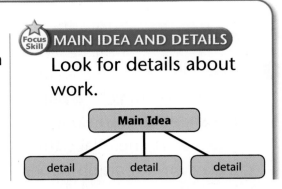

Different Types of Work

People use the word *work* all the time. Scientists do, too, but you might be surprised to find out what work means to a scientist.

Suppose your teacher asks you to solve a math problem in your head. You sit still and think hard. Then you get the answer. Your teacher says, "Good work!" However, a scientist would say that you did not do any work.

Suppose you want to open a jar. You twist the lid hard, but the lid doesn't move. You feel that you have done a lot of work. Again, however, a scientist would say you had done no work.

Is This Work?

◀ Solving a math problem isn't what scientists call work.

If the lid doesn't move, no work is being done. ▶

Now suppose you are playing soccer. A player passes the ball to you. As it rolls toward you, you pull your leg back. Then you kick hard. The ball flies up and forward. You make a goal!

As the crowd cheers, you say to yourself, "This is fun!" You do not think that you have done any work at all. Yet to a scientist, you have done work.

Which example shows what scientists call work?

Kicking a ball may seem like play, but scientists would say it was work.

One Type of Work

To a scientist, **work** is done only when a force is used to move an object. If you don't use a force, you don't do work. If nothing moves, you haven't done work.

Remember the math problem? You thought hard to solve it, but you didn't move anything. So to a scientist, you didn't do work.

Now think about opening the jar. You used a lot of force on the lid, but it didn't move. So to a scientist, you didn't do work on the jar.

What about the soccer game? Your muscles used force to lift your leg. Your foot used force to kick the ball. To a scientist, that was work.

 MAIN IDEA AND DETAILS
What must happen for work to be done?

How can you tell that work was done on the ball?

Measuring Work

Suppose you want to move a box full of toys across a room. To move such a heavy box, you will need to use a lot of force.

If you are not able to move the box alone, you might get a friend to help you. Then the force might be great enough to move it. You might also take out half of the toys. Then you might be able to move the box by yourself.

How could you measure how much work you had done? You would need to measure how much force you used and how far the object moved.

Insta-Lab

Move It!
Put a book on a desktop. Push gently on the side of the book. Slowly increase your force. Do the same thing with a stack of two books. Compare the force used to move two books with the force to move one book.

MAIN IDEA AND DETAILS
What two measurements tell you how much work is done?

Which picture shows the girl doing work?

full, not moving

half full, moving

Lesson Review

Essential Question
What is work?

In this lesson, you learned that scientists measure work by how much force is used to move an object and how far it is moved in the direction of the force.

1. **MAIN IDEA AND DETAILS**
   Draw and complete a graphic organizer for the main idea: Work is done when you do two things.

   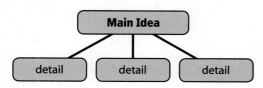

2. **SUMMARIZE** Use the word *work* to write a three-sentence summary about this lesson.

3. **DRAW CONCLUSIONS** Heather wants to carry her laundry basket to the washer. It is too heavy for her to pick up. What can she do to get the laundry to the washer?

4. **VOCABULARY** Write an example of work being done.

Test Prep

5. Which of these is **not** an example of work?
   - A. sitting and reading
   - B. playing fetch with a dog
   - C. pulling a chair across a room
   - D. pushing a box of books across the floor

Make Connections

 Writing

Expository
Think about community workers. List two or three workers who do things that scientists would call work. Write a **description** of how the workers' jobs include work.

 Art

Working Art
Make a drawing of someone doing work. Then, on the back of your paper, list the ways you did work while you were making the drawing.

639

LESSON 2

Essential Question

What Are Some Simple Machines?

Investigate to find out how levers can be used.

Read and Learn about different types of simple machines and how they are used.

Fast Fact

Important Levers

We use levers to do many things. Did you know that the nutcracker you use to crack a nut is a lever? In the investigate you will learn more about the importance of levers.

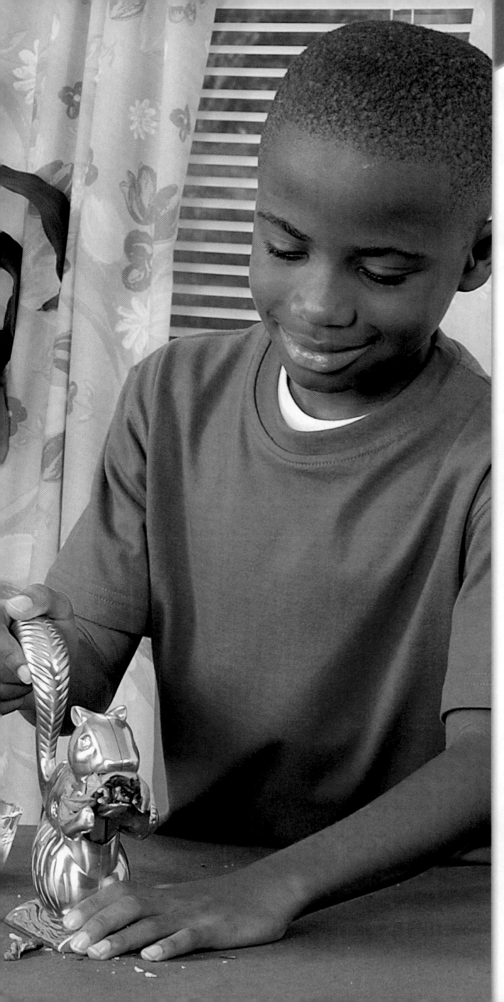

Vocabulary Preview

simple machine [SIM•puhl muh•SHEEN] A tool with few or no moving parts that helps people do work (p. 645)

lever [LEV•er] A simple machine made up of a bar that pivots, or turns, on a fixed point (p. 646)

fulcrum [FUL•kruhm] The fixed point on a lever (p. 646)

wheel-and-axle [weel•and•AK•suhl] A simple machine made up of an axle and a wheel that are connected and turn together (p. 648)

pulley [PUHL•ee] A simple machine made up of a wheel with a rope around it (p. 650)

641

Investigate

Help from Simple Machines

Guided Inquiry

Start with Questions

In the picture below, a crowbar is prying apart two things. A crowbar is a lever.

- How is the lever in this picture helping the person who uses it?
- What are other ways levers can be used to separate things?

Investigate to find out. Then read and learn to find out more.

Prepare to Investigate

Inquiry Skill Tip

To predict what might happen in an investigation, gather data from other similar investigations. Look at the data for clues about what might happen in your investigation.

Materials

- measuring spoons
- white rice and brown rice, uncooked
- jar lid
- forceps
- two paper plates

Make an Observation Chart

Method of Separation	Observation
Fingers	
Forceps	

642

Follow This Procedure

1. Measure out one tablespoon of white rice. Place it into the jar lid. Do the same with the brown rice.
2. Mix the two types of rice in the lid.
3. Use your fingers to separate the types of rice. **Record** your observations.
4. Put the rice you separated back into the lid. Mix the rice again.
5. This time, use forceps to separate the types of rice. **Record** your observations.

Draw Conclusions

1. Which way of separating the rice grains was easier? Why?
2. Which would be a safer way of handling food, using your fingers or using forceps?
3. **Inquiry Skill** Do you think using a spoon to separate the rice would be faster than using forceps? **Predict** which one you think might be faster. Then repeat the Investigate using a spoon and forceps to find out.

Step 3

Step 5

Independent Inquiry

You have measured how long it took to separate the rice. **Compare** times with four other classmates by making a bar graph. Were there differences? Why do you think so?

Read and Learn

VOCABULARY
simple machine p. 645
lever p. 646
fulcrum p. 646
wheel-and-axle p. 648
pulley p. 650

SCIENCE CONCEPTS
- what simple machines are
- how levers, wheel-and-axles, and pulleys make work easier

MAIN IDEA AND DETAILS
Look for details.

Simple Machines

Imagine that the lawn is covered with leaves. Your job is to clear them away. You grab a handful of leaves and put them in a trash bag. Then you pick up another handful of leaves, and another, and another. This is going to take a very long time!

The job might go faster if you used a machine. A machine is anything that changes the way work is done. For example, a leaf blower is a machine. This machine would make clearing up the leaves easy and fast.

Other examples of machines include cars, dishwashers, and bicycles. These machines have many parts. All the parts together make the machines work.

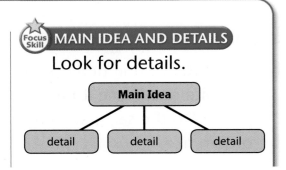

This leaf blower has an engine. The engine supplies the force to do the work.

Not all machines have a lot of parts. If you don't have a leaf blower, you could use a rake to help you clear up the leaves. A rake is a machine, even though it has no engine. A rake is an example of a simple machine.

A **simple machine** has few or no moving parts. The rake has no moving parts. To use a simple machine, you apply only one force. To use a rake, you pull on it with one hand.

MAIN IDEA AND DETAILS
What is a simple machine?

◀ With the rake, the boy provides the force that gets the work done.

Math in Science
Interpret Data

Gavin raked all the leaves in the yard. Then he scattered them across the yard and used a leaf blower. How much time could he have saved by using the leaf blower the first time?

Clearing Leaves	
Leaf Blower	🍃
Rake	🍃 🍃 🍃

Key: Each 🍃 = 1 hour

The Lever

A rake is a lever. A **lever** is a bar that pivots, or turns, on a fixed point. A fixed point is a point that doesn't move. The fixed point on a lever is called the **fulcrum** (FUHL•kruhm).

Think about how you hold a rake when you use it. One hand holds the end of the handle. That hand stays still. It is the fulcrum. The other hand pulls the middle of the handle. The end of the rake gathers the leaves.

Your hand moves the middle of the rake's handle a certain distance. The end of the rake moves a greater distance, so it gathers more leaves. That's what makes the work easier.

Do all levers work this way? No, but a broom and a fishing pole are levers that work this way. So is your arm. Your elbow joint is the fulcrum.

A broom is a lever that helps you clean up an area more easily. Where is the fulcrum? ▼

fulcrum

I Wonder

Where is the best place on a broom to put your hands? Try placing both your hands at the top of the stick and then at the bottom of the stick. What would happen if you separated your hands? Which way would allow you to sweep up more leaves?

A shovel is a lever that changes the direction and strength of a force. You push down on the handle with a certain force. A greater force moves the blade up, taking the rock with it.

fulcrum

Levers work in different ways, depending on where the fulcrum is and where you apply the force. In the Investigate, you used forceps. The fulcrum of the forceps was the hinge.

Suppose you want to move a large rock. You push the blade of a shovel under the rock. The ground is the fulcrum. You push down on the handle, and the blade comes up, bringing the rock with it. A crowbar works this way, too.

A nutcracker is also a lever. Its fulcrum is where its arms connect. When you push the other ends of the arms together, the force cracks the nut that is between them.

MAIN IDEA AND DETAILS

Why are a rake, a shovel, and a nutcracker all classified as levers?

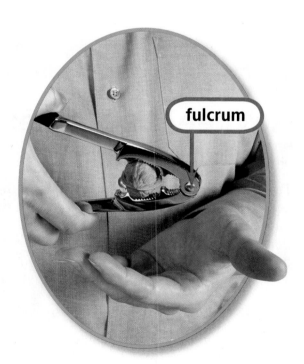

fulcrum

▲ A nutcracker doesn't change the direction of the force you apply. It increases the strength of that force.

647

The Wheel-and-Axle

Another type of simple machine is the wheel-and-axle. The **wheel-and-axle** is made up of a wheel and an axle that are connected so that they turn together.

You know that a bicycle has wheels, and you might know that it has axles. However, since the wheels and the axles of a bicycle aren't connected and don't turn together, they are not what a scientist would call a simple machine.

The doorknob shown on this page is an example of a wheel-and-axle. The wheel is the knob, and the rod that is connected to the knob is the axle. The knob and the rod form a simple machine, since they are connected and they turn together.

Science Up Close

For more links and animations, go to www.hspscience.com

A Wheel-and-Axle: The Doorknob

The knob part of a doorknob is a wheel. The rod connected to it is an axle. When you turn the knob, the axle turns, too. As it does, it pulls back the catch, and the door opens.

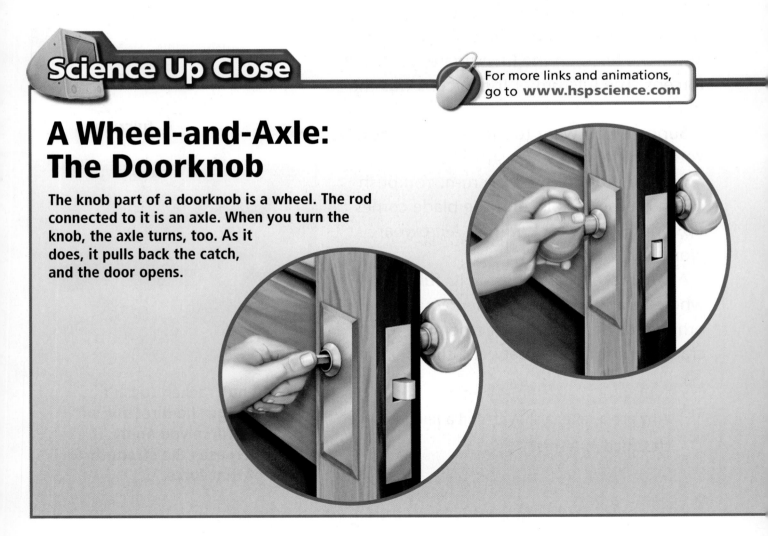

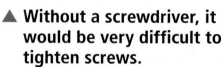

▲ Without a screwdriver, it would be very difficult to tighten screws.

The handle of the screwdriver is the wheel that turns the axle—the shank. ▼

A screwdriver is another example of a wheel-and-axle. The metal shank is the axle. The handle is the wheel. The shank would be very hard to turn by itself. You would have to apply a lot of force. The handle allows you to turn the shank with less force, making the work easier.

MAIN IDEA AND DETAILS

How does a wheel-and-axle make work easier?

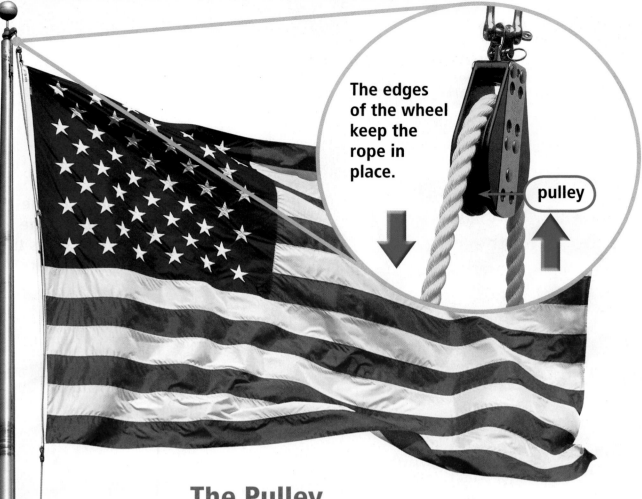

The edges of the wheel keep the rope in place.

pulley

◀ You pull down on one end of the rope, and the flag moves up.

The Pulley

Some old-fashioned wells have another type of simple machine. A bucket is attached to a rope. The rope is part of a simple machine called a pulley.

A **pulley** is a wheel with a rope around it. You pull one end of the rope one way, and the other end moves in the opposite direction.

The bucket in the well is attached to one end of the rope. It would be hard to raise the bucket by pulling up on the rope. With the pulley, you can pull down instead, which is easier, and the bucket comes up.

Some window curtains and blinds have pulleys. So do flagpoles. You wouldn't want to climb to the top of the pole in order to raise the flag. With a pulley, you don't have to.

How does a pulley work?

Lesson Review

Essential Question

What are some simple machines?

In this lesson, you learned that we use some simple machines every day to help us with work. A lever, wheel-and-axle, and pulley are all simple machines.

1. **MAIN IDEA AND DETAILS** Draw and complete a graphic organizer for the main idea: There are different types of simple machines.

   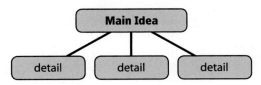

2. **SUMMARIZE** Write a sentence that tells the most important information in this lesson.

3. **DRAW CONCLUSIONS** What simple machine could you use to move a basket of fruit from the ground into a treehouse?

4. **VOCABULARY** Write one example of each kind of simple machine you learned about in this lesson.

Test Prep

5. Which simple machine would you use to open a paint can?
   - **A.** lever
   - **B.** lever and pulley
   - **C.** pulley
   - **D.** wheel-and-axle

Make Connections

 Writing

Expository
Write to a first grader explaining **how to** raise a flag by using a pulley. Draw pictures to go with your writing.

 Health

Body Levers
Find a diagram of a human arm, copy it, and label it to show how the arm works as a lever. What other body parts work that way?

Lesson 3

Investigate to find out how inclined planes are helpful.

Read and Learn about other types of simple machines.

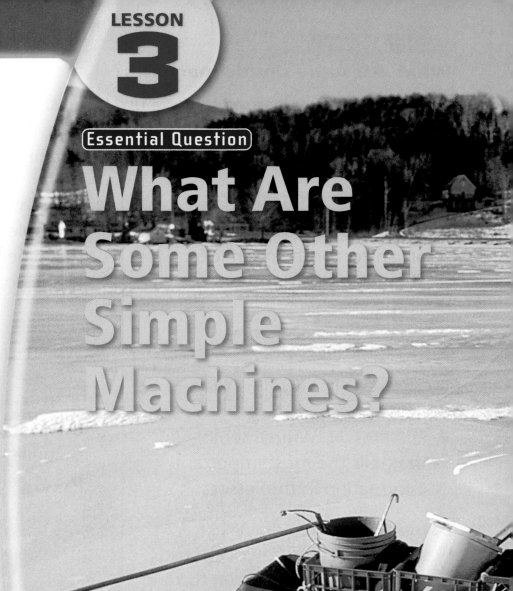

Essential Question

What Are Some Other Simple Machines?

Drilling the ice

Fast Fact

Watch Out!
An ice fisher must be careful when walking out onto the ice to drill using an auger. To hold the weight of one adult, the ice must be at least 5 centimeters (2 in.) thick. In the Investigate you will be measuring force using a spring scale.

Vocabulary Preview

inclined plane [in•KLYND PLAYN] A simple machine that makes moving or lifting things easier (p. 656)

wedge [WEJ] A simple machine that is made up of two inclined planes placed back to back (p. 658)

screw [SKROO] A simple machine you turn to lift an object or to hold two or more objects together (p. 660)

Investigate

Inclined to Help

Guided Inquiry

Start with Questions

Some mountain roads wind around and around like this one. It can take a long time to reach the top of a mountain.

- Have you ever been on a road that went straight up a hill or mountain?
- Why do you think this road was built with so many curves?

Investigate to find out. Then read and learn to find out more.

Prepare to Investigate

Inquiry Skill Tip

When you interpret data, you decide what the data means. Interpreting data involves other processes, such as inferring and drawing conclusions.

Materials

- safety goggles
- board
- chair
- tape measure
- string
- toy car
- spring scale

Make a Data Table

	Distance (cm)	Times	Force (newtons, N)	Equals	Work (N-cm)
No ramp		x		=	
Ramp		x		=	

Follow This Procedure

1. **CAUTION: Put on safety goggles.** Use a board to make a ramp from the floor to a chair seat. Measure to find the distance from the floor to the seat. **Measure** straight up and along the ramp. **Record** both distances.

2. Tie a loop of string to a toy car. Attach a spring scale to the string.

3. Hold on to the spring scale, and lift the car from the floor straight up to the chair seat. **Record** the force shown on the scale.

4. Hold on to the spring scale, and slowly pull the car up the ramp, from the floor to the chair seat. **Record** the force shown on the scale. Now complete the table.

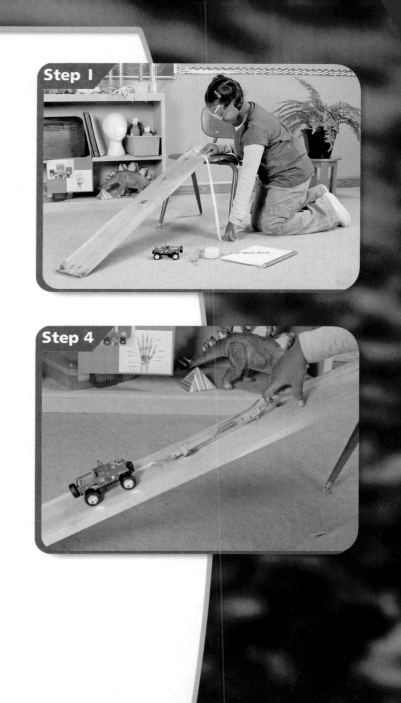

Draw Conclusions

1. How did the ramp affect the force needed to lift the car?

2. How did the ramp affect the distance?

3. **Inquiry Skill** Scientists interpret data to draw conclusions. What conclusions can you draw from your data?

Independent Inquiry

What might affect the force needed to lift the car? Plan and conduct a simple investigation to test some variables.

Read and Learn

VOCABULARY
inclined plane p. 656
wedge p. 658
screw p. 660

SCIENCE CONCEPTS
▶ how inclined planes, wedges, and screws are used

 COMPARE AND CONTRAST
Look for ways inclined planes, wedges, and screws are alike and different.

alike — different

The Inclined Plane

Suppose you wanted to get to the top of this hill. You could take a straight path to the top. That would be easier than climbing up a cliff, but you still might be very tired after riding up the straight path.

Instead, you could ride up a path that winds around the hill. The path is longer than the straight route, but it's easier to ride up a gentle slope.

Both routes are inclined planes. An **inclined plane** is a simple machine that makes moving and lifting things easier.

> Going up the straight path is faster, but it takes more effort. Going up the winding path takes longer, but it is easier.

▲ Pushing this chair up the ramp is easier than lifting it directly into the truck.

Suppose you need to put a box of books on a shelf. You try to lift the box, but it's too heavy. So you lean a board against the shelf. Then you drag the box to the end of the board. A hard push moves the box up the board. Soon it's on the shelf.

You have used the board as a ramp. A ramp is an example of an inclined plane. People use ramps to make it easier to get things out of basements or onto trucks.

COMPARE AND CONTRAST

How is a ramp like a path up a hill?

A ramp can make getting into buildings easier for people with physical disabilities. ▶

The Wedge

Suppose you want to put a slice of tomato on your sandwich. How can you get one thin slice off a whole tomato? That's easy—use a sharp knife.

To slice the tomato, you press down on the knife. The knife splits the tomato into two pieces. In a moment, they are separated from each other.

A knife is a simple machine known as a wedge. A **wedge** is made up of two inclined planes placed back to back. Wedges are used to force two things apart or to split one thing into two.

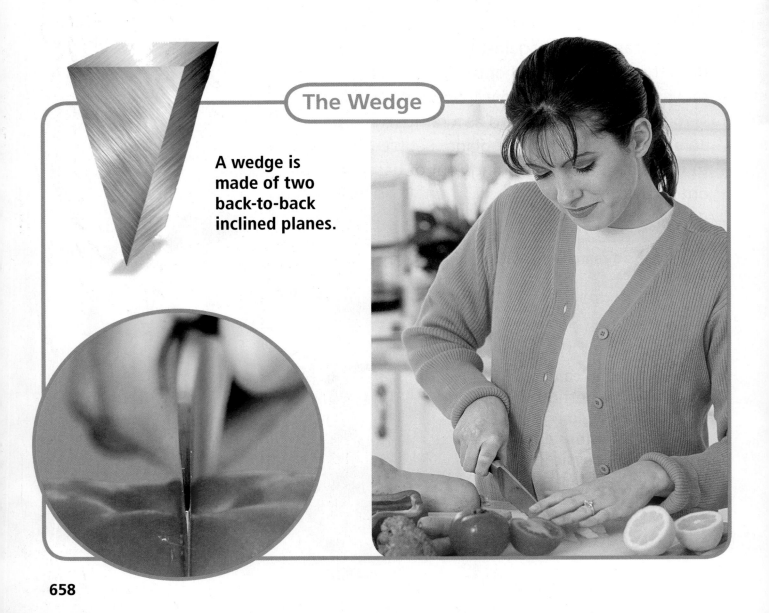

The Wedge

A wedge is made of two back-to-back inclined planes.

It's easy to see the wedge shape of this tool.

A chisel is a wedge, too. This man uses a chisel to separate small pieces of limestone from the big piece.

An ax is also a wedge. It can split a log into two pieces. The ax head looks like two ramps. To move something up a ramp, you push the object up. To move a wedge down through an object, you push the wedge down. In both cases, something slides along a ramp.

 COMPARE AND CONTRAST

How are a wedge and an inclined plane alike? How are they different?

The Screw

Suppose you want to hold two pieces of wood together. You could use a nail, but since a nail is smooth, it might slip out of the wood. It might be better to use a screw.

A **screw** is a simple machine you turn to lift an object or to hold two or more objects together. It is like a nail with threads around it. The threads make it hard to pull the screw out of the wood.

A nail goes straight into the wood. A screw moves around and around as it goes in. A screw travels farther than a nail does, but it takes much less force to get it in.

COMPARE AND CONTRAST

How are a screw and an inclined plane alike? How are they different?

Make a Model of a Screw

Cut out a paper triangle. How is the paper like an inclined plane? Now, wrap the paper around a pencil. How is the pencil now like a screw?

Some screws have threads that are far apart. Others have threads that are close together and wind around the screw more times. It's easier to turn these screws, but it takes longer. ▼

The screw's threads hold the pieces of wood together.

Lesson Review

Essential Question

What are some other simple machines?

In this lesson, you learned about more simple machines that help you do work. Inclined planes, wedges, and screws are all simple machines.

1. **COMPARE AND CONTRAST** Draw and complete a graphic organizer to show how inclined planes, screws, and wedges are alike and different.

2. **SUMMARIZE** Write a three-sentence summary of this lesson. Write one sentence about each simple machine described.

3. **DRAW CONCLUSIONS** Matthew is putting together a bench. Explain why it might be better for him to use screws than to use nails.

4. **VOCABULARY** Write a paragraph using the terms *inclined plane*, *wedge*, and *screw* correctly.

Test Prep

5. Your friend lives at the top of a hill. Which simple machine would you most likely use to go to your friend's house?
   - **A.** inclined plane
   - **B.** pulley
   - **C.** wedge
   - **D.** lever

Make Connections

Writing

Narrative

Write a **story** about an imaginary place that doesn't have simple machines. Your story can be funny or serious. Remember to give your place a name.

Social Studies

Egyptian Pyramids

Read about how the ancient Egyptians built the pyramids. Find out which simple machines scientists believe they used and how they probably used them.

Science Spin From Weekly Reader
TECHNOLOGY

Say Hello to ASIMO

How would you like having a robot around the house to help you with chores, such as cleaning your room? That reality may be closer than you think. Meet ASIMO, the world's most advanced humanoid robot.

This 4-foot, 115-pound robot is almost lifelike. Balancing on two legs, ASIMO can walk, climb stairs, and even dance. This robot has two arms and two hands, so it can shake hands, hold objects, turn on light switches, and open doors. It can also recognize faces and voices.

Engineers for a company in Japan spent more than 16 years developing ASIMO. Their goal was to build a robot that would improve people's lives. "ASIMO was created for the purpose of someday helping people in need," said Koichi Amemiya, president of the company.

Scientists hope *ASIMO* will be able to help out with chores around the house, such as taking out the garbage.

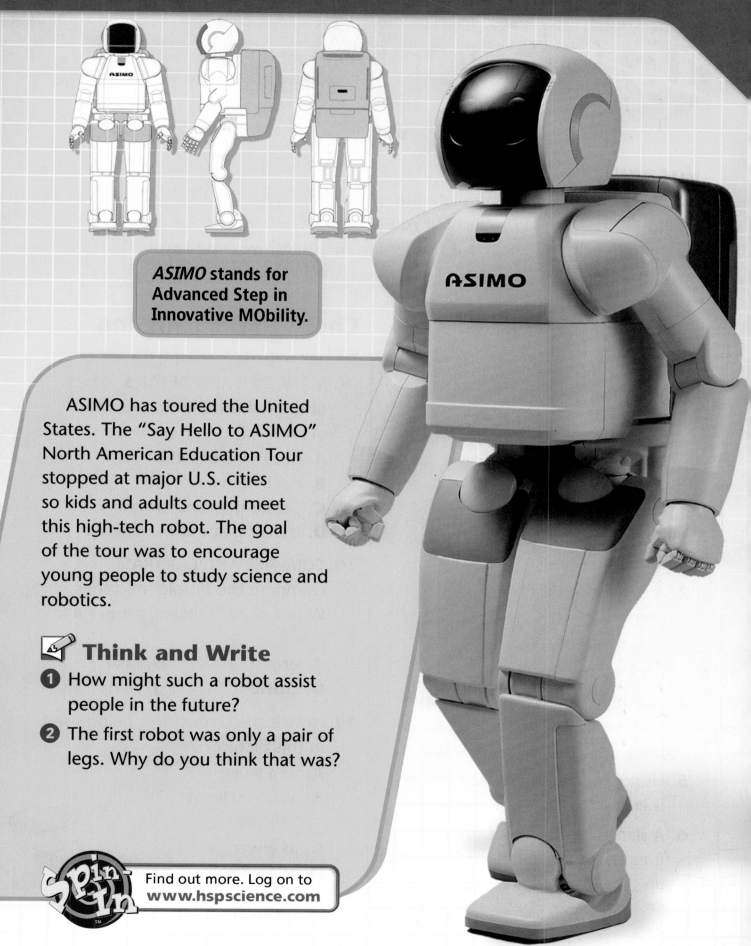

ASIMO stands for Advanced Step in Innovative MObility.

ASIMO has toured the United States. The "Say Hello to ASIMO" North American Education Tour stopped at major U.S. cities so kids and adults could meet this high-tech robot. The goal of the tour was to encourage young people to study science and robotics.

Think and Write

1. How might such a robot assist people in the future?
2. The first robot was only a pair of legs. Why do you think that was?

Find out more. Log on to **www.hspscience.com**

CHAPTER 16 Review and Test Preparation

Vocabulary Review

Use the terms below to complete the sentences. The page numbers tell you where to look in the chapter if you need help.

work   p. 636
simple machine   p. 645
lever   p. 646
fulcrum   p. 646
pulley   p. 650
inclined plane   p. 656
wedge   p. 658
screw   p. 660

1. A wheel with a rope that goes around it is a _____.
2. A bar that pivots on a fixed point is a _____.
3. Two inclined planes placed back to back form a _____.
4. Anything that makes work easier and has few or no moving parts is a _____.
5. Using a force to move an object is known as _____.
6. A slanted surface that makes it easier to move objects is an _____.
7. When you use a shovel to pry up an object, the ground acts as a _____.
8. Threads wrapped around a post form a _____.

Check Understanding

Write the letter of the best choice.

9. **MAIN IDEA AND DETAILS** Which of the following is an example of work?
   A. holding a baseball
   B. packing a lunch box
   C. pushing against a wall
   D. thinking about homework

10. **COMPARE AND CONTRAST** Compare the simple machines. Which of the following is **not** a lever?
    F. broom   H. rake
    G. crowbar   J. ramp

11. How is the screwdriver being used in this picture?
    A. as a lever
    B. as an inclined plane
    C. as a wedge
    D. as a wheel-and-axle

12. What type of simple machine is a nutcracker?
   F. axle
   G. inclined plane
   H. lever
   J. screw

13. What simple machine can you use to lift a can of paint to the second floor as you stand on the ground?
   A. pulley
   B. ramp
   C. wedge
   D. wheel-and-axle

14. What simple machine is made up of an inclined plane that winds around a post?
   F. fulcrum
   G. screw
   H. wedge
   J. wheel-and-axle

15. Leo is using a chisel to shape a piece of wood. What simple machine is he using?
   A. lever
   B. pulley
   C. ramp
   D. wedge

16. Which of these objects at a playground is an example of a lever?
   F. ladder
   G. seesaw
   H. slide
   J. swing

Inquiry Skills

17. A teacher asks her students to use classroom items to put together a simple machine. You see a ruler and a rubber eraser. Predict the two simple machines the students might make. Draw what they would look like.

18. You kicked a soccer ball as far as you could. Explain how you could use a jump rope to measure how far the ball was kicked.

Critical Thinking

19. Angie and her family have reached the airport late. They must hurry to catch their flight. Is Angie doing work as she pulls her suitcase up a ramp? Explain.

20. In this picture, Kyle is pushing a large bundle of newspapers up a ramp. **The Big Idea** What could he do to lessen the amount of force he needs to apply?

UNIT F PHYSICAL SCIENCE
Visual Summary

Tell how each picture shows the **Big Idea** for its chapter.

CHAPTER 15 Big Idea
Movement is caused by forces acting on an object.

CHAPTER 16 Big Idea
Simple machines allow people to change the direction or size of a force.

Glossary

Visit the Multimedia Science Glossary to see illustrations of these words and to hear them pronounced.
www.hspscience.com

Every entry in the glossary begins with a term and a *phonetic respelling*. A phonetic respelling writes the word the way it sounds, which can help you pronounce new or unfamiliar words.

The Pronunciation Key below will help you understand the respellings. Syllables are separated by a bullet (•). Small uppercase letters show stressed, or accented, syllables.

The definition of the term follows the respelling. An example of how to use the term in a sentence follows the definition.

The page number in ( ) at the end of the entry tells you where to find the term in your textbook. These terms are highlighted in yellow in the lessons. Every entry has an illustration to help you understand the term.

Pronunciation Key

Sound	As in	Phonetic Respelling	Sound	As in	Phonetic Respelling
a	bat	(BAT)	oh	over	(OH•ver)
ah	lock	(LAHK)	oo	pool	(POOL)
air	rare	(RAIR)	ow	out	(OWT)
ar	argue	(AR•gyoo)	oy	foil	(FOYL)
aw	law	(LAW)	s	cell	(SEL)
ay	face	(FAYS)		sit	(SIT)
ch	chapel	(CHAP•uhl)	sh	sheep	(SHEEP)
e	test	(TEST)	th	that	(THAT)
	metric	(MEH•trik)		thin	(THIN)
ee	eat	(EET)	u	pull	(PUL)
	feet	(FEET)	uh	medal	(MED•uhl)
	ski	(SKEE)		talent	(TAL•uhnt)
er	paper	(PAY•per)		pencil	(PEN•suhl)
	fern	(FERN)		onion	(UHN•yuhn)
eye	idea	(eye•DEE•uh)		playful	(PLAY•fuhl)
i	bit	(BIT)		dull	(DUHL)
ing	going	(GOH•ing)	y	yes	(YES)
k	card	(KARD)		ripe	(RYP)
	kite	(KYT)	z	bags	(BAGZ)
ngk	bank	(BANGK)	zh	treasure	(TREZH•er)

Multimedia Science Glossary: www.hspscience.com

A

absorbed
[ab•SAWRBD] Taken in by an object: Light is *absorbed* by the black pavement. (574)

adaptation
[ad•uhp•TAY•shuhn] Any trait that helps a plant or an animal survive: A giraffe's long neck is an *adaptation* for reaching leaves high on a tree. (178)

amphibian
[am•FIB•ee•uhn] A type of vertebrate that has moist skin, begins its life in water with gills, and develops lungs as an adult to live on land: Frogs are one type of *amphibian*. (131)

anemometer
[an•uh•MAHM•uht•er] A weather instrument that measures wind speed: An *anemometer* measures the speed of the wind. (393)

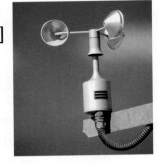

atmosphere
[AT•muhs•feer] The air around Earth: Earth's atmosphere has several layers. (390)

axis [AK•sis] A line—which you cannot see—that runs through the center of Earth from the North Pole to the South Pole: Earth rotates around its *axis*. (408)

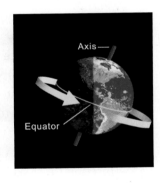

R2

Glossary

bird [BERD] **A type of vertebrate that has feathers:** *Birds* are the only type of animal that has feathers. (129)

camouflage [KAM•uh•flahzh] **Colors, patterns, and shapes that disguise an animal and help it hide:** The snow leopard's spots are *camouflage* to help keep it hidden from prey. (182)

canyon [KAN•yuhn] **A valley with steep sides:** A *canyon* is formed by running water. (288)

carnivore [KAHR•nuh•vawr] **An animal that eats other animals:** A *carnivore* has sharp teeth to help it eat meat. (205)

cell [SEL] **A tiny building block that makes up every part of an organism:** A plant *cell* has a cell wall. (56)

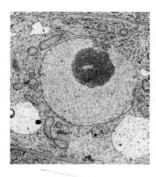

chlorophyll [KLAWR•uh•fil] **The green substance inside leaves that helps a plant use light energy to make food:** *Chlorophyll* is what makes leaves green. (103)

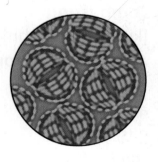

R3

circuit [SER•kuht] **A path that electricity follows:** Electricity flows through this *circuit*, which connects a light bulb with a battery. (529)

clay [KLAY] **Particles of rock so small you need a microscope to see them:** *Clay* soil holds a lot of water. (335)

combustion [kuhm•BUS•chuhn] **Another word for *burning*:** When a spark touches gas, you get *combustion*. (507)

community [kuh•MYOO•nuh•tee] **All the populations of organisms that live in an ecosystem at the same time:** A *community* has many kinds of living things. (160)

condensation [kahn•duhn•SAY•shuhn] **The process by which water vapor changes into liquid water:** Water on this glass results from *condensation*. (382, 474)

conduction [kuhn•DUK•shuhn] **The movement of heat between objects that are touching each other:** The metal is heating the liquid through *conduction*. (559)

Glossary

conductor
[kuhn•DUK•ter]
An object that heat can move through easily: Metal is a good *conductor*. (559)

conservation
[kahn•ser•VAY•shuhn]
The saving of resources by using them wisely: Using resources the right way is a part of *conservation*. (352)

constellation
[kahn•stuh•LAY•shuhn] A group of stars that appears to form the shape of an animal, a person, or an object: The Great Bear is a *constellation*. (434)

consumer
[kuhn•SOOM•er] A living thing that gets its energy by eating other living things as food: Animals are *consumers*. (203)

crest [KREST] The highest point of a wave: The *crest* of the wave is the highest part. (622)

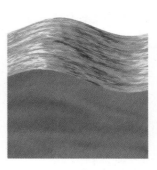

current electricity
[KER•uhnt ee•lek•TRIS•i•tee]
Electricity that moves through a wire: *Current electricity* travels through this circuit. (529)

deciduous [dee•SIJ•oo•uhs] **Relating to plants that lose all their leaves at about the same time every year:** Trees that lose leaves during winter are *deciduous* trees. (93)

decomposer [dee•kuhm•POHZ•er] **A living thing that breaks down dead organisms for food:** Mushrooms are one kind of *decomposer*. (203)

density [DEN•suh•tee] **The mass of matter compared to its volume:** *Density* is a property of matter. (463)

desert [DEZ•ert] **An ecosystem that is very dry:** A *desert* gets very little rain. (168)

distance [DIS•tuhns] **How far one location is from another:** The boys are measuring *distance* with a meterstick. (602)

earthquake [ERTH•kwayk] **The shaking of Earth's surface caused by movement in Earth's crust:** This seismograph is recording an *earthquake*. (306)

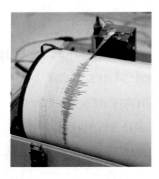

Glossary

ecosystem
[EE•koh•sis•tuhm]
The living and nonliving things that interact in an environment: A pond is one example of an *ecosystem*. (160)

energy [EN•er•jee]
The ability to make something move or change: *Energy* is needed to push a box across the floor. (496)

energy pyramid
[EN•er•jee PIR•uh•mid] A diagram that shows how energy gets used in a food chain: Producers are at the bottom of the *energy pyramid*. (214)

environment
[en•VY•ruhn•muhnt]
The things, both living and nonliving, that surround a living thing: Many kinds of living things can share the same *environment*. (159)

erosion
[ee•ROH•zhuhn]
The movement of weathered rock and soil: This gully was formed by *erosion*. (298)

evaporation
[ee•vap•uh•RAY•shuhn] The process by which liquid water changes into water vapor: These puddles are drying due to *evaporation* of the water by the sun. (383, 474)

evergreen [EV•er•green] A plant that stays green and makes food all year long: *Evergreen* trees keep their leaves all year. (93)

experiment [ek•SPAIR•uh•muhnt] A test done to see if a hypothesis is correct or not: In this *experiment*, the only difference between the cups is the type of soil the plants are growing in. (31)

F

fish [FISH] A type of vertebrate that breathes through gills and spends its life in water: *Fish* are animals that have backbones and live in the water. (132)

flood [FLUHD] A large amount of water that covers normally dry land: *Floods* can happen very suddenly. (310)

food chain [FOOD CHAYN] The path of food from one living thing to another: A producer is at the beginning of every *food chain*. (212)

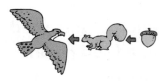

Glossary

food web [FOOD WEB] **Food chains that overlap:** *Food webs* show interdependence in ecosystems. (222)

force [FAWRS] **A push or a pull:** *Forces* affect the movement of objects. (610)

forceps [FAWR•seps] **A tool used to pick up and hold on to objects:** The student is using *forceps* to examine a seed. (7)

forest [FAHR•ist] **An ecosystem in which many trees grow:** Many kinds of animals live in a *forest*. (172)

formulate [FOR•myoo•layt] **To come up with a plan for something:** When you *formulate* a hypothesis, you will be ready to try an experiment. (23)

fossil [FAHS•uhl] **A trace or the remains of a living thing that died a long time ago:** We can learn about living things of long ago by studying *fossils*. (270)

fossil fuel [FAHS•uhl FYOO•uhl] **A resource that comes from the remains of plants and animals that lived long ago:** People burn *fossil fuels* for energy. (515)

R9

fresh water [FRESH WAWT•er] Water that has very little salt in it: The water that you drink is *fresh water*. (371)

fulcrum [FUL•kruhm] The fixed point on a lever: The *fulcrum* of this lever is the triangle-shaped base. (646)

G

gas [GAS] A form of matter that has no definite shape or volume: Air is the *gas* that is being used to fill this ball. (473)

generator [JEN•er•ayt•er] A device that uses a magnet to produce a current of electricity: People use portable *generators* when the power is out. (546)

glacier [GLAY•sher] A huge sheet or block of moving ice: Very high mountains often have *glaciers*. (300)

grassland [GRAS•land] An area of land that is generally hot in the summer and cold in the winter. The main plants found in this ecosystem are grasses: A *grassland* has fields of grass. (169)

Glossary

gravity [GRAV•ih•tee] **A force that pulls two objects toward each other:** If you fall, you experience the effects of *gravity*. (614)

groundwater [GROWND•wawt•er] **An underground supply of water:** The water that you drink usually comes from *groundwater*. (372)

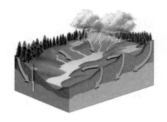

habitat [HAB•i•tat] **The place where an organism lives in an ecosystem:** In some grassland environments, prairie dogs live in an underground *habitat*. (161)

heat [HEET] **The movement of thermal energy from hotter to cooler objects:** *Heat* from a stove cooks food. (558)

herbivore [HER•buh•vawr] **An animal that eats only plants:** Cows are *herbivores*. (204)

hibernate [HY•ber•nayt] **To go into a deep, sleeplike state for winter:** When an animal *hibernates*, its breathing and heartbeat rate slow down and its body temperature drops. (180)

R11

humus
[HYOO•muhs] The part of some soil that is made up of broken-down parts of dead plants and animals: *Humus* helps plants grow. (332)

hypothesis
[hy•PAHTH•uh•sis] A possible answer to a question that can be tested to see if it is correct: The boy is writing down his *hypothesis* for a science experiment. (31)

I

igneous rock
[IG•nee•uhs RAHK] Rock that was once melted and then cooled and hardened: Granite is one kind of *igneous rock*. (258)

inclined plane
[in•KLYND PLAYN] A simple machine that makes moving or lifting things easier: A ramp is a type of *inclined plane*. (656)

infer [in•FER] To draw a conclusion about something: The girl can *infer* the differences between the balloons. (19)

inherit [in•HAIR•it] To have a trait passed on from parents: Young birds *inherit* traits from their parents. (68)

Glossary

inquiry [IN•kwer•ee] A question that is asked about something, or a close study of something: The students learned a lot from their *inquiry* about greenhouses. (6)

instinct [IN•stinkt] A behavior that an animal knows without being taught: Birds follow their parents by *instinct*. (178)

insulator [IN•suh•layt•er] An object that doesn't conduct heat well: Styrofoam is a common *insulator*. (560)

invertebrate [in•VER•tuh•brit] An animal without a backbone: A scorpion is an *invertebrate*. (138)

investigation [in•ves•tuh•GAY•shuhn] A study that a scientist does: The students performed an *investigation* to find out what substances dissolve in water. (30)

K

kinetic energy [kih•NET•ik EN•er•jee] The energy of motion: When something moves, it has *kinetic energy*. (498)

landform [LAND•fawrm] A natural shape on Earth's surface: Mountains, valleys, plains, and plateaus are all *landforms*. (286)

larva [LAHR•vuh] The stage of complete metamorphosis after an organism hatches from its egg: A caterpillar is the *larva* of a butterfly. (67)

leaf [LEEF] The part of a plant that grows out of the stem and is where a plant makes food: Sunlight, air, and water combine in a *leaf* to make food. (82)

lever [LEV•er] A simple machine made up of a bar that pivots, or turns, on a fixed point: A shovel is a useful *lever* for moving dirt. (646)

life cycle [LYF SY•kuhl] The changes that happen to an organism during its life: A frog begins its *life cycle* as a tadpole. (64)

liquid [LIK•wid] A form of matter that has a volume that stays the same but a shape that can change: Water is a *liquid*; its shape can change but its volume does not. (472)

Glossary

loam [LOHM] Soil that is a mixture of humus, sand, silt, and clay: *Loam* is the best kind of soil for growing fruits and vegetables. (336)

loudness [LOUD•nuhs] The amount of energy a sound has: You can tell that the *loudness* of the sound is great because the boy is covering his ears. (584)

lunar cycle [LOON•er SY•kuhl] The pattern of phases of the moon: The *lunar cycle* takes about 29 days. (418)

lunar eclipse [LOON•er i•KLIPS] An event in which Earth blocks sunlight from reaching the moon and Earth's shadow falls on the moon: A *lunar eclipse* can only occur during a full moon. (420)

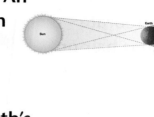

magnetic [mag•NET•ik] Able to attract objects that have iron in them: The *magnetic* force attracts the toy truck. (536)

R15

mammal
[MAM•uhl] A type of vertebrate that has hair or fur and feeds its young with milk from the mother; most mammals give birth to live young: A rabbit is a *mammal*. (128)

mass [MAS] The amount of matter in an object: A balance is used to measure how much *mass* an object has. (462)

matter [MAT•er] Anything that takes up space: *Matter* makes up everything in this picture. (458)

metamorphic rock
[met•uh•MAWR•fik RAHK] Rock that has been changed by heat and pressure: Gneiss is one kind of *metamorphic rock*. (259)

metamorphosis
[met•uh•MAWR•fuh•sis] A series of changes in appearance that some organisms go through: Caterpillars go through *metamorphosis* to become butterflies. (66)

migrate [MY•grayt] To travel from one place to another and back again: Some kinds of birds *migrate* south for the winter and north for the summer. (181)

Glossary

mimicry
[MIM•ik•ree] The imitating of the look of another animal: The flower fly is an example of *mimicry*. (182)

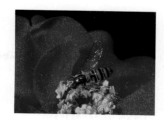

mineral
[MIN•er•uhl] A solid object found in nature that has never been alive: Amethyst is a *mineral*. (248)

mixture [MIKS•cher] A substance that has two or more different kinds of matter: Lemonade is a *mixture* of water, sugar, and lemon juice. (482)

moon phases
[MOON FAYZ•uhz] The different shapes that the moon seems to have in the sky when it is observed from Earth: *Moon phases* change as the moon orbits Earth. (418)

motion
[MOH•shuhn] A change of position: This student is in *motion*. (601)

mountain
[MOUNT•uhn] A place on Earth's surface that is much higher than the land around it: *Mountains* are one of Earth's natural landforms. (287)

R17

N

nonrenewable resource [nahn•ri•NOO•uh•buhl REE•sawrs] A resource that, when it is used up, will not exist again during a human lifetime: Coal and other fossil fuels are *nonrenewable resources*. (326, 516)

nutrients [noo•TREE•uhnts] The parts of the soil that help plants grow and stay healthy: Plants absorb *nutrients* from soil. (82)

O

omnivore [AHM•nih•vawr] A consumer that eats both plants and animals: A bear is an *omnivore*. (206)

opaque [oh•PAYK] Relating to objects that don't let light pass through: You cannot see through *opaque* objects. (575)

orbit [AWR•bit] The path a planet takes as it revolves around the sun or that a moon takes as it revolves around a planet: It takes about $365\frac{1}{4}$ days for Earth to *orbit* the sun. (428)

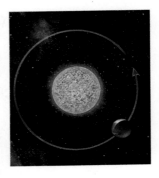

organism [AWR•guh•niz•uhm] Any living thing: Plants and animals are *organisms*. (54)

Glossary

oxygen [AHK•sih•juhn] A gas that all living things need and that plants give off into the air: Humans need *oxygen* to survive. (117, 390)

photosynthesis [foht•oh•SIN•thuh•sis] The process that plants use to make sugar: Plants need light and water to perform *photosynthesis*. (102)

physical property [FIZ•ih•kuhl PRAHP•er•tee] Anything that you can observe about an object by using one or more of your senses: Some of the *physical properties* of this lizard are its color and shape. (460)

pitch [PICH] The highness or lowness of a sound: Sounds with a very high *pitch* can hurt your ears. (585)

plain [PLAYN] A wide, flat area on Earth's surface: These giraffes live on a *plain* in Africa. (289)

planet [PLAN•it] A large body of rock or gas in space: Each *planet* in our solar system orbits the sun. (428)

plateau [PLA•toh] A flat area higher than the land around it: Most *plateaus* are wide. (290)

R19

pollution
[puh•LOO•shuhn]
Any harmful material in the environment: Oil spills are a kind of *pollution* that can kill many living things. (344)

population
[pahp•yuh•LAY•shuhn]
A group of organisms of the same kind that live in the same place: Available resources limit animal *populations*. (160)

potential energy
[poh•TEN•shuhl EN•er•jee] Energy of position or condition: Being on top of a hill increases *potential energy*. (498)

precipitation
[pree•sip•uh•TAY•shuhn] Rain, snow, sleet, or hail: *Precipitation* can be solid, like snow, or liquid, like rain. (384)

predator
[PRED•uh•ter] An animal that hunts another animal for food: Bobcats are *predators*. (216)

prey [PRAY] An animal that is hunted by a predator: Rabbits are one kind of *prey*. (216)

producer
[pruh•DOOS•er] A living thing that makes its own food: Grasses are *producers*. (203)

Glossary

pulley [PUHL•ee] A simple machine made up of a wheel with a rope around it: A *pulley* can help you lift something heavy. (650)

pupa [PYOO•puh] The stage of complete metamorphosis in which an organism is wrapped in a cocoon or a chrysalis: The *pupa* slowly changes inside the chrysalis to become a butterfly. (67)

R

recycle [ree•SY•kuhl] To reuse a resource by breaking it down and using it to make a new product: You can *recycle* your plastic bottles. (356)

reduce [ree•DOOS] To use less of a resource: To help save the environment, you should *reduce*, reuse, and recycle. (354)

reflection [ri•FLEK•shuhn] The bouncing of light off an object: You can see a *reflection* in the water. (566)

refraction [ri•FRAK•shuhn] The bending of light as it moves from one material to another: The straw appears broken because of *refraction*. (567)

R21

renewable resource [ri•NOO•uh•buhl REE•sawrs] A resource that can be replaced within a human lifetime: Wind is a *renewable resource*. (324, 516)

reptile [REP•tyl] A type of vertebrate that has dry skin covered with scales: Lizards and snakes are *reptiles*. (130)

resource [REE•sawrs] A material that is found in nature and that is used by living things: The amount of available *resources* changes with the seasons. (189, 322, 515)

reusable resource [ree•YOOZ•uh•buhl REE•sawrs] A resource that can be used again and again: Water is a *reusable resource*. (325)

reuse [ree•YOOZ] To use a resource again and again: If you *reuse* resources, you are helping the environment. (355)

revolution [rev•uh•LOO•shuhn] The movement of Earth one time around the sun: Earth's *revolution* takes one year to complete. (409)

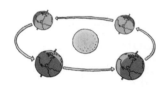

Glossary

rock [RAHK] A naturally formed solid made of one or more minerals: *Rocks* come in all shapes and sizes. (252)

root [ROOT] The part of a plant that grows underground and takes water and nutrients from the soil: The *root* holds the plant to the ground. (82)

rotation [roh•TAY•shuhn] The spinning of Earth on its axis: One *rotation* of Earth takes 24 hours. (408)

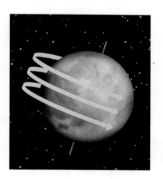

S

sand [SAND] Grains of rock that you can see with your eyes alone: When rocks are broken down into very small pieces, they become *sand*. (335)

scientific method [sy•uhn•TIF•ik METH•uhd] An organized plan that scientists use to conduct a study: This student is using the *scientific method* to answer a question. (30)

screw [SKROO] A simple machine you turn to lift an object or to hold two or more objects together: A *screw* looks like a nail with threads wrapped around it. (660)

sedimentary rock [sed•uh•MEN•ter•ee RAHK] Rock made when materials settle into layers and get squeezed until they harden into rock: *Sedimentary rock* often forms beneath lakes or rivers. (259)

seed [SEED] The first stage of life for many plants: A plant can grow from one *seed*. (91)

shadow [SHAD•oh] A dark area that forms when an object blocks the path of light: When you move, so does your *shadow*. (568)

silt [SILT] Grains of rock that are too small to see with your eyes alone: You can find *silt* at the bottom of a river. (335)

simple machine [SIM•puhl muh•SHEEN] A tool with few or no moving parts that helps people do work: Levers and pulleys are types of *simple machines*. (645)

Glossary

solar eclipse
[SOH•ler ih•KLIPS] An event in which the moon blocks sunlight from reaching Earth and the moon's shadow falls on Earth: You should not look at the sun during a *solar eclipse*. (422)

solar system
[SOH•ler SIS•tuhm] The sun, the planets and their moons, and the small objects that orbit the sun: Everything that orbits the sun is part of the *solar system*. (428)

solid [SAHL•id] A form of matter that has a volume and a shape that both stay the same: Marbles, pasta, and grains of sand are *solids*; their shapes and volumes do not change. (471)

solution
[suh•LOO•shuhn] A mixture in which the different kinds of matter mix evenly: If the girl mixes her ingredients evenly, she will have a *solution*. (483)

speed [SPEED] The distance that an object moves in a certain period of time: You can calculate the *speed* of an object if you know how far it traveled and the time it took. (604)

star [STAR] A hot ball of glowing gases that gives off energy: There are many *stars* in the night sky. (429)

static electricity [STAT•ik ee•lek•TRIS•uh•tee] An electric charge that builds up on an object: *Static electricity* might make your clothes stick to you. (528)

stem [STEM] The part of a plant that grows above ground and helps hold the plant up. It also carries water and nutrients up to the leaves: The *stem* carries food and water to the other parts of a plant. (82)

temperature [TEM•per•uh•cher] The measure of how hot or cold something is: The kids are wearing coats because of the *temperature*. (393, 508, 558)

thermal energy [THER•muhl EN•er•jee] The form of energy that moves particles of matter: Movement of *thermal energy* is heat. (558)

Glossary

translucent [tranz•LOO•suhnt] **Relating to an object that lets some light pass through:** You can see light through *translucent* objects, but you cannot see clearly through them. (575)

transparent [trans•PER•uhnt] **Relating to an object that lets most light pass through:** You can see clearly through *transparent* objects. (575)

trough [TRAWF] **The lowest point of a wave:** The *trough* is the lowest part of the wave. (622)

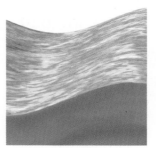

valley [VAL•ee] **A low area between higher land such as mountains:** *Valleys* are often protected from severe weather by the highlands around them. (288)

variable [VAIR•ee•uh•buhl] **The one thing that changes in a science inquiry or experiment:** The *variable* in this experiment is the color of the socks. (22)

vertebrate [VER•tuh•brit] **An animal with a backbone:** Antelope are one type of *vertebrate*. (127)

vibration
[vy•BRAY•shuhn] A series of back-and-forth movements: The *vibration* of a bell causes sound waves to form in the air. (584)

volcano
[vahl•KAY•noh] An opening in Earth's surface from which lava flows: The lava from the *volcano* flowed slowly. (308)

volume
[VAHL•yoom] The amount of space that matter takes up: You can measure how much *volume* a liquid has by using a graduated cylinder. (463)

water cycle
[WAWT•er SY•kuhl] The movement of water from Earth's land, through rivers toward the ocean, to the air, and back to the land: The *water cycle* includes evaporation and precipitation. (384)

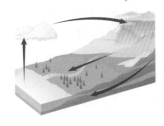

wave
[WAYV] A disturbance that travels through matter or space: Light travels in waves that are similar to *waves* in the ocean. (620)

Glossary

wavelength [WAYV•length] **The distance from one point of one wave to the same point on the next wave:** The *wavelength* of visible light determines its color. (622)

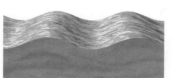

weather [WETH•er] **What is happening in the atmosphere at a certain place and time:** *Weather* can affect people's activities, especially if it is severe. (392)

weathering [WETH•er•ing] **The breaking down of rocks into smaller pieces:** *Weathering* has formed grooves in this rock. (296)

wedge [WEJ] **A simple machine that is made up of two inclined planes placed back to back:** An ax head is a *wedge*. (658)

weight [WAYT] **The measure of the force of gravity on an object:** You can learn your *weight* by stepping on a scale. (614)

wheel-and-axle [weel•and•AK•suhl] **A simple machine made up of an axle and a wheel that are connected and turn together:** The wheelbarrow has a *wheel-and-axle* that helps it move. (648)

work [WERK] The use of a force to move an object: Moving a chair is a kind of *work*. (636)

Index

AbioCor heart, 548–549
Absorbed light, 574
Adaptation, 178–179
  camouflage, 182
  to changes in ecosystems, 190
  hibernation, 180
  migration, 181
Agricultural extension agents, 71
Air, 390–391
  for animals, 117
  carbon dioxide in, 103
  as insulator, 560
  as matter, 459
  as mixture of gases, 473
  for plants, 80
  as reusable resource, 325
  sound movement through, 585
  vibrations through, 586
Air pollution, 345
Air pressure, 519
Alaskan brown bear, 112
Allergies to plants, 104
Alligator, 130
Amber, 270
Amemiya, Koichi, 662
Amethyst, 248
Amphibians, 127, 131
Anemometer, 393, 508
Anemone, 225
Animals
  adaptations of, 178, 180–182
  amphibians, 131
  camouflage, 182
  carnivores, 205
  cells of, 57
  clams, 142
  classifying, 126–132
  as consumers, 203
  defending themselves, 224–225
  in deserts, 168
  environments of, 158, 159
  fish, 132
  in food chains, 212–213
  food for, 55, 102
  in food webs, 222, 226
  in forests, 172
  fossils of, 266, 270–272, 274
  in freshwater ecosystems, 171
  in grasslands, 169
  growth and change in, 66–67
  herbivores, 204
  hibernation of, 180
  homes for, 114–115
  insects, 67, 140
  invertebrates, 138–144
  mammals, 128–129
  migration of, 181
  mimicry, 182
  needs of, 116–120
  omnivores, 206
  in polluted water, 346
  predators and prey, 216
  as renewable resources, 324
  reptiles, 130
  in saltwater ecosystems, 170
  snails, 142
  spiders and ticks, 141
  squids, 142
  vertebrates, 126–132
  water for, 370
Antarctica, fresh water in, 372
Anteaters, 216, 218
Ants, 216
Arches National Park, Utah, 168
Arctic hares, 178
Artificial heart, 548–549
Ash, volcanic, 309
ASIMO, 662–663
Asking questions, 30
Atlantic long-finned squid, 142
Atlantic Ocean, 374
Atmosphere, 390
  weather in, 392
Atomic bomb, 518
Attraction, magnetic, 536, 537
Ax, 659
Axis (Earth), 408, 411
Axle, 649. *See also* **Wheel-and-axle**

Backbone
  animals with, 126
  animals without, 138
Balances, 10
Banana spider, 141
Barometer, 519
Basic needs, 80
  of animals, 116–120

R31

in habitats, 162
of plants, 78–81
Basilisk lizard, 611
Bat, fruit, 122
Batteries, 499
Battery-operated artificial heart, 548–549
Bean plants, 64
Bears
grizzly, 206
homes for, 112
Beaver, 120, 174
Bedrock, 333
Bees, 144
Behaviors, as adaptations, 178
Big Bear constellation, 434
Binoculars, 58
Birds
camouflage of, 182
keeping warm, 560
migration of, 181
nests of, 159
in pond ecosystem, 161
as predators and prey, 216
shelters for, 120
as vertebrates, 126, 129
Bison, 169
Blizzards, 395
Blue crabs, 143
Bobcat, 214
Bobsledding, 606
Bones, as fossils, 270
Boston fern, 95
Botanists, 107
Bottlenose dolphin, 170
Breezes, sea and land, 392
Bromeliads, 179
Bubbles
colors of, 26, 29
longest, 14

making, 5
shapes of, 17
Bulbs, 65, 84
Burning
as chemical change, 484
combustion, 507
for fireworks, 492
Burrows, 120
Bushes, in grasslands, 169
Butterflies, 67
monarch, 134, 181
Buttes, 290

C

Cactus plants, 84, 168
Camouflage, 182
of cell towers, 193
Canada goose, 181
Cancer, microwave treatment of, 624
Cane toads, 226
Canyons, 288, 290
Carbon dioxide, in photosynthesis, 103
Careers
agricultural extension agent, 71
civil engineer, 625
energy manager, 519
environmental scientist, 229
materials scientist, 487
meteorologist, 399
nursery worker, 107
volcanologist, 312
Carnivores, 205
Carnivores, in energy pyramid, 215
Cast, 270, 271
Cat, 54
Caterpillars, 67, 204

Cell membrane, 56, 57
Cell phone towers, 192–193
Cells, 56
animal, 57
plant, 56
Cell walls, 56, 57
Centimeters, 602
Chameleon, 182
Chang Jiang, 358–359
Cheetah, 216, 225
Chemical changes in matter, 484, 492
Cherry tree, 94
Chicken, 206
Chimpanzee, 206
Chipmunk, 180
Chisel, 659
Chlorophyll, 103
Chloroplasts, 56, 57
Chrysalis, 67
Chu, Steven, 486
Cirrus clouds, 391
Civil engineer, 625
Clam fossils, 274
Clams, 142
Classifying (ordering)
animals, 126–132
as inquiry skill, 20
minerals, 247
plants, 90–91
Clay, in loam, 336
Clay soil, 335, 336
Cliff houses, 256
Clouds, 391
in atmosphere, 390
formation of, 384
of water vapor, 382
weather satellite pictures of, 396
Clownfish, 225

R32

Index

Coal, 515, 589
    as nonrenewable resource, 326
Coastal plain, 289
Coffer dam, 359
Colonies, bat, 122
Color
    of bubbles, 29
    of light, 576–577
    as physical property of matter, 461
    of rainbows, 570, 577
    separating, 578
Colorado River, 288
Combustion, 507
Communicating
    data on trash, 357
    as inquiry skill, 23
    mixtures, 479
    in scientific method, 33
    strength of magnets, 535
    temperature and matter, 469
    wind measurement, 389
Communities, 160–161
Comparing
    animal homes, 115
    colors of bubbles, 28
    as inquiry skill, 20
    kinds of waves, 619
    light for plants, 101
    light striking tilted surfaces, 407
    needs of plants, 79
    plants grown differently, 341
    size of bubbles, 5
    sounds, 583
Computers, 12
Condensation, 382
    as change of state, 474
    in terrariums, 379
    in water cycle, 384
Conducting experiments, 32
Conductors, 530, 559
Cones, 95, 172
Coniferous forests, 172
Conifers, 95
Conservation, 352, 353
Constellations, 434
    movement of, 435
Consumers, 203
    carnivores, 205
    in energy pyramid, 214, 215
    in food chains, 212
    herbivores, 204
    omnivores, 206
Controlling variables, 22
Cooking
    as chemical change, 484
    conduction of heat in, 559
    with microwaves, 624
    thermal energy for, 502
Copper, 249
Coquina clam, 142
Corals, 138
Core (Earth), 286
Creep, 298, 299
Crest (waves), 622
Crocodiles, 130
Crust (Earth), 286
    folds in, 285
    movement of, 306
Cumulus clouds, 391
Current electricity, 529
Cutting, 480
Cytoplasm, 56, 57

D

Daffodils, 65
Dam, Three Gorges, 358–359
Darden, Christine, 625
Day, 409, 412
    change in shadows during, 568
    on inner planets, 430, 431
    on outer planets, 432, 433
Deciduous forests, 172
Deciduous plants, 93
Decomposers, 203
    in food chains, 212
Deer tick, 141
Deltas, 298
Density, as property of matter, 463, 464
Deserts, 168
    plants in, 84
    sand dunes in, 299
Diamonds, 248
Dinosaur fossils, 274
Dinosaur tracks, 270, 271
Direction, 602
    changing, 611
    of forces, 612
Dirzo, Rodolfo, 229
Displaying data, 19
Distance, 602
Dogs
    life cycle of, 66
    speeds of, 602
Dolphin, bottlenose, 170
Doorknob, 648
Dragonflies, 205
Drawing conclusions
    growing lima beans, 88
    in scientific method, 33
    travel of thermal energy, 556
    worm farm, 136
Dropper, 7
Drought, 310, 394

Dunes, 299
Dwarf junipers, 95

E

Earth
core, 286
crust, 285, 286, 306
distance of sun from, 429
landforms on (See Landforms)
layers of, 286
mantle, 286
moon of, 402–403, 414
moon's revolution around, 418, 419
movement of, 408–409
rocks covering, 260
seasons on, 410–411
in solar system, 428, 430, 431
sunlight striking, 407
water on, 366, 371
Earthquakes
landforms changed by, 306–307
measuring, 313
Earthworms, 144, 203
Eclipse
lunar, 420–421
solar, 422
Ecosystems, 154, 160–161
changes to, 188–190
desert, 168
forest, 172
freshwater, 171
grassland, 169
saltwater, 170
Electricity, 510
current, 529
electromagnets, 540, 544–545
generated with magnets, 538
generating, 589
generators, 546
movement of, 530
static, 524, 527, 528
Electromagnets, 540, 544–545
Elephants, 118
Elf owl, 116
Energy
changes in, 500
from combustion, 507
from food, 55
in food chains, 212
forms of, 498–499
importance of, 514
kinetic, 498, 500
light, 566–567
measuring, 508
from muscles, 506
for plants, 102
from plants, 104
potential, 498–500
as renewable resource, 324
sound, 584
sources of, 515–516
from sun, 202, 490–491, 506
thermal, 557–560
using, 506–507
in waves, 620, 622
ways of saving, 515
wind, 318
Energy manager, 519
Energy pyramid, 214, 215
Environmental scientists, 229
Environments
changing, 187
ecosystems of, 160–161
of living things, 158–159
observing, 157
of plants, 84
protecting, 352–353
water in, 370
Erosion, 298–299
by earthquakes, 307
by glaciers, 300
Evaporation, 383
as change of state, 474
in water cycle, 384
Everglades National Park, Florida, 152–153
Evergreen plants, 92, 93
in coniferous forests, 172
Evidence, 24
Experiment, 31

F

Fact, 24
Farming, 342
drought and, 395
water for, 371
Feathers
of birds, 129
as insulators, 560
Feldspar, 252
Fermi, Enrico, 518
Ferns, 92
Fire, ecosystems changed by, 188
First-quarter moon, 419
Fish, 132
oxygen for, 117
in pond ecosystem, 160
schools of, 70
as vertebrates, 127, 132
Flamingos, 110–111
Floating, 464
Floods
in China, 359

Index

ecosystems changed by, 188
landforms changed by, 310
Florida coquina clam, 142
Flowers, 90, 91
in grasslands, 169
parts of, 83
plants with, 94
plants without, 95
Food
for animals, 102, 119
energy from, 55
invertebrates as, 144
making and getting, 202–206
plants as, 104
resources, 323
Food chains, 212–213
energy pyramid, 214
in food webs, 222
predators and prey in, 216
Food webs, 222
changes in, 226
making, 221
marsh, 222–223
Forceps, 7, 647
Forces
to do work, 636
energy in waves and, 622
friction, 610
gravity, 614
magnetic, 611
net force, 613
with simple machines, 645
that change motion, 612–613
types of, 610–611
Forests, 172
as renewable resources, 324
studying, 228

Formulating models
of folds in Earth's crust, 285
of food chains, 211
of food webs, 221
of fossils, 269
as inquiry skill, 22
of keeping warm, 125
of rocks, 257
of volcanoes, 305
of windmills, 512–513
Fossil fuels, 515, 516
Fossils, 270–271
of animals, 266
formation of, 272–273
in La Brea tar pit, 276–277
learning from, 274
making, 269
of starfish, *242–243*
Fresh water, 371–373
Freshwater ecosystems, 171
Friction, 610
Frogs, 131, 161
Fruit bat, 122
Fuels
fossil, 515
switchgrass and coal, 589
Fulcrum, 646, 647
Full moon, 419, 421
Fur, 128

G

g (grams), 462
Galápagos tortoise, 204
Garden mazes, 86
Garnet, 248
Gases, 473
in air pollution, 345
in atmosphere, 390
changes in state of, 474

condensation of, 382
on outer planets, 432
as state of matter, 470
from volcanoes, 309
water vapor, 380
Gasoline, 326, 507
Gathering data
as inquiry skill, 19
on weather, 393
Gazelles, 225
Generators, 546, 589
Geologists, 313
Geothermal energy, 516
Giant sequoia, 65
Gills, 132
Giraffes, 119
Glaciers, 277, 300
fresh water in, 372
Global warming, 345
Gneiss, 259
Gold, 248, 249
Graduated cylinder, 10
Grains, 104
Grain size (soils), 335
Grams (g), 462
Grand Canyon, 288, 290
Grand Tetons, Wyoming, 287
Granite, 252, 258, 264
Graphite, 248, 249
Grasses, 90–91
in grasslands, 169
roots of, 167
Grasshoppers, 67
Grasslands, 169
Gravity, 614
Gray whale, 181
Great Plains, 289
Great Salt Lake, Utah, 374
Great Wall of China, 358
Grizzly bear, 206

R35

Groundwater, 372, 384
Grouping. *See* Classifying
Growth and change (plants), 64–65
Gulf of Mexico, 374
Gulfs, salt water in, 374

Habitats, 161–162
    sharing, 190
Hair, 128
Halite, 249
Hand lens, 6, 58
Hardness (minerals), 244, 250
Harp, 584, 585
Hayden Planetarium, New York, 424
Hearing, 586
    observing matter by, 460
Heart, artificial, 548–549
Heat
    changes of state caused by, 474
    conductors of, 559
    definition of, 558
    insulators, 560
    keeping warm, 125
    observing use of, 505
    producing, 558
    See also Thermal energy
Helium, 473
Hematite, 249, 251
Herbivores, 204
    changes in food webs and, 226
    in energy pyramid, 214, 215
Heredity, 68
Hibernation, 180

Hippos, 117
Homes
    for animals, 114–115
    for bears, 112
    for living things, 50, 53
Honeybee, 144
Hoover Dam, Colorado River, 184
Horse, 68, 154, 204
Horseshoe crabs, 139
Hubble Space Telescope, 438–439
Human body, water in, 118, 370
Hummingbirds, 196–197, 204, 604
Humus, 332, 333, 336
Hurricanes, 394–395
Hypothesizing
    in scientific method, 31
    about static electricity, 527

Ice, 380, 381
    glaciers, 300, 372
    rock cracked by, 297
Ice age, 277
Ice storms, 395
Identifying variables, 22
Igneous rock, 258
    formation of, 260
    fossils in, 272
    in rock cycle, 262–263
    uses of, 264
Iguanas, 130
Imprints, 270
Inches, 602
Inclined plane, 655–657
Inferring
    about condensation in terrariums, 379

about grass roots, 167
about homes for living things, 53
how fast seeds grow, 63
how insects hide, 177
as inquiry skill, 19
about living and nonliving things, 157
about objects placed in sunlight, 505
about the path of light, 565
about phases of the moon, 417
about speed on ramps, 609
about uses of electromagnets, 543
Inherited traits, 68
Inner planets, 430–431
Inquiries, 6
    skills for, 18–23
    tools for. *See* Science tools
Insects, 139, 140
    how they hide, 177
    metamorphosis of, 67
    oxygen for, 117
    as prey, 225
Insta-Lab
    Amazing Paper Clips, 611
    Big Ears, 586
    Blow a Super Bubble! 32
    Chemical Change, 484
    Classify Animals, 131
    Compare Densities, 463
    Compare Leaves, 93
    Cover Up, 575
    Dimples, 68
    Disappearing Act, 383
    Ecosystems Around You, 161
    Energy from Food, 507

R36

Index

Energy in Motion, 499
Energy Pyramid, 214
Feeling the Heat, 560
Finding Constellations, 434
Fizzy Rock, 259
Float Levels, 171
Fossil Find, 274
Fulcrum of Broom, 646
Getting There Fast, 604
Growing Ice, 299
Hiding from Predators, 225
Horseshoe Magnets and Bar Magnets, 537
How Much Water? 372
Hunting for Resources, 325
Is It Solid? 471
Jobs for Teeth, 205
Make a Graph, 104
Make a Landform, 290
Make a Model, 23
Make a Soil Model, 335
Make It Light, 529
Model Animal Homes, 117
Modeling an Eclipse, 420
Modeling Motion, 409
Model Ladybird Beetles, 140
Model of a Screw, 660
Model Plant Stems and Leaves, 81
Move It! 621, 638
Ready for an Emergency, 546
Save Fossil Fuels, 515
Seeing Air Pollution, 345
Streak It, 251
Thumbs Down, 179
Trashy Items, 189
Use a Magnifying Box, 7
Use a Rain Gauge, 393
What Can Be Reused? 355
What Do You See? 567
What Is a Feather? 58
Where's the Energy? 307

Instincts, 178
Insulators, 530, 560
Interpreting data
  as inquiry skill, 19
  making motion, 599
  water at work, 295
Invertebrates, 138
  clams, 142
  insects, 140
  snails, 142
  spiders and ticks, 141
  squids, 142
Investigate
  Animal Homes, 114–115
  Bubble Colors, 28–29
  Changing the Environment, 186–187
  Checking Teeth, 200–201
  Condensation in a Terrarium, 378–379
  Folds in Earth's Crust, 284–285
  Getting Warmer? 556–557
  Grass Roots, 166–167
  Growing Lima Beans, 88–89
  The Heat Is On, 504–505
  Help from Simple Machines, 642–643
  Homes for Living things, 52–53
  How Fast Do Seeds Grow? 62–63
  How Insects Hide, 176–177
  How Sunlight Strikes Earth, 406–407
  Inclined to Help, 654–655
  Keeping Warm, 124–125
  Lights, Plants, Action! 100–101
  Looking for Static Electricity, 526–527
  Make a Maraca, 582–583
  Make a Model Fossil, 268–269
  Make a Model Rock, 256–257
  Make a Paper Windmill, 512–513
  Make It Move, 598–599
  Making Bubbles, 4–5
  Making a Food Chain, 210–211
  Making a Food Web, 220–221
  Making Rainbows, 572–573
  Measuring Volume, 456–457
  Measuring Wind, 388–389
  Mining Resources, 320–321
  Model Volcano, 304–305
  Moon's Phases, 416–417
  Needs of Plants, 78–79
  Observe an Environment, 156–157
  Observing Soil, 330–331
  Observing Temperature, 494–495
  Planets, 426–427
  Pollution and Plants, 340–341
  Shapes of Bubbles, 16–17
  Simple Sorting, 542–543
  Speed Ramp, 608–609
  Taking a Look at Trash, 350–351

Temperature and Matter, 468–469
Testing Minerals, 246–247
Two Kinds of Waves, 618–619
Water at Work, 294–295
Where in the World Is Water? 368–369
Where's the Light? 564–565
Which Magnet Is Stronger? 534–535
Will It Mix? 478–479
Work with Me, 632–633
A Worm Farm, 136–137
Iron, 249
Islands, 287
Ivy plants, 82

Jellyfish, 138
Jupiter, 424, 428, 432

Kangaroos, 128
Keeping warm, 125
Kelp, 170
Kilograms (kg), 462
Kilometers (km), 602
Kinetic energy, 498, 500
Knives, 658
Koala, 159

La Brea tar pit, California, 276–277
Lake Powell, 373
Lakes made by dams, 359
Land
  changes people make to, 342

pollution of, 344
resources from, 342
uses of, 342–343
Land breezes, 392
Landfills, 344
Landforms, 286–290
  caused by erosion, 298, 299
  changed by earthquakes, 306–307
  changed by floods, 310
  changed by glaciers, 300
  changed by volcanoes, 308–309
  mountains, 287
  plains, 289
  plateaus, 290
  valleys, 288
Landslides, 307
Larva, 67
Lava, 308, 309
Lava lamp, 554
Leaf blower, 644
Leaves, 82, 83, 92–93, 179
Leopard, 205
Lever, 640, 646–647
Life cycle, 64
  of animals, 66–67
  of butterflies and moths, 67
  of grasshoppers, 67
  of plants, 64
Light. *See also* Sunlight
  colors of, 576–577
  as energy from sun, 429
  movement of, 566–567
  path of, 565
  from planets, 430
  for plants, 80, 81, 100–101
  reflection of, 566
  refraction of, 567
  shadows and, 568
  stopping, 574–575

Lighthouses, 562
Lightning, 528
Light waves, 620
Lilies, 94
Lima bean, 88–89
Limestone, 259, 272
Liquids, 472
  changes in state of, 474
  condensation of, 382
  sound movement through, 585
  as state of matter, 470
  water, 380
Living things (organisms), 54–55
  See also Animals; Plants
  animal life cycles, 66
  basic needs of, 80
  in ecosystems, 160–161
  ecosystems changed by, 188
  energy needed by, 514
  environments of, 158–159
  food for, 202–206
  fossils of, 270–271
  habitats of, 161–162
  heredity in, 68
  homes for, 50, 53
  observing, 58
  parts of, 56–57
  plant life cycles, 64
  survival of, 178–179
  water for, 370
Lizards, 611
Loam, 336
Locations, 601
  distance between, 602
Lodges, beaver, 120
Loggerhead sea turtles, 204
Long-eared owls, 172

Index

Loudness, 584
Lunar cycle, 419
Lunar eclipse, 420–421
Lungs
   of amphibians, 131
   of birds, 129
   of mammals, 128
   of reptiles, 130

M

Machines, simple. *See* **Simple machines**
Magma, 308
Magnetic force, 611
Magnetite, 536
Magnets, 536
   electromagnets, 540, 544–545
   in generators, 546
   magnetic force, 611
   poles of, 537
   strength of, 532, 535
   uses of, 538
Magnifying box, 7
Magnolia trees, 93
Making models. *See* **Formulating models**
Mammals, 126, 128–129
Mantle (Earth), 286
Maple trees, 92
Maps, weather, 396
Maracas, making, 583, 584
Marble, 259, 264
Mars, 428
   moon of, 398
   in solar system, 430, 431
Marsh food web, 222–223
Mass, 462–463
   change in motion and, 612
   gravity and, 614
   as property of matter, 462
Materials scientist, 487
Matter, 458–459
   chemical changes in, 484
   gases, 473
   liquids, 472
   measuring, 462–463
   mixtures of, 482
   physical changes in, 480–481
   physical properties of, 460–461
   solids, 471
   states of, 470–471
   temperature and, 469
Mazes, 86
McHugh, Tara, 487
Measuring
   with body parts, 2
   distances, 602
   earthquakes, 313
   energy, 508
   with help from simple machines, 643
   as inquiry skill, 19
   motion, 602, 603
   time, 409
   tools for, 8–10
   volume, 457
   waves, 622
   work, 633, 638
Measuring cup, 9
Measuring of matter, 462–463
Measuring spoons, 10
Measuring tape, 8
Mercury, 428, 430
Mesas, 290
Metals
   as conductors, 559
   as nonrenewable resources, 326
Metamorphic rock, 259
   formation of, 260, 261
   fossils in, 272
   in rock cycle, 262–263
   uses of, 264
Metamorphosis, 66, 67
Meteorologists, 399
Meters, 602
Mica, 248, 252
Mice, 216, 224
Microscope, 10, 56–58
Microwave oven, 624
Microwaves, 620, 624
Migration, 181
Miles, 602
Mimicry, 182
Minerals, 248–249
   hardness of, 244, 250
   identifying, 250–251
   in rocks, 252
   streak of, 251
   testing, 247
Mining resources, 321
Mixtures, 479, 482
Models, making. *See* **Formulating models**
Mohs scale, 250
Mold, 270, 271
Moles, 120
Monarch butterfly, 134, 181
Monkeys, 126
Moon (Earth's), 402–403
   distance of, 414
   eclipses of, 420–421
   phases of, 416–419
Moon phases, 416–419
Moons (of other planets)
   inner planets, 430, 431

outer planets, 432, 433
researching, 398
Moonseed, 107
Moths, 67, 182
Motion (movement)
caused by waves, 620
of constellations, 435
creating, 599
definition of, 601
direction, 602
distance, 602
of Earth, 408–409
in Earth's crust, 306
of electricity, 530
forces causing, 610
forces changing, 612–613
of light, 566–567
of rock pieces, 298–299
of sound, 585
speed of, 602–604
of stars, 435, 436
of thermal energy, 557
types of, 600–601
Mountain goat, 162
Mountains, 287
earthquakes in, 307
folded, 282
in oceans, 286–287
volcanoes, 308–309
Mount Etna, 302
Mount Rushmore, 264
Mount St. Helens, 188
Movement. See **Motion**
Muscle cells, 57
Muscles
energy from, 506
work done by, 636
Musical instruments
percussion, 580
sound from, 584–585

Nails, 660
National Aeronautics and Space Administration (NASA), 625
Natural gas, 515
Natural resources. See **Resources**
Nectar, 204
Needs. See **Basic needs**
Neptune, 429, 432, 433
Net force, 613
New moon, 419
Niagara Falls, 316–317
Nice, Margaret Morse, 228
Night, 412
movement of stars in, 436
Nonliving things, 54, 159
Nonrenewable resources, 326, 352, 516
Nucleus, 56, 57
Nursery worker, 107
Nutcrackers, 647
Nutrients
from photosynthesis, 102, 103
for plants, 80
from plants, 104

Oak trees, 90, 92
Observing
colors of bubbles, 26
with computers, 12
as inquiry skill, 23
living things, 58
mining resources, 321
in scientific method, 30
soil, 331
teeth, 201
temperature, 495
Obsidian, 258
Ocampo, Adriana, 398
Oceanographer, 70
Oceans
mountain ranges in, 287
mountains and valleys in, 286
oil in, 346
salt water in, 374
saltwater ecosystems in, 170
valleys in, 288
water temperature in, 466
waves in, 616
Octopus, 225
Oil, 515
as nonrenewable resource, 323, 326
in oceans, 346
searching for, 312
Okubo, Akira, 70
Omnivores, 206
Opaque objects, 574, 575
Opinion, 24
Opossum, 224
Oranges, 96
Orange trees, 338
Orangutan, 198
Orbits, 428
Orb spiders, 141
Ordering. See **Classifying**
Organisms, 54. See also **Living Things**
Ornithologist, 228
Ortiz, Rosa, 107
Ospreys, 208
Outer planets, 432–433

Index

Owls
  elf, 116
  long-eared, 172
Oxygen
  in air, 390
  for animals, 117
  for fish, 132
  in photosynthesis, 103

P

Pacific Ocean, 374
  weather of, 399
Palmetto trees, 93
Pan balance, 10
Pandas, 116
Panthers, 153
Paper, recycled, 356
Paperwhites, 84
Particle physics, 486
Patterns
  of stars, 434
  weather, 392
Peas, 96
Pelicans, 119
Penguins, 129
People
  ecosystems changed by, 189
  land changed by, 342
People in Science
  Akira Akubo, 70
  Steven Chu, 486
  Christine Darden, 625
  Rodolfo Dirzo, 229
  Enrico Fermi, 518
  Tara McHugh, 487
  Margaret Morse Nice, 228
  Adriana Ocampo, 398
  Rosa Ortiz, 107
  Marisa Quinones, 312
  Charles Richter, 313
  Percy Spencer, 624
  Marie Clark Taylor, 106
  Evangelista Torricelli, 519
  Charles Henry Turner, 71
  Bin Wang, 399
Percussion instruments, 580
Petrified wood, 271
Pets, needs of, 116
Phlox, 94
Phobos, 398
Photosynthesis, 102–103
Physical changes in matter, 480–481
  mixtures, 482
  solutions, 483
Physical properties of matter, 460–461
  density, 463, 464
  floating, 464
  mass, 462
  volume, 463
Pigs, 128
Pinball games, 596
Pine trees, 92
Pitch, 585
Plains, 289
Planets, 428–429
  inner, 430–431
  newly discovered, 438–439
  ordering, 427
  origin of, 439
  outer, 432–433
  researching, 398
Planning experiments
  as inquiry skill, 23
  in scientific method, 31
Plants
  adaptations of, 179
  as animal shelter, 120
  cells of, 56
  in deserts, 168
  energy for, 102
  environments of, 84, 158, 159
  with flowers, 90, 91, 94
  without flowers, 95
  in food chains, 212
  in food webs, 226
  fossils of, 270–271, 274
  in freshwater ecosystems, 171
  grasses, 90–91
  in grasslands, 169
  growth and change of, 64–65
  helpful and harmful, 104
  leaves, 92–93
  light for, 100–101
  as living things, 54, 55
  motion of, 604
  needs of, 78–81
  parts of, 82–83
  photosynthesis, 102–103
  pollution and, 341
  as producers, 203
  as renewable resources, 324
  in saltwater ecosystems, 170
  seeds of, 91, 96
  shrubs, 90–91
  structure of, 179
  trees, 90–91
  water for, 370
  weathering by, 296
Plastic, recycled, 356
Plateaus, 290
Pluto, 427
Poison ivy, 104
Poisons, from plants, 104
Poles, of magnets, 537
Pollen, 104

R41

Pollution
  air, 345
  ecosystems changed by, 189
  from electricity generation, 589
  land, 344
  plants and, 341
  from using resources, 353
  water, 346
Pond ecosystem, 160–161
Ponds, 171
Population, 160
Position, 601
Potential energy, 498–500
Power companies, 589
Power plants, 589
Prairie dogs, 158–159
Precipitation, 384, 391
Predators, 216
  in food webs, 222
Predicting
  changes in environment, 187
  as inquiry skill, 20
  rainbows, 573
  shapes of bubbles, 17
  volume of water, 457
  weather, 396
Prey, 216
  defending themselves, 224–225
  in food webs, 222
  insects as, 225
Prisms, 576, 577
Producers, 203
  in energy pyramids, 214, 215
  in food chains, 212
  in food webs, 222

Properties of matter
  density, 463
  mass, 462
  minerals, 250
  physical, 460–461
  volume, 463
Prop roots, 179
Puffins, 216
Pupa, 67

Q

Quartz, 248–250, 252
Quinones, Marisa, 312

R

Rabbits, 120, 216
Raccoons, 206
Radar, 624
Rain
  floods from, 310
  fresh water from, 372
  for rainbows, 577
  in storms, 394
  in water cycle, 384
  water pollution from, 346
Rainbow Bridge, Utah, 280–281
Rainbows
  colors of, 570
  formation of, 577
  making, 573
  sun and rain for, 577
Rakes, 645, 646
Ramps, 655, 657
Raytheon Company, 624
Recording data, as inquiry skill, 19
Recycling, 348, 356, 540
Red-tailed hawk, 205

Reducing use of resources, 354
Reflection, 566
Refraction, 567
Renewable resources, 324, 516
Repeating experiments, 34
Repelling, magnetic, 537
Reproduction, 54
Reptiles, 127, 130
Reservoirs, 184
Resources, 322–323
  competition for, 190
  conserving, 352, 353
  ecosystems and use of, 189
  energy, 515
  in environments, 159
  land, 342–344
  mining, 321
  nonrenewable, 326, 352, 516
  protecting, 352–356
  renewable, 324, 516
  reusable, 325
  soil, 332–336
Reusable resources, 325
Reusing resources, 355
Revolutions
  of Earth around sun, 409
  of moon around Earth, 418, 419
Richter, Charles, 313
Richter scale, 313
Rivers, 171
  erosion by, 298
  valleys formed by, 288, 290
Robins, 120
Robot, ASIMO, 662–663
Rock cycle, 262–263

Index

Rocks
  bedrock, 333
  erosion of, 298–300
  formation of, 260–261
  minerals in, 252
  models of, 257
  movement of pieces, 298–299
  in soils, 334, 335
  types of, 258–259
  uses of, 264
  weathering of, 296–297
Roots, 82, 83, 179
  adaptations of, 179
  of cactus plants, 168
  of grasses, 167
Rotation of Earth, 408, 409, 412
Rotting, 484
Rulers, 8
Rusting, 484

S

Safety in science, 38
Saguaro cactus, 116
Salamanders, 131
Salt, 249
Salt water, 371, 374
Saltwater ecosystems, 170
Salvage yards, 544–545
Sand
  in loam, 336
  in soils, 334, 335
Sand castles, 476
Sand dunes, 299
Sandstone, 259
San Francisco earthquake, 306
Satellites, weather, 396
Saturn, 429, 432
Saw palmettos, 93

Scales
  of fish, 132
  of reptiles, 130
Scallops, 138
Scarlet macaw, 162
Science tools, 6–12
  computer, 12
  dropper, 7
  forceps, 7
  graduated cylinder, 10
  hand lens, 6
  magnifying box, 7
  measuring cup, 9
  measuring spoons, 10
  measuring tape, 8
  microscope, 10
  pan balance, 10
  rulers, 8
  spring scales, 9
  thermometers, 8
Science Up Close
  doorknobs, 648
  energy pyramid, 215
  forces at work, 611
  how a fossil forms, 272–273
  how a battery works, 499
  how ice cracks rock, 297
  lunar eclipses, 421
  magnets, 537
  measuring mass, 462–463
  metamorphosis, 67
  photosynthesis, 103
  pond ecosystem, 160–161
  shadows throughout the day, 568
  soil layers, 333
  spider webs, 141
  water cycle, 384
Scientific method, 30–34
Screw, 660
Screwdriver, 649

Sea anemones, 143
Sea breezes, 392
Seals, 50
Seas, salt water in, 374
Sea slugs, 142
Seasons, 410–411
Sea turtles, 130
Seaweed, 170
Sedimentary rock, 259
  formation of, 260, 261
  fossils in, 272
  in rock cycle, 262–263
  uses of, 264
Seeds, 91, 96
  growth of, 63
  in life cycle, 64
  production of, 94
Seismologists, 312
Senses, observing with, 460
Severe weather, 394–395
Shadows, 420, 568
Shale, 272
Shape
  of bubbles, 17
  of gases, 473
  of liquids, 472
  as physical property of matter, 461
  of solids, 471
Shaw, Chris, 277
Shelter, for animals, 120
Shovels, 647
Shrubs, 90–91
Sight, observing with, 460
Silt, 335, 336
Simple machines, 630, 644–645, 656–657
  help from, 643
  inclined plane, 656-657
  lever, 646–647
  pulley, 650

R43

screw, 660
wedge, 658–659
wheel-and-axle, 648–649
Size, as physical property of matter, 461
Skin, of amphibians, 131
Skin cells, 57
Skunks, 224
Slate, 264
Sloths, 48–49
Smell, observing with, 460
Snails, 142, 161
Snakes, 127, 130, 224
camouflage of, 182
Snow
blizzards, 395
fresh water from, 372
in storms, 394
Soil
carried by floods, 310
with earthworms, 144
erosion of, 298–300
importance of, 336
layers of, 332–333
as nonrenewable resource, 326
observing, 331
for plants, 80, 81
types of, 334–335
Solar eclipses, 422
Solar energy, 202, 324, 506, 516
Solar system, 428 See also Earth; Sun
inner planets, 430–431
outer planets, 432–433
sun, 429
Solids, 471
changes in state of, 474
ice, 380, 381

sound movement through, 585
as state of matter, 470, 471
Solutions, 483
Sound, 584–585
Sounds, hearing, 586
Sound waves, 621
Space
taken up by gases, 473
volume as measure of, 463
Speed, 602, 604
differences in, 609
of Earth's motion, 408
height of starting point and, 606
of motion, 603
Spencer, Percy, 624
Spiders, 139, 141
Spider webs, 141
Spores, 95
Spring scale, 9
Spring toys, 621
Spruce trees, 95
Squids, 142
Squirrels, 172
Star
definition of, 429
movement of, 435, 436
in patterns, 434
sun as, 429
Starfish, 143
Starfish fossil, 242–243
States of matter, 470
changes in, 474
gases, 473
liquids, 472
solids, 471
water, 380–384
Static electricity, 528
looking for, 527

Van de Graaff generator, 524
Steel, 536
Stems, 82, 83, 179
adaptations of, 179
of cactus plants, 84
growing new plants from, 65
Stonefish, 182
Storms, 394
Stratus clouds, 391
Streak (minerals), 251
Streams, 171
valleys formed by, 290
Strength, of forces, 612
Subsoil, 333
Sugar maples, 96
Summer, 410, 411
Sun
Earth's revolution around, 409
eclipse of, 422
energy from, 202, 324, 490–491, 506
planets orbiting, 428
in solar system, 429
water cycle and, 384
Sunflowers, 203
Sunlight
as energy, 429
as mixture of colors, 576
for rainbows, 577
striking Earth, 407, 410, 411
Switchgrass fuel, 589
Synthetic materials, 147

T

Tadpoles, 131
Taste, observing with, 460
Taylor, Marie Clark, 106

Index

Technology
  AbioCor heart, 548–549
  ASIMO robot, 662–663
  cell phone tower camouflage, 192–193
  Hubble Space Telescope, 438–439
  La Brea tar pit fossils, 276–277
  switchgrass fuel, 588–589
  synthetic waterproof material, 146–147
  Three Gorges Dam, 358–359

Teeth
  fossils of, 270, 274
  observing, 201
  of omnivores, 206

Telescopes
  Hubble Space Telescope, 438–439

Temperature
  definition of, 393, 558
  in deserts, 168
  of hibernating animals, 180
  matter and, 469
  measuring, 508
  observing, 495
  of ocean water, 466
  state of water and, 381–383

Tent bats, 120
Terrarium, 379
Texas State House, 252
Texture
  friction and, 610
  as physical property of matter, 461

Thermal energy
  conduction, 559
  for cooking, 502
  definition of, 558
  insulators, 560
  movement of, 557. *See also* Heat
  states of water and, 382, 383

Thermometers, 8, 393, 508
Third-quarter moon, 419
Thistle, 96
Three Gorges Dam, China, 358–359
Thunderstorms, 395
Ticks, 141
Tigers, 128
Time
  as measure of motion, 603
  measuring, 409

Time/space relationships, 21
Time zones, 412
Topsoil, 333
Tornadoes, 394
Torricelli, Evangelista, 519
Toucans, 126
Touch, observing with, 460
Traits, 68
Translucent objects, 574, 575
Transparent objects, 575
Trash
  gathering data on, 350–351
  land pollution from, 344

Tree frogs, 127
Trees, 90–91
  in forests, 172
  in grasslands, 169
  leaves of, 92–93
  as renewable resource, 324

Triceratops, 273
Trilobites, 271
Tropical ecologists, 229
Tropical rain forest, 172
Trough (wave), 622
Tubers, 65
Tulips, 65, 76
Tundra, 164
Turbine, 358
Turner, Charles Henry, 71
Turtles, 160, 204
Twisters, 395

Uranus, 429, 432, 433
U-shaped valleys, 288
Using numbers, 19
  to compare land area to water area on Earth's surface, 369
  to order planets, 427

Vacuoles, 56, 57
Valleys, 288
  made by glaciers, 300
  in oceans, 286

Van de Graaff generator, 524
Variables, 22
Venus, 428, 430
Vertebrates, 126–132
  amphibians, 131
  birds, 129
  fish, 132
  mammals, 128
  reptiles, 130

Vibrations, 584–585
  in objects, 586
  waves from, 621

Vines, 179
Volcanoes
  igneous rock from, 260

landforms changed by, 308–309
model of, 305
Mount Etna, 302
Mount St. Helens, 188
Volcanologists, 312
Volume
of gases, 473
of liquids, 472
measuring, 457
as property of matter, 463
of solids, 471
V-shaped valleys, 288

W

Walking stick, 182
Wang, Bin, 399
Water
for animals, 118
change in brick from, 295
changes in state of, 382–383, 474
condensation of, 383
in deserts, 168
on Earth, 366, 371
erosion by, 298–299
evaporation of, 383
floating in, 464
floods, 310
fresh, 371–373
groundwater, 372
for horses, 154
importance of, 370–371
locations of, 369
need for, 454
in photosynthesis, 103
for plants, 80, 81
plateaus formed by, 290
as resource, 323
as reusable resource, 323
rock broken down by, 261

in rock cycle, 262, 263
salt, 371, 374
in soils, 334
states of, 380–381
weathering by, 296, 297
Water cycle, 384
Waterlilies, 84, 147, 161
Water pollution, 346
Waterproof material, 147
Water treatment plants, 346
Water vapor, 380–382
in atmosphere, 390
Water waves, 616
Wavelength, 622
Waves
definition of, 620
kinds of, 619
measuring, 622
sound, 621
water, 616
Weather
barometers, 519
blizzards, 395
cloud types and, 391
definition of, 392
in different seasons, 410, 411
droughts, 394
gathering data on, 393
hurricanes, 394–395
ice storms, 395
meteorologists, 398
predicting, 396
severe, 394–395
tornadoes, 394
Weathering, 296–297
Weather instruments, 393, 396
Weather maps, 396
Weather patterns, 392
Weather satellites, 396

Weather vanes, 386
Wedge, 658–659
Weight, 614
Whales
gray, 181
oxygen for, 117, 128
Wheel-and-axle, 648–649
White light
bending, 576
making, 578
Wildebeests, 225
Wind
erosion by, 298–299
measuring, 389, 393
plateaus formed by, 290
as renewable resource, 516
rock broken down by, 261
in rock cycle, 262, 263
in tornadoes, 395
weathering by, 296
Wind generators, 318
Windmills, making, 512–513
Wind speed, 508
Wings, of insects, 140
Winter, 410, 411
Wolves, 205, 216
Wood
burning, 484
petrified, 271
from plants, 104
Wool, 480, 481
Woolly mammoth, 276
Work, 633
done by simple machines, 644–645
forces used in, 636
meaning of, 634–635
measuring, 638
types of, 634–635
Worm farm, 136–137
Worms, 143, 144

Index

X-rays, 620
Xylophone, 621

Yards, 602
Year, 409
   on inner planets, 430, 431
   on outer planets, 432, 433
   seasons of, 410–411
Young animals
   amphibians, 131
   birds, 129
   fish, 132
   mammals, 128
   reptiles, 130

Photo Credits

Page Placement Key: (t) top, (b) bottom, (l) left, (r) right, (c) center, (bg) background, (fg) foreground

Table of Contents:

Introduction:
12 (tr) John Madere/CORBIS;

Unit A:
39 BRUCE COLEMAN INC./Alamy; 47 Michael T. Sedam/Corbis; 48 Getty Images; 50 Fred Bruemmer/Peter Arnold, Inc.; 52 Frans Lanting/Minden Pictures; 54 Photolibrary.com pty.ltd/Index Stock Imagery; 55 Larry Lefever/Grant Heilman Photography; 56 (t) Carolina Biological/Visuals Unlimited; 56 (b) Kevin R. Morris/Bohemian Nomad Picturemakers/Corbis; 57 (t) Carolina Biological/Visuals Unlimited; 57 (b) Gary Meszaros/Bruce Coleman, Inc.; 58 (b) Ed Reschke/Peter Arnold, Inc.; 58 (cr) Konrad Wothe/Minden Pictures; 58 (tr) Mark Moffett/Minden Pictures; 58 (bl) Michael S. Yamashita/Corbis; 60 Roy Morsch/Corbis; 62 David R. Frazier Photolibrary, Inc./Alamy; 64 J. Douillet/Peter Arnold, Inc.; 65 (l) Ferrer & Sostoa/Age Fotostock America; 65 (r) Royalty-free/Corbis; 66 (br) E.A. Janes/Age Fotostock America; 66 (tl) Jane Burton/Bruce Coleman, Inc.; 66 (tr) Photodisc Green (Royalty-free)/Getty Images; 66 (bl) Photodisc Red (Royalty-free)/Getty Images; 68 (br) D. Robert & Lorri Franz/Corbis; 68 (tr) Kit Houghton/Corbis; 68 (l) Lindy Craig/Grant Heilman Photography; 71 (tl) Courtesy of Terri Small-Turner, Charles Henry Turner II and Charles I. Abramson; 71 (br) John Lamb/Getty Images; 71 (bg) KONRAD WOTHE/Minden Pictures; 74-75 Zoran Milich/Masterfile; 76 John McAnulty/Corbis; 78 RICHARD SHIELL/Animals Animals - Earth Scenes84 (t) Royalty-Free/Corbis; 84 (b) David Young-Wolff/PhotoEdit/PictureQuest; 86 Estock Photo; 88 Jon Arnold Images/Alamy; 90 (inset) George Harrison/Grant Heilman Photography; 90 (b) Kent Foster/Bruce Coleman, Inc.; 91 Martin Fox/Index Stock Imagery/PictureQuest; 92 (t) Martin B. Withers; Frank Lane Picture Agency/Corbis; 92 (cr) Michael P. Gadomski/Dembinsky Photo Associates; 92 (cl) Phil Degginger/Bruce Coleman, Inc.; 92 (b) Randy M. Ury/Corbis; 93 (t) Ed Kanze/Dembinsky Photo Associates; 94 (l) R-R/S/Grant Heilman Photography; 94 (c) Skip Moody/Dembinsky Photo Associates; 94 (r) C-JG/Grant Heilman Photography; 95 (l) Grant Heilman Photography; 95 (r) Bruce Coleman, Inc.; 96 (t) C-MON/Grant Heilman Photograpy; 96 (cl) Michael P. Gadomski/Dembinsky Photo Associates; 96 (cr) Michelle Garrett/Corbis; 96 (bl) Randall B. Henne/Dembinsky Photo Associates; 98 Royalty-Free/Corbis; 100 David C Tomlinson/Getty Images; 102 Patti McConville/Dembinsky Photo Associates; 104 (b) C-TB/Grant Heilman Photography; 104 (t) Getty Images; 104 (cl) John Robinson/Peter Arnold; 104 (cr) Larry West/Bruce Coleman, Inc.; 106 (bg) Clouds Hill Imaging Ltd./CORBIS; 106 (tr) Courtesy of Howard University; 107 (tl) Courtesy of Rosa Ortiz; 107 (bg) Mills Tandy; 114 Colin Hawkins/Getty Images; 110-111 Mark J. Thomas/Dembinsky Photo Associates; 112 Getty Images; 114 Colin Hawkins/Getty Images; 116 (r) John Cancalosi/Nature Picture Library; 116 (l) John Giustina/Bruce Coleman, Inc.; 117 ZSSD/Minden Pictures; 118 (t) BIOS/Peter Arnold; 118 (b) Royalty-Free/Corbis; 119 Getty Images; 119 (inset) Steve Kaufman/Peter Arnold; 120 (t) Michael & Patricia Fogden/Corbis; 122 Theo Allofs/Corbis; 124 Thomas Mangelsen/Minden Pictures; 126-127 (bg) Norman O. Tomalin/Bruce Coleman, Inc.; 126 (l) Staffan Widstrand/Corbis; 126 (r) Wolfgang Kaehler/Corbis; 127 (t) Getty Images; 127 (br) Bruce Coleman, Inc.; 127 (bl) Joe McDonald/Bruce Coleman, Inc.; 128 (t) Bruce Coleman, Inc.; 128 (bl) David Fritts/Animals Animals; 128 (br) Thomas Mangelsen/Minden Pictures; 129 (tl) Thomas D. Mangelsen/Peter Arnold; 129 (tr) Tim Davis/Corbis; 129 (c) Tom Vezo/Minden Pictures; 129 (b) Diane Miller/Monsoon Images/PictureQuest; 130 (t) Zigmund Leszczynski/Animals Animals; 130 (c) Getty Images; 130 (b) George McCarthy/Corbis; 131 (r) John Burnley/Bruce Coleman, Inc.; 131 (l) T. Young/Tom Stack & Associates; 132 (t) Getty Images; 132 (c) Getty Images; 132 (b) Scott Kerrigan/Corbis; 134 Frans Lanting/Minden Pictures; 136 Chris Close/Getty Images; 138 (t) Jesse Cancelmo/Dembinsky Photo Associates; 138 (t) Jesse Cancelmo/Dembinsky Photo Associates; 138 (bl) Getty Images; 138 (br) Sue Scott/Peter Arnold; 139 (t) Dwight Kuhn; 139 (bl) Douglas P. Wilson; Frank Lane Picture Agency/Corbis; 139 (br) Larry West/Bruce Coleman, Inc.; 140 (tl) Thomas Boyden/Dembinsky Photo Associates; 140 (r) CISCA CASTELIJNS/FOTO NATURA/Minden Pictures; 140 (bg) Craig Aurness/Corbis; 140 (bl) Skip Moody/Dembinsky Photo Associates; 141 (t) Donald Specker/Animals Animals; 141 (b) Larry West/Bruce Coleman, Inc.; 142 (tr) Gordon R. Williamson/Bruce Coleman, Inc.; 142 (cr) Getty Images; 142 (bl) Gail M. Shumway/Bruce Coleman, Inc.; 142 (br) Priscilla Connell/Index Stock Imagery; 143 (tr) Getty Images; 143 (cl) Fred Bavendam/Minden Pictures; 143 (bl) E. R. Degginger/Bruce Coleman, Inc.; 144 (t) ABPL/Roger De La Harpe/Animals Animals; 144 (b) David M. Dennis/Animals Animals; 119 (l) Michael & Patricia Fogden/Minden Pictures; 119 (r) Theo Allofs/Corbis.

Unit B:
151 © OKLAHOMA TOURISM; 152-153 Gail M. Shumway/Bruce Coleman, Inc.; 154 Richard T. Nowitz/Corbis; 156 Tom Payne/Alamy; 158-159 Claudia Adams/Dembinsky Photo Associates; 158 (inset) R-GH/Grant Heilman Photography; 159 (t) John Cancalosi/Nature Picture Library; 159 (inset) Tim Fitzharris/Minden Pictures; 162 (b) Mary Clay/Dembinsky Photo Associates; 162 (t) Royalty-Free/Corbis; 164 Stefan Meyers/Animals Animals/Earth Scenes; 166 JOHN LEMKER/Animals Animals - Earth Scenes; 168 Photodisc Green (Royalty-free)/Getty Images; 169 Jim Brandenburg/Minden Pictures; 170 Flip Nicklin/Minden Pictures; 171 Getty Images; 172 (b) Nancy Rotenberg/Animals Animals/Earth Scenes; 172 (inset) OSF/M. Hamblin/Animals Animals/Earth Scenes; 172 (t) Roger Wilmshurst; Frank Lane Picture Agency/Corbis; 174 Yva Momatiuk/John Eastcott/Minden Pictures; 176 Mitsuhiko Imamori/Minden Pictures; 178 (b) Barbara Von Hoffmann/Animals Animals/Earth Scenes; 178 (inset) Steve Kaufman/Corbis; 179 (c) Gary Braasch/Corbis; 179 (l) I-JG/Grant Heilman Photography; 179 (r) Nancy Rotenberg/Animals Animals/Earth Scenes; 180 Breck P. Kent/Animals Animals/Earth Scenes; 181 (l) Alan G. Nelson/Animals Animals/Earth Scenes; 181 (r) Stephen Dalton/Animals Animals/Earth Scenes; 182 (tl) Carl Roessler/Bruce Coleman, Inc.; 182 (tr) Anthony Bannister/Gallo Images/Corbis; 182 (bl) Martin Withers/Dembinsky Photo Associates; 182 (br) Breck P. Kent/Animals Animals/Earth Scenes; 184 Michael E. Lubiarz/Dembinsky Photo Associates; 186 Carlo Allegri/Getty Images; 188 (t) Gary Braasch/Corbis; 188 (b) David Muench/Corbis; 188 (inset) Dennis MacDonald/Index Stock Imagery/PictureQuest; 189 (r) Smith Aerial Photos; 189 (l) Smith Aerial Photos; 190 (t) Getty Images; 190 (c) Raymond Gehman/Corbis; 190 (bl) Royalty-Free/Corbis; 190 (br) Dominique Braud/Animals Animals/Earth Scenes; 192 Index Stock/Picture Quest; 193 Marianna Day Massey/Corbis; 196-197 Michael & Patricia Fogden/Minden Pictures; 198 Getty Images; 200 Yva Momatiuk/John Eastcott/Minden Pictures; 202 Ed Reschke/Peter Arnold; 204 (t) Fred Unverhau/Animals Animals; 204 (c) Yva Momatiuk/John Eastcott/Minden Pictures; 204 (b) Stephen Frink/Corbis; 205 (t) Bruce Coleman, Inc.; 205 (b) Bruce Coleman, Inc.; 206 (tl) Bruce Coleman, Inc.; 206 (inset) Getty Images; 206 (tr) Konrad Wothe/Minden Pictures; 208 Getty Images; 210 Sumio Harada/Minden Pictures; 212-213 (bg) I-R/S/Grant Heilman Photography, Inc.; 212 (l) R-GH/Grant Heilman Photography, Inc.; 213 (r) Bruce Coleman, Inc.; 213 (l) Barry Mansell/npl/Minden Pictures; 214 Kevin Schafer/Peter Arnold; 216 (t) Frans Lanting/Minden Pictures; 216 (b) Gerard Lacz/Animals Animals; 218 Tom Brakefield/Corbis; 224-225 (t) Kennan Ward/Corbis; 220 SuperStock; 224 (c) Joe McDonald/Corbis; 224 (bl) Pete Oxford/Nature Picture Library; 224 (br) Tom Brakefield/Corbis; 225 (tr) Fred Bavendam/Peter Arnold; 225 (b) Sea Images, Inc./Animals Animals; 226 Getty Images; 228 (tr) AP Images; 228 (bg) B. Renevey/Peter Arnold, Inc.; 229 (tl) Courtesy of Stanford News Service; 229 (bg) Dirk Anschutz/Getty Images; 233 Emma Lee/Life File/Getty Images.

Unit C:
192 Lee Foster/Lonely Planet; 242-243 Layne Kennedy/Corbis; 246 Wayne Scherr/Photo Researchers, Inc.; 248 (t) Breck P. Kent Photography; 248 (cl) Barry Runk/Stan/Grant Heilman Photography; 248 (cr) Breck P. Kent/Smithsonian Institute; 248 (bl) Th Foto-Werbung/Science Photo Library/Photo Researchers; 248 (bc) Albert Copley/Visuals Unlimited; 248 (br) Edward R. Degginger/Bruce Coleman, Inc.; 249 (tl) Breck P. Kent Photography; 249 (tc) John James Wood/Index Stock Imagery; 249 (tr) Mark Schneider/Visuals Unlimited; 249 (bl) Jose Manuel Sanchis Calvete/Corbis; 250 (tr) Getty Images; 252 (t) Bob Daemmrich Photography; 252 (cl) Mark A. Schneider/Visuals Unlimited; 252 (c) Breck P. Kent Photography; 252 (cr) Breck P. Kent Photography; 252 (bl) Sinclair Stammers/Science Photo Library/Photo Researchers; 252 (br) Wally Eberhart/Visuals Unlimited; 254 Tom Bean/Corbis; 256 DAVID R. FRAZIER Photolibrary, Inc.; 258 (r) Doug Sokell/Visuals Unlimited; 259 (t) Wally Eberhart/Visuals Unlimited; 259 (cl) Barry Runk/Stan/Grant Heilman Photography; 259 (cr) Barry Runk/Stan/Grant Heilman Photography; 260 Jim Sugar/Corbis; 261 (t) Grant Heilman/Grant Heilman Photography; 261 (b) Ben S. Kwiatkowski/Fundamental Photographs; 262 Rob C. Williamson/Index Stock Imagery; 2264 (t) Bob Rashid/Brand X Pictures/Alamy Images; 266 Albert Copley/Visuals Unlimited; 268 Gunter Marx Photography/CORBIS; 270 (t) Layne Kennedy/Corbis; 270 (b) Dick Roberts/Visuals Unlimited; 271 (b) Francesc Muntada/Corbis; 271 (t) Ken Lucas/Visuals Unlimited; 271 (c) Sylvester Allred/Fundamental Photographs; 273 (t) Kevin Schafer/Corbis; 274 (b) Richard T. Nowitz/Corbis; 274 (t) Sternberg Museum of Natural History; 276 Philip James Corwin/Corbis; 277 AP/Wide World Photos; 280-281 David Muench/Corbis; 282 Mark Turner/Turner Photographics; 284 DAJ/Getty Images; 288 (inset) ArtPhoto/Diomedia/Alamy Images; 289 Photodisc Red (Royalty-free)/Getty Images; 290 Getty Images; 292 Gavriel Jecan/Corbis; 294 Comstock Images/Jupiter Images; 296 (b) Chris Bell/Lonely Planet Images; 296 (inset) E.R. Degginger/Color-Pic; 298 (t) Dick Roberts/Visuals Unlimited; 298 (b) Gordana Uzelac/Domedia/Alamy Images; 299 (t) Annie Griffiths Belt/Corbis; 299 (inset) Science VU/Visuals Unlimited; 300 (b) Bernhard Edmaier/ Science Photo Library/Photo Researchers; 300 (inset) George Wilder/Visuals Unlimited; 302 Alfio Scigliano/Sygma/Corbis; 304 Jean-Pierre Pieuchot/Getty Images; 306 Roger Ressmeyer/Corbis; 307 Roger Ressmeyer/Corbis; 308 Amos Nachoum/Corbis; 309 (t) Michael S. Yamashita/Corbis; 309 (b) Roger Ressmeyer/Corbis; 310 (t) Bill Ross/Corbis; 310 (b) Philip Wallick/Corbis; 312 (bg) imagebroker/Alamy; 313 (tl) Bettmann/CORBIS; 313 (bg) Tom McHugh/Photo Researchers, Inc.; 316-317 Jan Butchofsky-Houser/Corbis; 318 ML Sinibaldi/Corbis; 320 Karen Kasmauski/Corbis; 322 (inset) Fred Bavendam/Minden Pictures; 322 (b) Larry Lefever/Grant Heilman Photography; 323 (inset) Lou Jacobs, Jr./Grant Heilman Photography; 323 (r) Tom Campbell/Index Stock Imagery; 324 David Whitten/Index Stock Imagery; 325 Jean-Michel Bertrand/Index Stock Imagery; 326 (inset) Getty Images; 326 (b) Mike Dobel/Masterfile; 328 PhotoDisc Blue (Royalty-Free)/Getty; 330 Jan Van Arkel/Foto Natura; 334 (b) Getty Images; 335 Royalty-Free/Corbis; 336 Joel W. Rogers/Corbis; 338 Eunice Harris/Index Stock Imagery; 340 Oliver Strewe/Getty Images; 342-343 David R. Frazier Photolibrary, Inc.; 343 (inset) PhotoDisc Green(Royalty-Free)/Getty Images; 344 Getty Images; 345 (l) Chase Swift/Corbis; 345 (r) Douglas Slone/Corbis; 346 (b) Bettmann/Corbis; 346 (inset) J. Watney/Photo Researchers; 350 Stephen Wilkes/Getty Images; 352-353 Norbert Wu/Minden Pictures; 354 (c) Getty Images.

Unit D:
363 (b) www.niagarafallslive.com; 364-365 Guy Motil/Corbis; 366 Volvox/Index Stock Imagery; 368 Peter Schulz/Index Stock Imagery; 370 Francisco Erizel/Bruce Coleman, Inc./PictureQuest; 371 Ray Ellis/Photo Researchers; 372-373 Steve Vidler/Superstock; 373 (inset) Getty Images; 374 Corbis; 376 Getty Images; 378 Jeremy Walker/Getty Images; 380 J. David Andrews/Masterfile; 381 Royalty-Free/Corbis; 382 (bg) J.A. Kraulis/Masterfile; 386 Jeff Greenberg/PhotoEdit; 388 Frithjof Hirdes/zefa/Corbis; 390 Willie Holdman/Imagestate; 393 (l) Peter West/National Science Foundation; 393 (r) Simon Fraser/CGBAPS/Photo Researchers; 394 (cr) Eric Nguyen/Jim Reed Photography/Photo Researchers, Inc.; 394 (bl) Grady Harrison/Alamy; 394 (bg) Jochen Schlenker/Masterfile; 394 (br) Paul Glendell/Alamy; 394 (cl) Valley Times, Jim Ketsdever/AP Images; 395 (br) David Hancock/Alamy; 395 (bl) Dennis MacDonald/Alamy; 395 (cl) RICK WILKING/Reuters/Corbis; 395 (cr) Scott Olson/Getty Images; 396 (c) 2004 www.ACCUWEATHER.com; 398 (t) International Pacific Research Ctr. Univ of Hawaii; 399 (bg) J. Pat Carter/AP Images; 400-401 NASA-HQ-GRIN; 404 George H.H. Huey/Corbis; 406 Antonio M Rosario/Getty Images; 412 (l) Jeff Greenberg/Index Stock Imagery; 412 (r) Thomas Craig/Index Stock Imagery; 414 David Nunuk/Photo Researchers; 416 Rob Matheson/CORBIS; 419 (t) Dr. Fred Espenak/Photo Researchers; 419 (cl) Eckhard Slawik/Photo Researchers; 419 (cr) Eckhard Slawik/Photo Researchers; 419 (b) J. Sanford/Photo Researchers; 420 John Sanford/Photo Researchers; 422 Dr. Fred Espenak/Photo Researchers; 424 Geray Sweeney/Corbis; 426 Eckhard Slawik/Photo Researchers, Inc.; 429 NASA-JPL; 430 (t) NASA-JPL; 430 (b) USGS/Photo Researchers; 431 (b) NASA-JPL; 432 (b) NASA-JPL; 432 (t) NASA-JPL; 433 (tr) NASA-HQ-GRIN; 433 (b) NASA-JPL; 433 (tl) STSI/NASA/Photo Researchers; 438-439 NASA; 443 Clive Sawyer/Alamy;

Unit E:
456 Wayne Linden/Alamy; 460 (br) Getty Images; 460 (bl) Getty Images; 461 (r) Min Roman/Masterfile; 463 (r) Min Roman/Masterfile; 466 Getty Images; 468 Doug Young Travel/Alamy; 470 Getty Images; 471 (tr) Alamy Images; 471 (l) Getty Images; 471 (br) Joseph Sohm/ChromoSohm Inc./Corbis; 472 (b) Dennis Degnan/Corbis; 476 Zoran Milich/Masterfile; 478 Geoffrey Kidd/Alamy; 480 (b) Klaus-Peter Wolf/Animals Animals; 480 (c) Tim Wright/Corbis; 482 (b) Getty Images; 484 (b) Owen Franken/Corbis; 486 (bg) Alfred Pasieka/Photo Researchers, Inc.; 486 (tr) SZENES JASON/CORBIS SYGMA; 487 (tl) Courtesy of USDA; 487 (bg) Courtesy of USDA;492 Getty Images; 496 (l) Hubert Stadler/Corbis; 496 (c) Getty Images; 496-497 Getty Images; 497 (t) Brand X Pictures/Creatas Royalty Free Stock Resources; 497 (b) Robert Harding World Imagery/Getty Images; 500 Ed Bock/Corbis; ; 504 Swerve/Alamy; 506-507 Lloyd Sutton/Masterfile; 507 (inset) Colin Garratt; Milepost 92 1/2 /Corbis; 508 (l) David Young-Wolff/PhotoEdit; 510 Getty Images; 512 GRANT HEILMAN/Grant Heilman Photography; 514 Getty Images; 516 Bob Krist/Corbis; 518 (tr) Science Source/Photo Researchers Inc.; 518 (bg) SuperStock, Inc./SuperStock; 519 (cl) Chris Pancewicz/Alamy; 519 (bg) Russell Monk/Masterfile; 519 (br) Tony Freeman/PhotoEdit; 522 Getty Images; 524 Paul A. Souders/Corbis; 526 John Gilmore; 528 William Koplitz/Index Stock Imagery; 534 Paul Silverman/Fundamental Photographs, NYC; 536 (t) Sinclair Stammers/Science Photo Library/Photo Researchers; 540 Jeremy Walker/Science Photo Library/Photo Researchers; 542 imagebroker/Alamy; 552 Mary Kate Denny/PhotoEdit; 556 Heath Robbins/Jupiter Images; 562 Getty Images; 564 Sandro

Vannini/CORBIS; 566 (b) Getty Images; 570 Getty Images; 572 Elinor Osborn; 582 Vasina Natalia Vladimirovna/Shutterstock; 576 ThinkStock LLC/Index Stock Imagery; 584 Getty Images.

Unit F:
593 Grace Davies/Omni-Photo Communications; 574-575 Getty Images; 596 Getty Images; 598 Phil Schermeister/CORBIS; 600 (l) Bob Daemmrich/PhotoEdit; 602 Martin Rugner/Age Fotostock America; 603 (t) Getty Images; 604 (l) Maryann Frazier/Photo Researchers; 604 (r) Joe McDonald/Corbis; 606 Ariel Skelley/Corbis; 608 ThinkStock/Age Fotostock; 610 (r) NASA; 611 McDonald Wildlife Photography/Animals Animals/Earth Scenes; 509 Bill Varie/Corbis; 614 (l) David Ball/Corbis; 614 (r) Getty Images; 616 RF/Getty Images; 618 Fred Hirschmann/Getty Images; 622 Walter Bibikow/Age Fotostock America; 624 (tr) Courtesy of the Raytheon Company; 624 (bg) Judd Pilossof/Jupiter Images; 625 (tl) Courtesy of Nasa; 625 (br) Gareth Brown/Corbis; 625 (bg) Koichi Kamoshida/Getty Images; 628-629 Kevin Smith/Alaska Stock Images/PictureQuest; 632 Gary Gerovac/Masterfile; 634 (l) Comstock Images; 635 (r) Royalty-Free/Corbis; 635 (inset) Royalty-Free/Corbis; 642 Spencer Grant/Photo Researchers, Inc.; 644 (inset) David Young-Wolff/Photo Edit; 649 (r) Getty Images; 652 Alden Pellett/The Image Works; 654 Tom Bushey/The Image Works; 657 (b) Superstock; 658 (r) Index Stock Imagery, Inc.; 659 (inset) Annie Griffiths Belt/Corbis; 659 (t) Index Stock Imagery, Inc.; 662-663 © Honda Corp.

All other photos © Harcourt School Publishers. Harcourt Photos provided by the Harcourt Index, Harcourt IPR, and Harcourt photographers; Weronica Ankarorn, Victoria Bowen, Eric Camden, Doug Dukane, Ken Kinzie, April Riehm, and Steve Williams.

LIFE SPAN Slider turtles can live up to 50 years.

CHARACTERISTIC Some slider turtles spend so much time in the water that they're covered in algae.

YOUNG Slider hatchlings have a high death rate; about three-fourths of hatchlings die before they grow up.

CLAWS Male slider turtles have very long front claws.

CHARACTERISTIC The turtle is the only reptile with a shell.